D0487641

THE MYSTERIOUS UNIVERSE

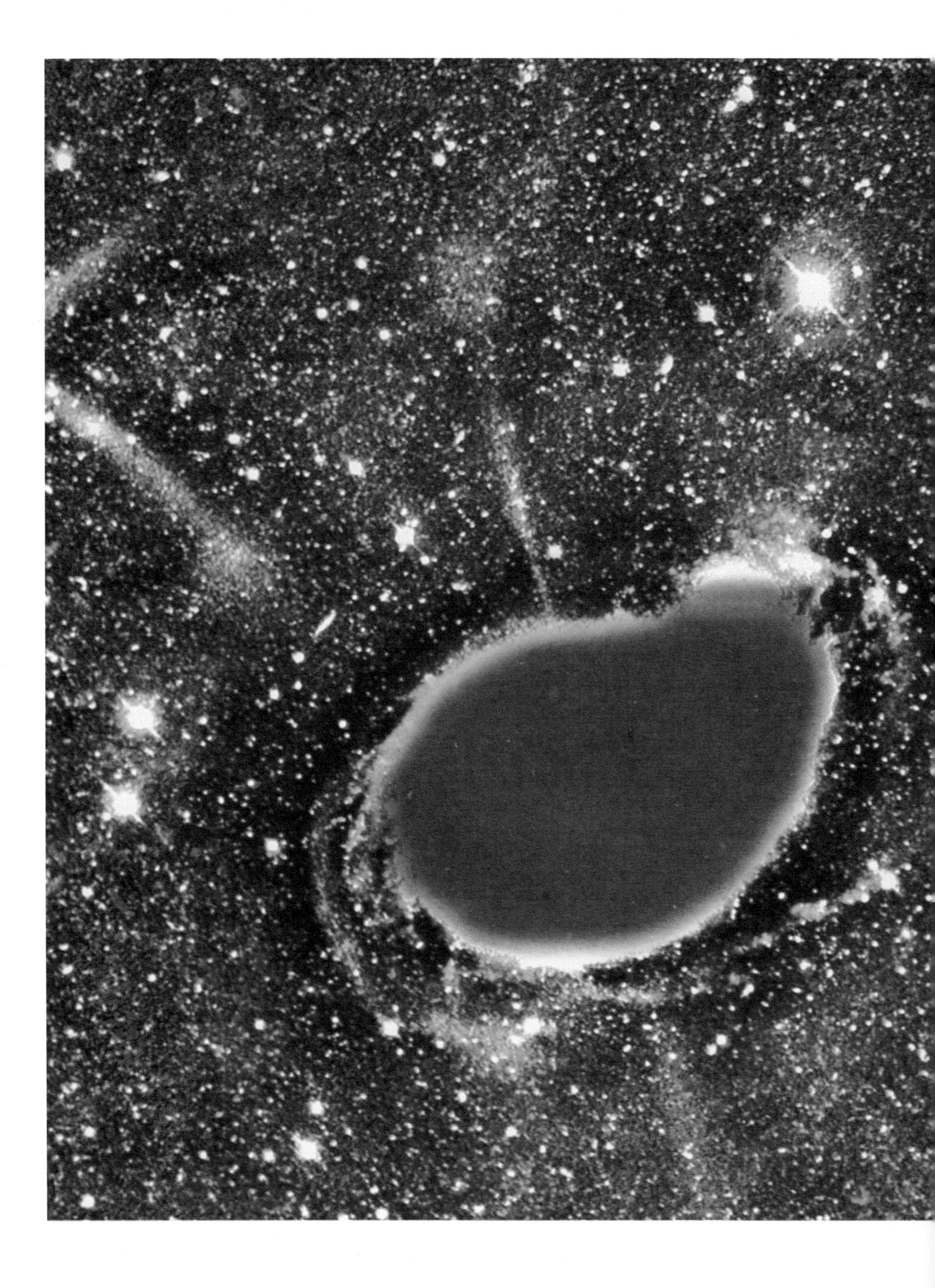

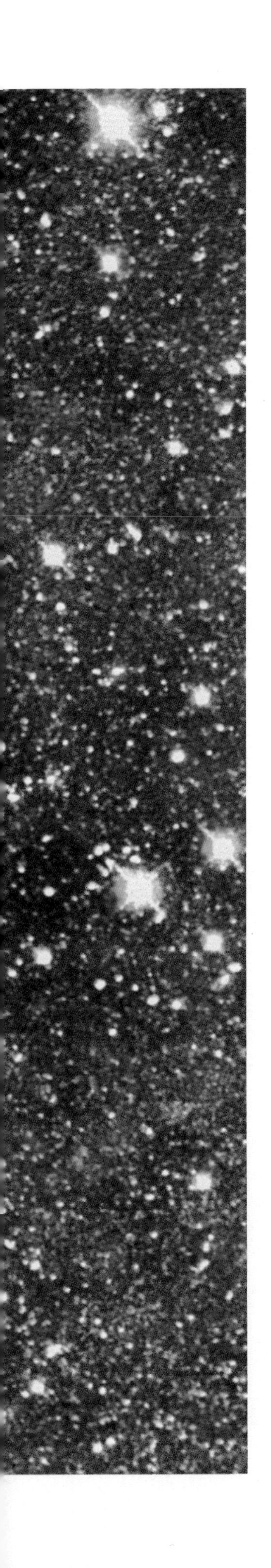

THE MYSTERIOUS UNIVERSE

NIGEL HENBEST

EBURY PRESS
London

Endpapers : Computer-enhanced photograph of the galaxy M82 shows violent activity at its centre. Streamers of hydrogen gas (red) reach out from the centre of the galaxy's disc of stars, seen here edge-on as a glowing band (green). Astronomers dispute whether M82 is suffering a major explosion at its core, or if there is an implosion as the galaxy draws in gas from an intergalactic gas cloud.

Title page : Computer-enhancement brings out the very faint coloured 'jets' which stretch out 100,000 light years into space from the galaxy NGC 1097. The process has emphasized both the colour of the jets and their brightness, to the extent that no detail can be seen in NGC 1097 itself. The original photographs show it is a spiral-shaped galaxy, like the Milky Way Galaxy in which we live, but its mysterious jets are more reminiscent of those found in the distant and extremely powerful quasars.

Contents page : Spectacular close-up view of the planet Saturn shows two of its fifteen icy moons, Tethys (*above*) and Dione, as well as the famous rings. Sunlight is casting shadows of the rings and Tethys onto the globe of Saturn. This picture is one of thousands sent back by the highly successful Voyager 1 spaceprobe as it swept past Saturn in November 1980.

Published by Ebury Press Limited
National Magazine House
72 Broadwick Street
LONDON W1V 2BP

First impression 1981

ISBN 0 85223 212 8

Produced by
Marshall Cavendish Books Limited
58 Old Compton Street
LONDON W1V 5PA

Edited by Isabel Moore
Designed by Eddie Pitcher

Printed by L.E.G.O., Vicenza, Italy

INTRODUCTION

From Earth, we look out at a vast Universe around us. Space stretches out unimaginable distances in every direction, and it is inhabited by many strange objects.

It is the astronomer's job to make sense of the Universe, to confirm understood facts and to interpret new knowledge and theories. We now know as facts, for instance, that Earth is one of nine planets, endlessly circling the much larger Sun and that the stars in the sky are other suns, dimmed by their great distance. And we also now know that stars congregate together in enormous islands called galaxies, which are speeding apart from one another – a legacy of the explosive force of the Big Bang with which our Universe began. Theory allows us to add a time dimension, and therefore to estimate changes in the sky which cannot be seen because they take far longer to reach us than the recorded history of astronomy. The result is an understanding not only of what stars *are*, but also of how they are born and die.

And there is always new information to be sought, and made sense of, because our Universe is still, even with today's knowledge, a mysterious place. There are uncharted reaches of space, strange objects to be discovered, and even unrecognized regions of time. We still do not know, for example, how the planets of the solar system formed. Observations can give only indirect clues about this long-past event, so we must rely heavily on theory and here, as in most mysteries of the Universe, there are many ideas, each staunchly supported by its own proponents. It is a 'controversial theoretical topic, with inadequate observational data' – or, more simply, a mystery.

Other 'mysteries' in this book are individual astronomical objects. When something new and unexpected is discovered, the first instinct is not to call it a mystery, but to look for a rational explanation: there will undoubtedly be several conflicting ones at first. So when Cambridge radio astronomers detected regularly-pulsing sources (now called pulsars) they did not announce the discovery of a 'mysterious' new object. After presenting their observations in a scientific paper, they suggested two possible explanations – both wrong, as it turned out. Their only concession to the weirdness of the radio sources was a reference to the 'remarkable nature' of the pulses.

Mysteries in astronomy are not permanent. Further observations, and refined theories, will eventually lead to a satisfactory explanation of even the strangest objects in the Universe, and of the least-understood episodes in its history. But the effort of explaining the inexplicable is an exciting process: it is a detective story, with the astronomer as sleuth, combing the Universe in space and time for clues. Each investigation carries us closer to understanding the Universe as a whole; it is a faltering step, turning to a confident stride – until we are brought up short by the next unknown.

This book is a round-up of some mysterious aspects of the Universe, as I see them. Inevitably, the choice has been subjective. To do full justice to the subject would require a whole book on each section of each chapter in this volume, and it would be never-ending as new discoveries are made. I must therefore apologise to those astronomers whose work is not mentioned here. There are, for example, many more types of interesting star than I have had room for; but the net has been cast wide: from cosmic influences on the Earth to the search for life on Mars and the possibility of life elsewhere in the Universe; from the embarrassing lack of neutrinos from the Sun to Voyager's epic journey to Saturn and then out to Pluto; from pulsars to black holes; from spiral galaxies with their hidden 'missing mass' to distant quasars, the most violent of all the denizens of the Universe. And, finally, we consider the beginning of the Universe itself, and its ultimate fate.

Every few months sees our understanding of some of these mysteries advanced. The most puzzling questions, however, will undoubtedly succumb only after many years of dedicated onslaught. A good mystery story is above all an exciting read, and this account is an attempt, not just to describe the frontiers of present-day knowledge, but also to convey the flavour and excitement of man's continuing quest to understand the Universe about him.

Nigel Henbest

CONTENTS

CHAPTER 1

BIRTH OF THE PLANETS

One summer morning in 1908 a second sun flashed into the skies of Siberia. The blue-white column of blinding light burned to death reindeer grazing the tundra. Under the fireball, trees were felled like matchsticks. Torn from their roots, they lay pointing outwards, spokes in a huge devastated circle some 50 kilometres (30 miles) across. A power from space had smitten the remote Tunguska valley. An explosion high in the air had shattered the eternal stillness of Siberian life with the force of a twelve megatonne bomb—600 Hiroshima-type atom bombs in one.

Whatever it was, no solid pieces from the exploding fireball hit the ground. In 1927 Leonid Kulik, the first scientist to investigate it, reached the remote site halfway between Mongolia and the Arctic Ocean, and to his surprise the centre of the ring of devastation was not marked by a blast-crater. The fireball must have shattered into microscopic fragments some 8500 metres (28,000 feet) up.

Science fiction writers have claimed that the Tunguska explosion was caused by a nuclear spaceship manned by beings from another planet—perhaps stopping off to refill its tanks from Lake Baikal, the largest freshwater lake on Earth. But all the supposed 'evidence' for this alien craft has either been dreamed up, or can be explained quite readily by natural phenomena. It is now certain that what exploded over the Tunguska valley was a natural visitor from beyond the realm of the planets—a comet. It must have approached Earth from the direction of the Sun and burned up, a rival to the Sun in the Russian skies.

But the scientific explanation belies its own mystery. What *is* a comet, and where does it come from? We cannot study directly the solid nucleus from which the comet's woolly head

Comet nuclei form a huge swarm far beyond Pluto's orbit. Occasionally, one of these snowballs (*first inset*) takes an orbit in towards the Sun. Its evaporating gases make a planet-sized 'head' (*second inset*); they may stream away in two tails up to a billion kilometres long (*third inset*) at the comet's closest approach to the Sun.

and spectacular tail grow, because it is just too small to be seen even with powerful telescopes. Only by inference, by scientific detective work, can we surmise what gives rise to a visible comet. Even more mysterious is its origin. Comets' orbits bring them in towards the Sun from out beyond the Sun's family of planets, from a region further off than even remote Pluto. Is there a huge cloud of them out there, invisible because of their distance? Does the Sun pick up comets as it circles around in the Milky Way? Or did they form at the same time as the planets?

Most astronomers subscribe to the latter view. To understand the comets, we must first, however, know more about the Sun's family of planets and how it began. And, strangely enough, we now know more about the birth of the entire Universe than about the formation of the planets in our small pocket of space. Earth is just one of the Sun's family of planets, and the Sun is just an average star, similar to thousands of very remote stars we can see in the night sky. Understanding the birth of our planetary system is vital, not just because Earth is a planet itself, but because debris from the solar system's birth can affect us dramatically—as Tunguska valley saw in 1908.

PLANETARY NEIGHBOURS

Our viewpoint in space is Earth, third planet out, which circles the Sun once a year. The massive Sun controls all nine planets in the solar system, its immense gravitational pull keeping them in almost circular orbits about itself. Within the Sun there is a thousand times as much matter as in all its attendant planets put together—the Sun would outweigh the rest of the solar system a thousand times over.

Just as important to us, the Sun is the only body in the solar system to shine in its own right. Light and heat pour from its incandescent surface—at a temperature of 5500°C—illuminating and warming its planetary family. The Sun has many mysteries of its own, but it does not hold many clues for the scientist seeking to solve the mystery of the planets' formation. The planets themselves, and the debris leftover from their birth, must be the lodestone in this quest.

Leaving aside individual quirks for the moment, it is clear that the planets fall into two categories. The four nearest to the Sun are Earth-like: astronomers call them the terrestrial planets. They are comparatively small (on the planetary scale), have solid surfaces, and are made of rock throughout, apart from a central core of iron-rich metal. Closest to the Sun, truly in the hot seat, is Mercury, smallest of the four. Next out is Venus, roughly Earth's size, whose cloud-laden atmosphere veiled its surface from astronomers until pierced by spaceprobes and radar beams in recent years. Third planet Earth is unique in many ways—including having a species of intelligent life indigenous to it. Further out is Mars, a rusty-red desert of a world.

In the outer parts of the Sun's system are four very different planets. All are much larger than Earth, and they do not have solid surfaces, their bulk being composed of substances we find as gases on Earth: hydrogen, helium, ammonia and methane. But towards the centre, the weight of overlaying layers crushes the gases until they behave more like a liquid. Only right at the centre may there be a solid core, no larger than Earth. Greatest of these planets is Jupiter, which contains more matter than all the other planets put together. Its strong gravity controls 15 circling moons. Further from the Sun is Saturn, with its beautiful rings; the quartet is completed by Uranus and Neptune, two more 'gas giants', so distant that little is known about them.

Beyond the giant planets, and patrolling the boundaries of the solar system, is another small planet, Pluto. Farther out still, somewhere towards the nearest stars, are the comets. Coming in nearer to the Sun again, we come across the final members of the solar system. The asteroids (or minor planets) are chunks of rock smaller than any planet, and they infest space between the major planets, interplanetary debris posing a threat to spacecraft wending the paths of the solar system. Most asteroids, however, are well-behaved enough to confine their wanderings to a belt which lies between the orbits of Mars and Jupiter.

The Sun's systematic arrangement of planets demands explanation. For a start there is the sharp division into the four inner rocky planets and the four enormous gas giants in the outer reaches of the solar system. But there is also a plan, a mathematical blueprint, to their arrangement and motion. All the planets necessarily orbit the Sun under the influence of its gravity, the time for one round increasing from 88 days for Mercury to 248 years in the case of Pluto. Yet they all orbit in the same plane—draw a flat sheet in space which contains the Earth's orbit, and the orbits of the other planets will lie practically on the same sheet, without looping much either above or below. They all also travel around the Sun in the same direction: counterclockwise if we view them from above the Sun's north pole. And the orbits are all nearly circular, despite the fact that the law of gravity permits elongated orbits, too.

We are so used to seeing diagrams of the solar system on flat posters and the pages of books that this arrangement seems obvious. But is it? In clusters of stars, for example, each star pursues its own elongated orbit, and these orbits are orientated at random in space. As a result, stars travel from the centre of the cluster to the outside, mixing with other stars to emerge quite possibly at a totally different part of the cluster, and their orbits fill a spherical volume of space instead of being confined to one plane.

The Titius-Bode Law

The orderly arrangement of the solar system appealed irresistibly to the mechanically-minded scientists of the eighteenth and nineteenth centuries, and they built beautiful mechanical models of the planets' motions. On turning the handle of one of these orreries, the planets would move about the Sun, geared in such a way that they travelled at the correct relative rates.

And while astronomers were looking at the arrangement of the planets as a simple 'mechanical' system, they discovered that the sense of order in the solar system is so highly developed that the sizes of the planets' orbits follow a rule. The first man to notice this regularity was the German mathematician, Johann Daniel Titius. He pointed out that the sizes of the orbits could be set down as a series: call the distance of the Earth from the Sun 10, and the distances of the planets known at this time—Mercury to Saturn—gave the following reading (roughly) 3.9, 7.2, 10, 15, 52, 95.

Is there any meaning to the series? In 1766 Titius found one, which enabled him to predict other numbers in the series. Take the series 3, 6, 12, 24 . . . (where each number is twice its predecessor), put 0 at the beginning, and add 4 to each number. The series now becomes 4, 7, 10, 16, 28, 52, 100. . . . Leaving aside that 28 in the middle, it's a fairly close approximation of the planets' distances from the sun. Titius did not publicize his discovery—that was left to his contemporary, Johann Bode, director of the Berlin Observatory. So for a long time it was known, rather unfairly, as Bode's Law; only recently has the name been changed to the Titius–Bode Law, fairer even if more cumbersome.

It was rather a mathematical curiosity at the time. You could have taken any number to begin with—why 3?—and that odd zero at the beginning rather spoils the neatness of the series—it ought to be $1\frac{1}{2}$ (half of 3). And then there was 28, which did not fit any planet known at that time. Given a string of six numbers, a mathematician would not have had too much difficulty in finding a series to fit them if he had had as much leeway as Titius had.

But things changed in 1781. In that year William Herschel discovered Uranus, the first new planet to be found since the dawn of civilization. Its distance from the Sun was 192 (in the units we have been using); the next number in Titius' series was 196. The coincidence was just too strong. Many astronomers felt that the Law was a real feature of the solar system, and a team banded together as the 'celestial police' to search for a faint planet between the orbits of Mars and Jupiter, where the number 28 fell in the series.

Despite their efforts, it was an outsider to their force, Giuseppe Piazzi at Palermo in Sicily, who discovered the minor planet Ceres on 1 January, 1801. The celestial police continued their investigations, however, and turned in a further three minor planets. It became evident that a whole belt of minor planets (asteroids) lay between the orbits of

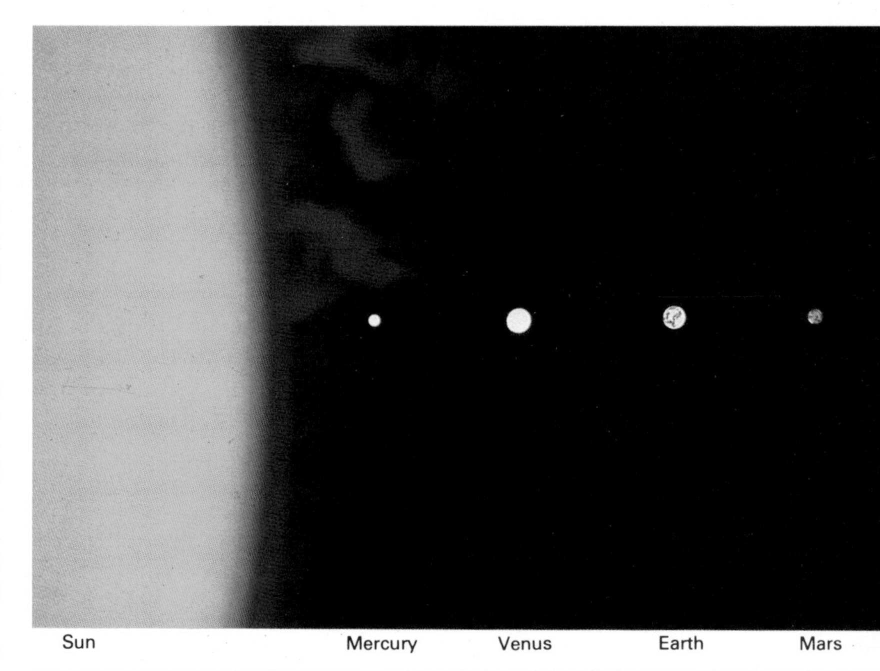

Solar system table

		Mercury	Venus
Average distance from Sun (millions of kilometres)		58	108
Period of revolution about Sun		88 days	225 days
Rotation period relative to stars (W to E, unless otherwise stated)		59 days	243 days (E to W)
Tilt of axis		2°	2°
Day length (sunrise to sunrise)		176 days	120 days
Diameter at equator (kilometres)		4878	12,103
Mass (relative to Earth)		0.055	0.81
Density (relative to water)		5.5	5.2
Temperature (°C)	surface:	350 (day) −170 (night)	465
	cloud tops:		
Number of known satellites (as of 1980)		0	0

Earth	Mars	Jupiter	Saturn	Uranus	Neptune	Pluto
150	228	778	1427	2870	4497	5900
365 days	687 days	11.9 years	29.5 years	84 years	165 years	248 years
23 hr 56 min	24hr 37min	9hr 50 min (clouds) 9hr 55min (core)	10hr 14min (clouds) 10hr 39min (core)	16 hours (?) (E to W)	22 hours (?)	6 days 9 hours (E to W)
23½°	25°	3°	27°	82°	29°	50°
24 hours	24hr 39min	9hr 50min (clouds)	10hr 14min (clouds)	16 hours (?)	22 hours (?)	6 days 9 hours
12,756	6794	142,796	120,000	50,800	48,600	3600
1.00	0.11	318	95	15	17	0.002
5.5	3.9	1.3	0.7	1.3	1.7	0.5
15	−23					−230
		−150	−180	−210	−220	
1	2	15	15	5	2	1

Mars and Jupiter. On the scale we have been using, the largest of them, Ceres, lies 28 units from the Sun.

In spite of these successes, the Titius–Bode Law does not hold for the outermost two planets, Neptune and Pluto. They were discovered later, and instead of lying at 388 and 772 units as predicted by the Law, they fell in at 300 and 394. Pluto is virtually where Neptune 'should' be, while Neptune does not fit the sequence at all.

The Titius–Bode Law is a mystery. There is still argument today over whether it is simply coincidence that most of the planets follow the Law—it could be, for instance, that the spacing of the planets' orbits increases with distance from the Sun, in which case some sort of mathematical series should produce a rough fit anyway. Some astronomers have pointed out that it is possible to choose other series which are simpler, in mathematical terms: for instance, say each planet's orbit is 73 percent larger than that of the planet next nearest to the Sun. This series gives a better agreement for the orbits of Neptune and Pluto, although it doesn't work quite so well for the innermost planets.

The mysteries of rotation

The reasons for the solar system's orderliness must lie in the time when the planets formed, from a cloud of gas and dust surrounding the embryonic Sun. And there are two further problems that require solution, both concerning *rotation* in the solar system.

The first is the astonishing fact that virtually

The distances of successive planets from the Sun increase in a fairly regular way. Two mathematical series which fit these distances reasonably well are here compared to the actual spacings. One series is derived simply by increasing the previous distance by 73 percent; the other is the Titius–Bode Law, explained on page 10.

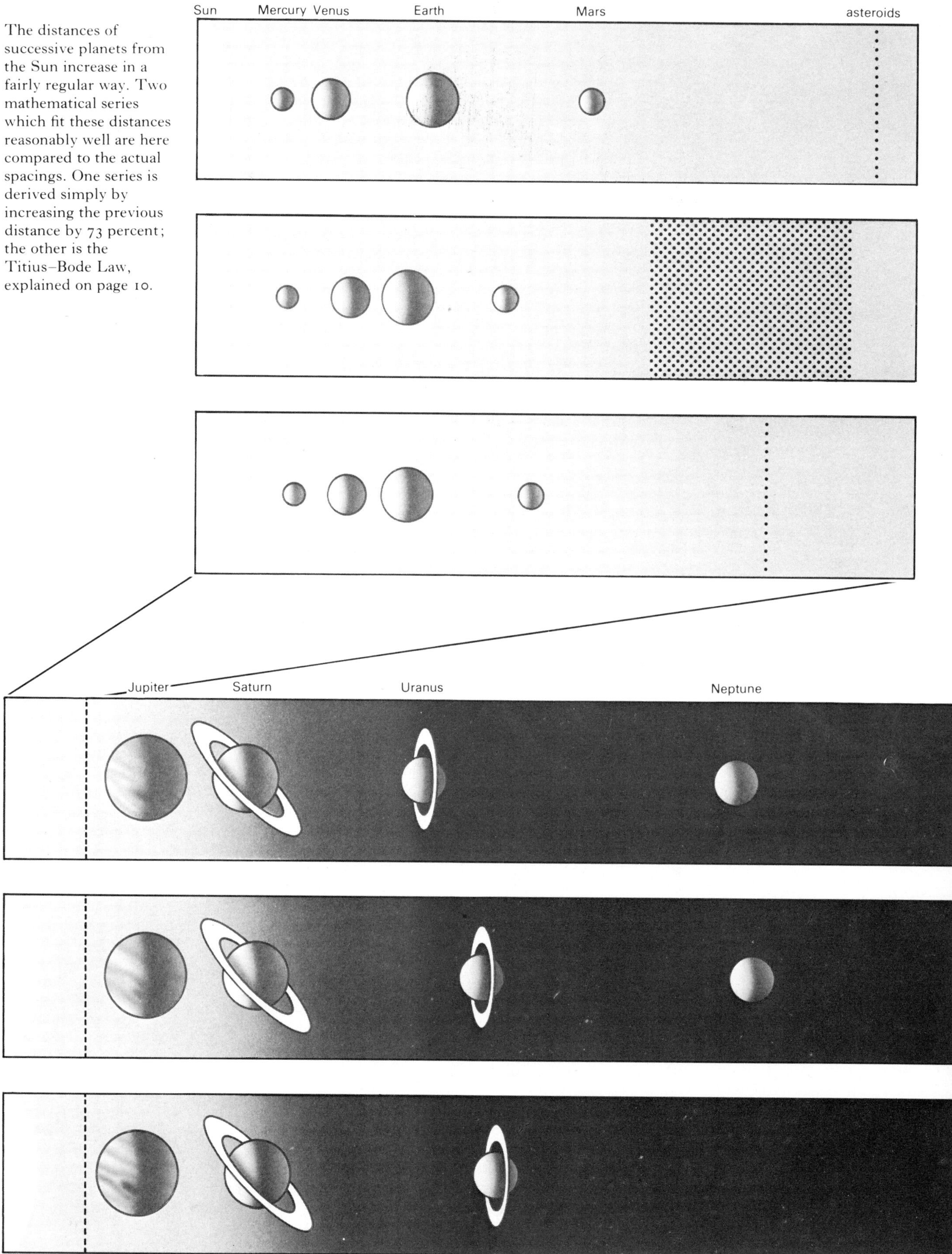

all the planets—from the minutest asteroid to giant Jupiter—rotate in approximately the same time, about 10 hours. Jupiter rotates in 9 hours 55 minutes, and Saturn in 10 hours 39 minutes. At the other end of the scale, the average asteroid spins once in about 9 hours. There is some variation on this average—tiny Icarus, for example, turns in only 2¼ hours—but that is not too surprising. We should expect many of this host of small objects to have had their spin-rates changed as they encountered the gravity of larger bodies over the billions of years of the solar system's history.

There are some exceptions to the rule, even among the planets, but generally these have also been caused by forces from outside. Mercury and Venus turn far more slowly, but only because they have been 'braked' by the Sun's strong gravitational pull in the inner part of the solar system. And Pluto turns only once in 6⅓ days, oddball in this as in many other ways. Earth does actually fit the cosmic rule—although our day is all of 24 hours long, there is a lot of 'rotation' (astronomers call it *angular momentum*) bound up in the Moon's orbit around us. If all that were added to the Earth's spin, our planet would also turn in a few hours.

Since their rotation periods are so similar, we might expect the planets all to turn in the same direction. They move around the Sun counterclockwise as seen from the north, and it seems natural they should rotate on their axes in the same way. Most do, but Venus and Uranus take a perverse pleasure in spinning clockwise. Uranus actually lies virtually on its side, with its equator almost at right angles to its orbit.

The other rotation problem is more subtle. Their orbital motion around the Sun gives the planets a tremendous quantity of 'rotation'—far more than the Sun itself possesses by its own turning. This may not seem too remarkable at first sight, but when it comes to calculating the birth pangs of the solar system, it is an enormous problem to explain. However you calculate it, the answer always comes out the same: the Sun should end up with most of the angular momentum of the solar system,

	Mercury	Venus	Earth	Mars	Asteroids	Jupiter	Saturn	Uranus	Neptune	Pluto
'73% Increase' Law	3.3	5.8	10	17.3	29.9	52	90	155	268	464
Actual (mean) orbit	3.9	7.2	10	15.2	27.7	52	95.4	192	300	394 (ranges 279–491)
Titius–Bode Law	4	7	10	16	28	52	100	196	388	772

Measurements in units of distance from the Sun.

Pluto

'73% Increase' Law

Actual (mean) orbit

Titius–Bode Law

The Horsehead Nebula in Orion is a huge cloud of gas and dust. The dust grains block off the light from the glowing gases behind, to produce the striking silhouette. The solar system condensed from a similar dense cloud some 4600 million years ago.

and the planetary motions with very little. As a result the Sun should rotate, not at its actual leisurely rate of once every 25 days, but whip around in only a few hours.

Most of these facts have been known for many years, although that does not mean that they have been understood. They could, however, provide clues to understanding how the solar system came about.

If we could travel back in time and see the solar system form, we would know exactly how all these regularities—and oddities—have worked out. But doing the opposite is not so easy. And despite the man-centuries of mind-stretching work which have been devoted to the problem, the answer is still not clear. By fitting together the clues in different ways, we have merely ended up with a plethora of theories.

BIRTHPANGS OF THE SOLAR SYSTEM

Astronomers, whatever their favourite theory, do agree on one fact: the Sun and planets formed at the same time, some 4600 million years ago. The naturally-occurring 'clock' which tells us this, is the gradual decay of unstable atoms, such as uranium, into lead. In principle, geologists can estimate the age of any rock by comparing the amounts of uranium and lead in it. But the rocks of the Earth and Moon do not tell us their *true* ages, for their 'clocks' have been reset by the later impact of meteorites, and by geological activity. The oldest known rocks are a type of dark crumbly meteorite that occasionally falls to Earth. These meteorites do not show any sign of traumatic events which could have 'reset' their clocks, so scientists believe they reveal their true ages. And they tell us that the solar system's birth took place 4600 million years before our own time.

The Sun and its planets were born from the gas and dust which litter space. Astronomers dispute exactly how this happened. There is however a 'standard picture'—most likely to be somewhere near the truth—but even here there is disagreement about the details.

In this standard picture, the gas falling inwards to make up our Sun was a rotating eddy, like the whorls in a river near rapids. The gas near the edge felt the tug of competing forces: gravity pulled it inwards, but the centrifugal force of its rotation tugged it away from the centre. In this merry celestial tug-of-war the result was roughly a draw. The rotating gas flattened into a thin rotating disc, of gramophone record proportions, with a higher concentration of gas at the centre. Although the disc was to become the planets, and the centre the Sun, at this stage there was no real division between the two.

But a gas which is compressed becomes hot. Ardent cyclists know that a bicycle pump heats up when a tyre is inflated, and that that has nothing to do with friction in the pump barrel. It happens simply because the air in the pump is being compressed by the force of the piston. Similarly, the spinning gas of the 'solar nebula' was being compressed by its own gravity, forcing it into a smaller volume.

This was not particularly important at the outermost edge of the disc, where the gases stayed near the frigid temperatures of interstellar space, only a few degrees above the absolute zero of temperature. Here, substances we know as liquids or gases—water, methane and ammonia—were frozen out as tiny solid crystals of ices, interstellar snowflakes whirling round the Sun in a disc consisting of hydrogen and helium gases. Planets forming here would naturally consist of these substances, and could attract large amounts of the gases around to build up huge gas giants. Thus were formed the outer planets, huge balls of hydrogen, helium and other volatile substances, with a small central rocky core which has inherited the small amounts of dust mixed in with the original gaseous disc.

Closer in to the Sun, it was hotter. Where the Earth is now, the temperature was a sizzling 300°C (500°F). Here the ices must have evaporated into gases, and only the rocky and metallic grains of the original cloud could survive as solids. As these grains came together in this part of the disc, they naturally made up rocky planets with iron cores—the terrestrial planets Mercury, Venus, Earth and Mars.

So far the standard theory works quite well—it certainly explains why the inner and outer planets became so different from one another. But what has happened to all the gases that should be hanging around in the inner parts of the solar system after the dust grains separated out into the planets? There turns out to be a simple answer, which also neatly ties up the problem of the system's excessive amount of rotation—for the picture so far would leave the Sun turning in only a few hours.

Other stars in the sky can provide the clue, for among the huge number visible in telescopes there must be some which have only recently been born. If we can identify these, they should tell us something about the Sun's condition when it was young. Some stars draw attention to themselves by varying in brightness. These *variable stars* can be inconstant for a number of reasons: some rhythmically expand and contract, others have their light blocked off as a companion star periodically moves across in front of them. But some change irregularly, suddenly dimming, and then without warning brightening again. Astronomers believe this flickering to be caused by veils of dust moving around the star and unpredictably blocking its light. The best-known of these stars is called T Tauri, and lies in the constellation of Taurus, the bull. The whole class of 'T Tauri stars' seems to be young on the solar system timescale; they are also unusually bright and are shedding gases out into space. In the T Tauri stars we are probably seeing a re-enactment of the early days of the Sun's history, just after the planets had formed, and when the remaining dust in the rotating disc formed irregular veils sweeping across its face.

If our Sun was indeed a T Tauri star in its extreme youth, then the gases ripping from its surface out into space would have swept out the gas remaining between the orbits of the planets. Any of the original disc's matter which had not by now been incorporated into the planets would have been pushed out into further space by the force of the 'T Tauri wind' blowing outwards from the young and superenergetic Sun. Left behind would be the planets we see now, with virtually empty space between them. Indeed, interplanetary space is even now occupied by a gentle outflow of gases from the Sun's surface, a mild breeze compared to the gales of its violent youth.

In addition, the gases blowing outwards as the T Tauri wind would have carried away some of its rotation. In robbing the Sun of its angular momentum, the gusty wind would have left it as the sedate, languidly turning star we see today.

The formation of the planets

Even those who would lay their bets on the theory described above are in disagreement about the finer details. The broad outlines seem reasonable, but the details must also be filled in—and must work as expected. For example, the planets must have somehow accumulated into individual globes from the thinly spread haze of dust grains, 'snowflakes' and gas in the disc. How?

The simplest answer is that the dust grains—and in the outer solar system, the icy snowflakes—jostled together and stuck, building up fist-sized dustballs and snowballs. Now their gravity was strong enough to make them bunch together, in *'planetesimals'*, minor planets perhaps a dozen kilometres across. In this version of events, the asteroids circling the Sun between Mars and Jupiter are simply planetesimals that have survived since the solar system's birth. But the vast majority ended up in the planets for, as they wheeled around the Sun, they cannoned into one another and, in the smash, would often split into smaller bodies, but sometimes unite as larger ones. In the end, the planetesimals which had started largest would inherit the advantages of their birth: they would tend to accumulate the smaller ones and the broken-up fragments, and grow into the planets we know today.

Such a picture, of course, does not tell us why the planets are spaced as they are according to the Titius–Bode Law. A. J. R. Prentice of Monash University in Australia explains the Titius–Bode Law by an updated version of the first serious attempt to explain the solar system. The great French mathematician, Pierre Simon de Laplace, suggested in 1796 that the solar system could have formed as the early Sun contracted from an enormous size and threw off rings of gas from its equator. Each ring ultimately became a planet. His theory is in many ways like the rotating disc of

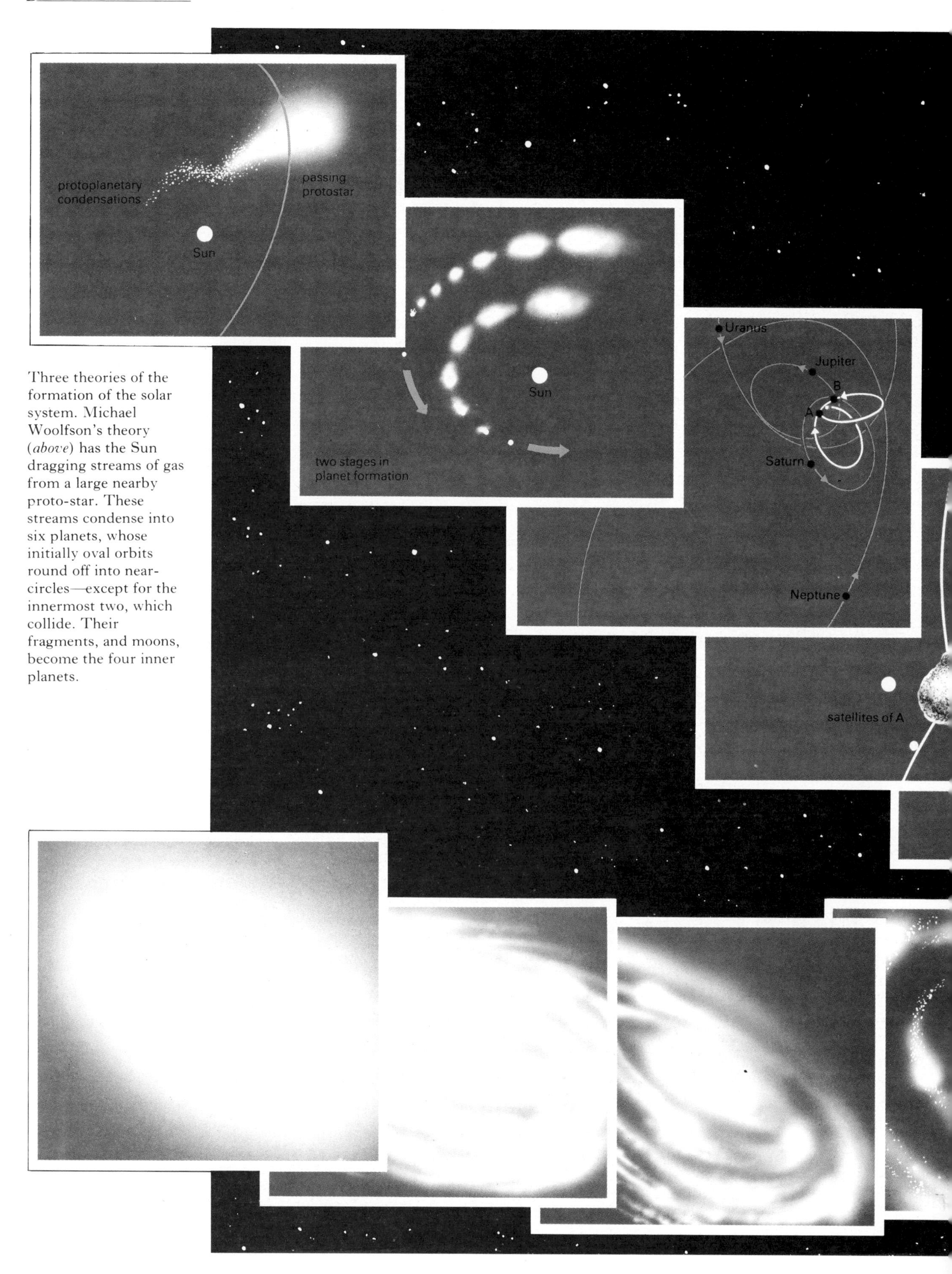

Three theories of the formation of the solar system. Michael Woolfson's theory (*above*) has the Sun dragging streams of gas from a large nearby proto-star. These streams condense into six planets, whose initially oval orbits round off into near-circles—except for the innermost two, which collide. Their fragments, and moons, become the four inner planets.

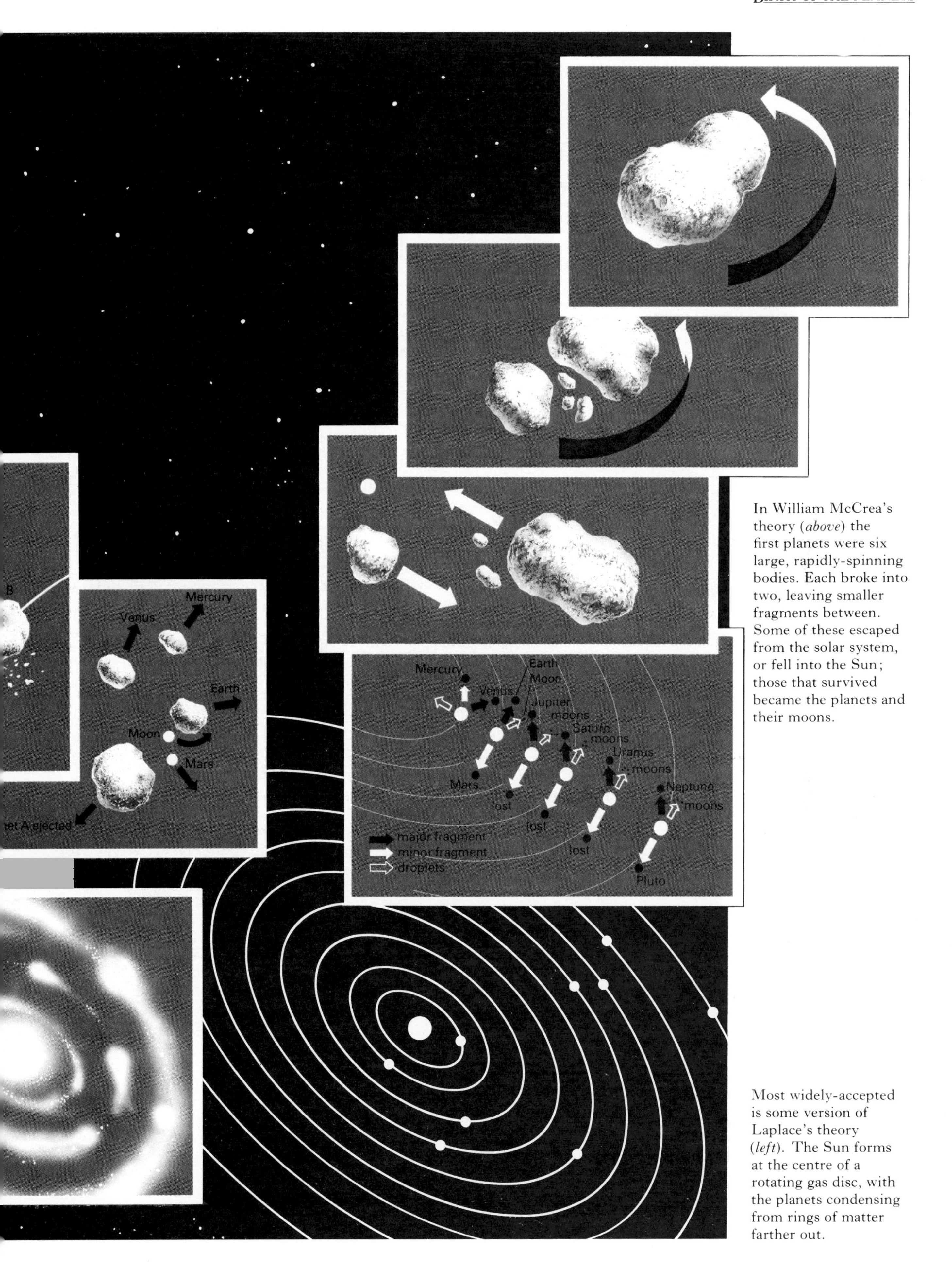

In William McCrea's theory (*above*) the first planets were six large, rapidly-spinning bodies. Each broke into two, leaving smaller fragments between. Some of these escaped from the solar system, or fell into the Sun; those that survived became the planets and their moons.

Most widely-accepted is some version of Laplace's theory (*left*). The Sun forms at the centre of a rotating gas disc, with the planets condensing from rings of matter farther out.

today's favourite theory, but it ran into problems with the rotation bugbear—most of the angular momentum should have ended up in the Sun. Laplace did not have the advantage of knowing about T Tauri stars, so there seemed no alternative but to ditch the theory.

Prentice resuscitates it, by using modern knowledge of how gas moves, quite beyond the understanding of Laplace's time. Rejecting the T Tauri wind explanation for the rotation puzzle, he shows that the gas *could* have ended up with most of the angular momentum in the outer reaches of the disc. As long as the central core collapses fast enough into the proto-Sun, and the gas moves in the way Prentice describes, then Laplace's long-rejected idea does work. As the disc contracts, the rings thrown off become more and more closely spaced, just as the planets' orbits are. And the calculations show that the orbits should increase in size outwards in a ratio close to the 73 percent actually found in the alternative form of the Titius–Bode Law.

This theory says that the planets did not all form at exactly the same time but, over a fairly short interval, were born in succession. The outermost formed first, then the inner ones as successive rings were cast off by the shrinking disc. A. G. W. Cameron of Harvard College Observatory also believes that the planets formed successively—but from the inside of the solar system outwards! Unlike Prentice, he thinks that the disc of the solar nebula had very little matter in the centre to begin with. As the planets formed, gas was all the time falling inwards past them towards the centre of the disc and as the central core grew to become the Sun, its increasing amount of matter gave it a continually stronger gravitational influence. Thus its increasing tug would have made the planets follow orbits which gradually decreased in size until the Sun was complete.

Cameron contends that Mercury was the first planet to form, at about the position where Jupiter's orbit is now. Rather than simply forming at its present size, Mercury became far larger according to this theory, a Jupiter-type world. But as it gradually came closer to the Sun, the Sun's gravitational pull stripped off its huge gaseous body, to leave only the rocky core which is Mercury today. At about this time, another planet would have started to condense further out, a giant world whose core would eventually become Venus. And so successive planets were produced, in order outwards from the Sun. According to these calculations, each orbit would have been a definite amount larger than the previous planet's, and the increase would have been close to 73 percent.

So two totally different ways of forming the planets from a rotating disc can—according to their proponents—naturally produce the properties of the solar system.

Swedish astronomer Hannes Alfvén puts forward a third theory. He believes that magnetic fields were as important as gravity in forming the planets. Since magnetism is produced by electric currents, his ideas have been dubbed those of the 'electricians', as contrasted with the 'dustmen' whose calculations concern only the coalescence of dust and ices under the influence of gravity.

Alfvén believes that the early Sun turned very quickly, had a strong magnetic field—but did not have planets. Within the first hundred million years of its life, however, it gradually drew in gas, and this gas was picked up by the whirling magnetic field around the rapidly-spinning Sun. As a result, the gas—which eventually coalesced into planets—acquired a rapid rotation around the Sun, while the Sun itself was slowed down. Thus Alfvén explains the unequal distribution of rotation in the solar system: unlike the dustmen who see the Sun's rotation lost in a gale of outblowing wind, the electricians believe it was transferred to matter falling *inwards* by the Sun's magnetic field.

It is fair to say that only a few astronomers subscribe to the electricians' point of view. But it is not easy to prove them wrong—and Alfvén has been very successful in proving the importance of magnetic fields elsewhere in the Universe. Only further painstaking research will show if magnetism is indeed the key to unlocking the riddle of the solar system's birth.

Theories on the fringe

There are smaller bodies in our solar system which may hold clues to its formation out of all proportion to their minor status. These are the satellites of the giant planets. The larger planets are not satisfied with just one moon, like Earth, or even two, as Mars has. Both Jupiter and Saturn have at least 15—it is impossible to give an exact number because faint new ones are found almost every year.

These families seem, at first sight, to be miniature solar systems. Like the planets orbiting the Sun, the satellites go around their controlling planet in the plane of its equator—even Uranus's five moons do this, though it is tipped over at right angles to its orbit about the Sun.

Those who believe that the satellite families really are mini-solar systems point out that they even have their own version of the Titius–Bode Law. In the case of the four main satellites of Jupiter, each orbit is larger than the preceding satellite by 65 percent; and Saturn has five consecutive moons, the orbits of which increase in size by 30 percent as one moves outwards.

Representatives of both the 'dustmen' and

the 'electricians' have proved (to their own satisfaction at any rate) that the formation of a gas giant planet would have been like the formation of the solar system on a smaller scale. And so each side argues that the processes which formed the planets would have worked here as well, to leave a Sun orbited by the mini-solar systems of the giant planets and their families.

Not everyone agrees. For a start, there is angular momentum. The Sun's spinning accounts for only about 1 percent of the 'rotation' of the solar system, with the rest being carried by the motion of nine planets around it. But in Jupiter's system the central body, rapidly-spinning Jupiter itself, accounts for fully 99 percent of the 'rotation'. The analogy between the two families obviously is not exact. Related to this is the fact that Jupiter's family is relatively much closer in. If we scaled Jupiter up to the size of the Sun, its satellites would become similar in size to the planets, but they would all be circling within the present orbit of Mercury.

As a result, some astronomers say the formation of the satellites was completely different. One of these is William McCrea of the University of Sussex, who puts forward one of the more unusual theories of the solar system's birth. He dispenses altogether with the idea that the forming Sun was surrounded by a spinning disc. Instead, the gas cloud from which the Sun and its companion stars formed would have been full of very small eddies, and the gas in these could have combined to make either stars or, when less of them coalesced, planets. If the planets happened to form near a young star they would be captured and end up as its family, otherwise they would end up undetectably wandering about between the stars.

The accumulations of eddies started off as large and rapidly rotating 'proto-planets'. As they contracted, they spun faster and faster until the proto-planet stretched so much it broke in two. According to the calculations, the smaller fragment would end up with roughly one-tenth as much matter as the larger; and in the outer part of the solar system where the Sun's pull is weaker, the smaller fragment would have shot off to escape from the solar system altogether.

Today's family of planets would then have resulted from six proto-planets. The innermost split into Venus and Mercury, the latter with one-fifteenth as much matter as Venus. The major portion of the second became Earth, the smaller part (one-ninth), Mars. The smaller fragments of the outer planets would have escaped—with the possible exception of Neptune, where the smaller fragment could be Pluto. One or more much smaller droplets should have been left over from the split-up, and these could have gone into orbit around the larger fragments. Hence the Earth and the four giant planets have comparatively large satellite companions—Venus's droplet must have fallen into the Sun.

It is fair to say that very few astronomers set any store by this version of events, but it does go to show that the solar system *could* have evolved in a way very different from the standard picture. A canvas where huge, rapidly-whirling proto-planets are rupturing into fragments is certainly exciting, with an appeal to the imagination far stronger than the placidly turning gas disc of the standard picture.

And while on the outer fringes of orthodoxy—though still remaining with ideas which are scientifically sound—there is another theory which conflicts with the standard story in an unusual and appealing way. Michael Wolfson of York University in England has returned to an old idea that the planets could have formed from a filament of gas pulled from a star—but not, in this case, from the Sun itself. When the Sun formed, it was in a cluster of stars, all comparatively closely packed compared to the present distances between the Sun and its neighbouring stars. Although the stars in the clusters formed almost simultaneously, there must have been a short interval when the Sun was fully formed—and in Woolfson's picture without a surrounding gas disc or planets—but some of its neighbours were still large contracting balls of gas. There is something like a one-in-a-hundred chance that the Sun would have approached one of these fluffy proto-stars so closely that it could have pulled off the proto-star's outermost shreds of gas.

The long filament of gas would then have fallen in towards the Sun, breaking up into six blobs as it did so. And because the Sun and the proto-star were passing, the gas would have had a sideways motion which would have prevented it from just falling straight in and ending up at the Sun's surface. Instead, the blobs would have gone into orbit around the Sun and formed into six large planets; because they were spinning fast, their outer matter would have broken off and condensed into satellites.

In this unconventional picture, then, the similarity between satellite families and the solar system itself is mere coincidence. Planets are *not* a natural part of star formation; they only occur in a small proportion of cases, when a newly-formed star disrupts a nearby neighbour.

Most astronomers do not take this 'filament' theory very seriously, however. It has problems accounting for the inner planets, including Earth. Calculations show that the innermost of the blobs from the condensed filament would lie at about Mars's distance from the Sun. Woolfson ingeniously gets around this by putting a planet five times heavier than the Earth where Mars now lies, and another 30 times heavier than Earth in the region of the present asteroid belt. By letting them have oval orbits, he can let them collide in a cosmic pile-up which splits the smaller planet in two. These fragments end up closer to the Sun, as Venus and Earth. The larger planet was speeded up so much that it shot out of the solar system altogether. Mercury is then a still smaller chunk of the broken-up planet, while the Moon and Mars are former satellites of the larger planet.

It seems rather too much like jiggery-pokery to explain the orderly arrangement of the inner planets as the end result of a messy pile-up. Yet our solar system hides its past so well that it is impossible to rule this idea out at the moment.

THE RUBBLE OF SPACE

To discover the most important clues to our solar system's birth, astronomers now think we must turn to the smallest members of the system. Not to the planets themselves, nor even to their satellites, but to rock fragments that orbit the Sun between Mars and Jupiter as the asteroids, and to the distant icebergs of the comets. The smallest of these bodies are only a few kilometres across, and they should have changed least since the solar system formed: their composition should therefore reflect the make-up of the original whirling gas-and-dust disc.

The asteroids number some 50,000, according to latest estimates, although only 2000 have had their orbits accurately calculated. They are probably planetesimals left over from the formation of the planets (although one or two astronomers still believe an older theory that they are the remains of an exploded planet). The gravitational pull of neighbouring giant Jupiter prevented the planetesimals here from coming together as a planet, but even if they had, the planet would have been minute; the amount of matter in the asteroid belt would build a planet only half the size of Earth's Moon.

Comet West was discovered by R. M. West of the European Southern Observatory, Chile, in November 1975. As it neared the Sun, Comet West grew two tails: in the true-colour picture (*below*) the gas tail appears blue and the dust tail yellowish. The false-colour view (*bottom left*) is computer-processed to show the brightness levels more clearly. After passing the Sun in February 1976, Comet West was a brilliant sight in the morning sky, and as it retreated its nucleus broke into four pieces.

Controversy currently centres around whether some asteroids are double—twins orbiting each other in perpetual gravitational embrace as they journey round the Sun together. When asteroid Herculina passed in front of a distant star in June 1978, astronomers at the right place on Earth measured for how long the star disappeared. They then calculated that Herculina is an oval body some 240 by 210 kilometres (150 by 130 miles) in size. But the star also disappeared briefly a few minutes before the true occultation, which suggests that it must have a smaller companion some 1000 kilometres (620 miles) from it: at one-quarter Herculina's size, should it be called a satellite of the asteroid, or should Herculina be classified as a slightly unequal 'double asteroid'?

Not everyone is convinced of the reality of the reported secondary disappearance of the star, even though the asteroid Melpomene produced a similar event the following year. Even more bizarre, though, is the suggestion that some very elongated asteroids are the end result of a double asteroid system, where the two bodies have spiralled in towards each other until they have touched together as a cosmic dumb-bell.

Until 1977 the most far-ranging asteroid known was Hidalgo, which strays from Mars's orbit right out past Jupiter to reach Saturn's orbit. In that year, Charles Kowal of the Hale Observatories in California noticed a faint streak on a photographic plate—a streak that turned out to be the track of an asteroid circling in the farther reaches of the solar system. Some 200 kilometres (120 miles) across, the new asteroid orbits the Sun between Saturn and Uranus. Kowal named it Chiron, after the wise centaur, half-man half-horse of Greek mythology.

What Chiron is remains a mystery. Does a second asteroid belt exist in the outer regions of our planetary system, with Chiron its brightest member? Or is it an ordinary asteroid whirled out from the asteroid belt to a new habitat by the gravitational pull of a neighbouring planet? Recent calculations have proved that wherever Chiron came from, it cannot remain in its present orbit for ever. Within a few million years, the gravitational tug of one of the giant planets will swing it into a new path taking it out of our solar system altogether, into the vastness of interstellar space.

Asteroids

The asteroids—or minor planets—circle the Sun between the orbits of Mars and Jupiter. They are probably planetesimals which never formed into a planet; indeed, collisions have now broken most of them up again. Composition depends on which part of the disrupted planetesimal they once formed: iron asteroids are from the core region, rocky ones from farther out, while those that are dark and carbon-rich came from the original surfaces.

Asteroids farther from the Sun are generally darker in colour; they are less easy to spot from Earth and so some were discovered comparatively late.

A few asteroids have unusual orbits. Some cross the orbit of Earth, and so constitute a potential hazard to our planet. They may be the dead cores of old comets, rather than true asteroids from the main belt. The Trojan asteroids follow Jupiter's orbit about the Sun, both 60° ahead and 60° behind the planet. Chiron, discovered in 1977, is unique in orbiting so far from the Sun—on average, farther out than Saturn.

Number (in order of discovery)	Name	Average distance from Sun (astronomical units)	Diameter (km)	Type
The largest main belt asteroids				
1	Ceres	2.79	1025	carbon-rich
2	Pallas	2.77	583	rocky
4	Vesta	2.36	555	rocky
10	Hygeia	3.14	443	carbon-rich
704	Interamnia	3.06	338	rocky
511	Davida	3.18	335	carbon-rich
Earth-crossing asteroids (35 known)				
1566	Icarus	1.08	2	rocky
2062	Aten	0.97	1	iron/rocky
Trojan asteroids (22 known)				
624	Hektor	5.15	300 × 150	uncertain
1437	Diomedes	5.08	191	carbon-rich
Chiron (unique)				
2060	Chiron	13.7	100 *or* 320	icy-surface carbon-rich

The comets

Beyond the confines of our planetary system set by the orbit of Pluto, the furthest planet now known, are other dwarf denizens of the Sun's family. The comets, like the asteroids, have probably remained unaltered from the earliest days of the solar system. Yet they differ, for they were formed in the outer reaches where ices of water, methane and ammonia abound; unlike the rocky asteroids, they are lumps of ice, 'dirty snowballs' in astronomers' jargon.

Although a comet appears as a fuzzy head wielding a long tail when it is close to the Sun, this is all extravagant show concealing its true identity. In its original habitat beyond the planets, a comet is a mere speck on the astronomical scene, a small 'iceberg' some 10 kilometres (6 miles) across—smaller than the tiny moons of Mars. Being so minute, there is not much light reflected from a comet's nucleus and it is not possible for us to see it directly.

If comets stayed out here, in large circular orbits roughly halfway to the nearest star, we would not know much about them. But occasionally the gravitational pull of other stars will perturb one from its course: some may wander away from the solar system altogether; others will begin to topple in towards the Sun. In the latter case, its orbit will be an elongated oval, taking it close to the Sun and then back into the remote depths of space.

Once a comet gets within the orbit of Jupiter, the Sun's warmth begins to melt the ices. In the near vacuum of interplanetary space they evaporate off the comet's nucleus spreading out into a huge cloud around it—a cloud which sometimes grows as large as the Sun, but contains only a tiny amount of gas thinly spread. It is as a round fuzzy ball that a new comet swims into human view.

Many comets remain like this as they swing round the Sun and head back into the depths of space. But the spectacular ones enliven our skies by growing a tail, sometimes as long as the distance from the Earth to the Sun, the popular hallmark of the comet. In fact, they usually come equipped not with a single tail, but with two. The force of sunlight itself can push on the tiny dust particles in the comet's head and drive them out in an elongated dusty tail; while the stream of electrically charged particles constantly flowing from the Sun's surface out through the solar system (the 'solar wind') picks up gas from the comet's head and drives this out as a second tail. The tails point in slightly different directions, but both must be directed more or less away from the Sun, since it is energy from the Sun which pushes them out from the comet's head. As it retreats from the Sun, the tails are thus sticking out in front of it, and a comet chases its own tails as it heads back to beyond the planets.

It is now virtually agreed that comets do indeed grow from such a small dirty snowball of a nucleus, but it is difficult to prove since the

nucleus is too small to see directly. Raymond Lyttleton of Cambridge University still upholds an older theory, in which the fuzzy cometary head is a crowd of small dust and ice particles linked only by their own gravity: a huge flying sandbank orbiting the Sun. This controversy may be resolved once and for all in 1986, when spaceprobes are scheduled to fly by Halley's comet at close range.

Some astronomers do not believe that comets come from a permanent reservoir in the outer parts of the solar system. If they do, there must be something like 100,000 million of them out there, a huge number, but because each is so small they do not total any more matter than Jupiter. William McCrea, whose planetary theory we have already considered, holds that comets are swept up from the clouds of gas between the stars when the Sun passes through one. We may be privileged to be living in one of those rare times when the Sun has just swept up a new batch, and bright comets are a comparatively common sight.

In close-up, an asteroid would probably resemble Mars's moon Phobos, seen right in a Viking Orbiter photograph. Phobos is similar in size to the smaller asteroids and its surface reflects light in a similar way to some of the darker asteroids.

Halley's comet at its last passage by the Sun in 1910 (*below*—the illuminated squares are superimposed to aid measurement). At its next return, in early 1986, the comet will be comparatively faint, and not easily visible to the unaided eye.

Once a comet is in its elongated orbit, it will continue to follow that path. But it takes so long that its spectacular returns to the Sun will be separated by millions of years. There are two things that can change this leisurely pace of life, though, and make an orbit so small that it stays within the realm of the planets. If a comet happens to pass too close to a giant planet, the latter's gravitational pull can swing it into a new orbit whose furthest point from the Sun is roughly where that planet's orbit lies. Jupiter is master at this game, with a 'family' of some 45 comets which it has perturbed in this way.

The other is more subtle. As the Sun warms a comet's nucleus, the evaporating gases push back against it. The effect of this is only small, but it builds up dramatically over time. When you take into account the fact that the nucleus must be spinning (one recent estimate concluded that it turns a thousand times a second), the comet is either speeded up or slowed down in its orbit in the process. In the first case it can acquire enough speed to leave the solar system; if, instead, it is slowed, its orbit will rapidly shrink until its furthest point is reduced from 100,000 to only 100 times larger than the Earth's orbit.

Thus, comets eventually become members of the inner solar system, and their days are then numbered. At each passage of the Sun, they lose 1 percent of their matter, evaporated off and distributed through interplanetary space. So each time they approach the Sun, they become smaller and fainter. The most famous comet of all, Halley's, has suffered this fate. Edmund Halley did not actually discover 'his' comet, but he did realize that several which had blazed in the skies during the previous four centuries were actually reappearances of the same object every 76 years. In Chinese historical records, Halley's comet has now been traced back to 1057 BC, when it appeared during a winter campaign of King Wu—it has a good sense of military timing, in fact, for it also put in an appearance in AD 1066 in England, just before the Battle of Hastings. At the first recorded sighting, it was extremely bright, outshining all the planets, but it has now faded. Partly because of its gradual dissipation, and partly because it will be approaching the Sun from an awkward angle

Debris from an old disrupted comet appears as a shower of meteors (shooting stars) (*above left*) when the Earth crosses the comet's orbit.

A 250,000 tonne iron meteorite survived the fall through our atmosphere and blasted out the Arizona Meteorite Crater (*above right*) some 50,000 years ago. Small pieces of meteoritic iron are still found nearby (*right*).

A polarized-light view of a meteorite section (*below*) reveals the small rock grains which accumulated to make its parent body in the asteroid belt.

for viewers on Earth, its next scheduled appearance in 1986 will be a disappointing affair. Rather than a glorious sky-stretching spectacle, it will appear only as a faint patch of light, not even as bright as the brightest stars—in many ways a disappointment, for its 76-year orbit means that the 1986 appearance will be the only one visible for most people living today.

We can all expect to see at least one bright comet, though, for those on long orbits constantly appear unexpectedly. Around 600 comets have been catalogued, of which about 100 are 'new', making their first expedition to the solar system's warmer realms after being perturbed from their home cloud. Another hundred comets circle in the inner solar system, dim objects on the verge of extinction after losing most of their gases. The others are in between: their furthest points lie in the region of the outermost planets, and they have yet to experience the planetary pull which will embark them on the still smaller orbits which will herald their demise.

In its final passages past the Sun, a comet

sheds huge amounts of gas and dust into space—comet Kohoutek of Christmas 1973, for example, lost four million tonnes of matter every second. The dust spreads around the orbit, making an oval ring of grains in the inner solar system. The Earth plunges through many of these rings—some are the mortal remains of comets which are now totally dissipated—on its yearly journey around the Sun. The dust grains spray into our atmosphere, and burn up as a 'shower' of meteors.

Meteor showers are a regular fixture in the astronomer's calendar. They always occur on the same day, and the meteors appear to spread out from the same constellation each year. But the actual number cannot be predicted. The grains are not spread evenly around the orbit, and if the Earth encounters a bunch of them, the effect can be stupendous. On 17 November 1966 Earth passed through the debris ring from comet Tempel–Tuttle, and astronomers turned out to watch the meteors expected. But no one foresaw so many: astonished observers in the United States saw a veritable storm, shooting stars raining down at a rate of 40 *every second*.

Eventually, however, a comet must die. Some may disperse entirely into dust; others may leave a rocky core devoid of gases and ices.

Astronomers know of some 20 'asteroids' which live in the inner part of the solar system, following elliptical paths which cross the Earth's orbit, and it is most likely that these are the skeletons of dead comets.

One comet that is well on the way to joining them is Encke's. In 1819 Johann Encke calculated that a comet seen the previous year had the shortest orbit of any known, and it still holds this record. It returns to the Sun every $3\frac{1}{3}$ years, transgressing inside Mercury's orbit after reaching a far point near Jupiter. Since its discovery, the amount of gas and dust it has lost at each return has dwindled from around 10 million tonnes to under one million. Its reservoirs are reaching exhaustion point, and within the next century it will probably lose its final reserves to become a dark, asteroid-like object, only a couple of kilometres across—and without its glowing gaseous head, it will be an inconspicuous speck.

Because its passages by the Sun are so frequent, Encke's comet still deposits a lot of dust in the inner solar system. Dust from Encke reaches the Earth as the Taurid meteor shower on 8 November each year. Interestingly enough, the Tunguska fireball which devastated the frozen Russian forests in 1908 appeared to come from a similar direction. As we have seen, this object exploded above the Earth, breaking up in the atmosphere. It must have been a fragile object, despite a size of 100 metres (300 feet) and a weight of almost a million tonnes (estimated from the brightness of the explosion). Its fragility suggests it was a small cometary nucleus. Lubor Kresák from Czechoslovakia suggests that the Tunguska 'comet' was actually a small fragment of Encke's comet, broken off in its final demise, and still following the same path through the inner solar system.

The devastation of Tunguska was thus a visitation from the earliest days of the solar system; comet Encke was born right at the birth of the Sun's family. It circled in the deep-freeze of the outer solar system for aeons, until the unkind nudge of a passing star's gravity sent it in towards the Sun. First by its ice's evaporation, then by the pull of Jupiter, it was forced to follow an ever smaller orbit. Passing the Sun so frequently its doom was sealed. Final demise is imminent. And in its break-up, a small fragment ended up on collision course with the Earth.

The original form of our solar system is not just a question for academics to pursue over coffee and busy computers. The debris which for some reason did not succeed in becoming part of a planet still threatens us on Earth today: the Tunguska valley in 1908, where tomorrow?

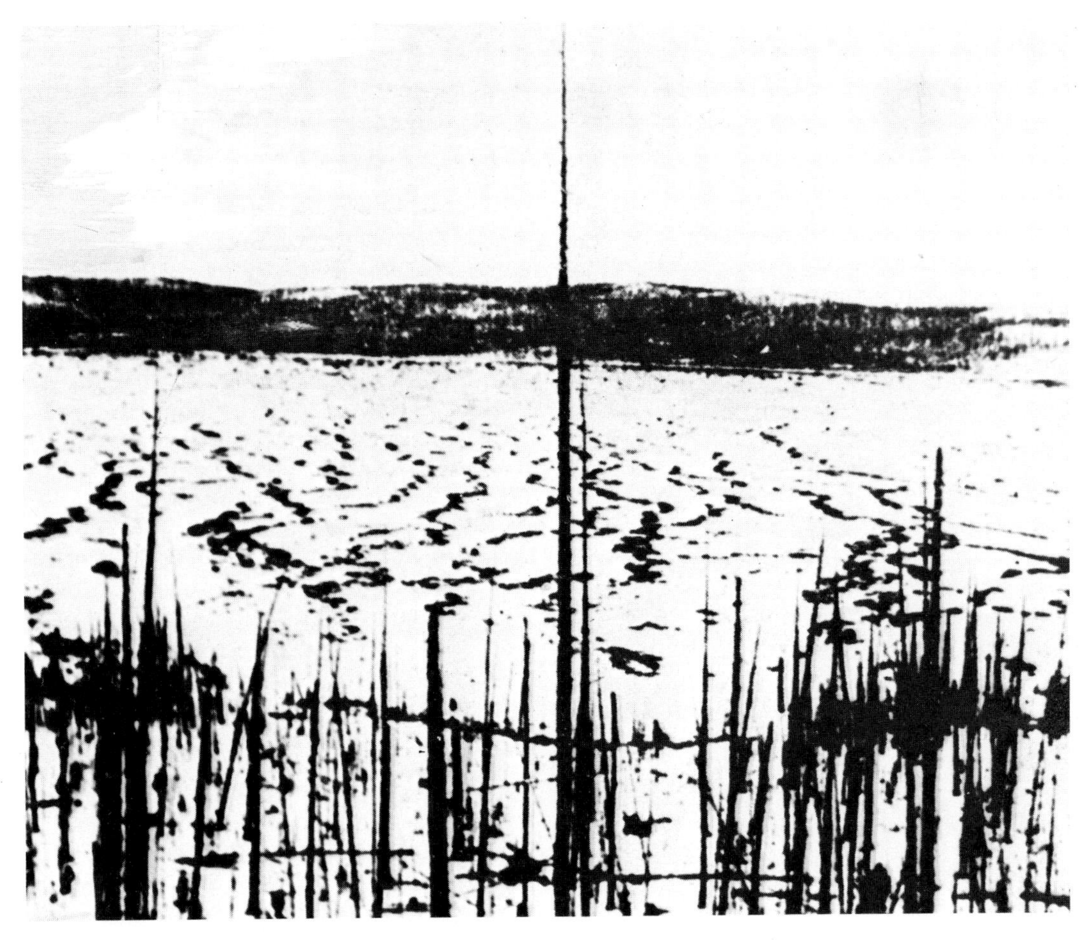

The Tunguska Valley, Siberia, at the point immediately below the explosion of June 1908. Trees have been totally destroyed over a region of several square kilometres, while those around the edge are devoid of bark and twigs—even in this photograph, taken in 1927. There is no major crater here, only a number of small holes a few metres across.

CHAPTER 2

THE OUTER PLANETS

Our knowledge of the other planets in our solar system has exploded in the past 20 years. Spaceprobes have hurtled out past Saturn, and dropped in as close to the Sun as Mercury, scanning them for details of their surfaces and clouds, and testing nearby space for their influence. We now take for granted the technological miracle of receiving television pictures from a billion kilometres away in space: radio signals from the Pioneer 11 and Voyager spaceprobes as they swept past Saturn appeared on the World's TV screens after a journey of almost an hour and a half travelling at the speed of light.

These and other automatic spaceprobes have revealed all the planets to be real worlds, with as many individual quirks as our own. The view from Earth is limited. The world's largest telescopes could, in theory, show a tremendous amount of detail on the nearest planets, but their performance is degraded, not by their own shortcomings but by the Earth's turbulent and ever-changing atmosphere.

Light coming down from space is bent slightly from its true path by eddies in the Earth's covering of air. As winds blow the eddies along, the light is bent back and forth. The stars' twinkling is one result; another is that it is impossible to see or photograph the finest details that should be visible on a planet's disc. Telescope views of the planets are always disappointing for this reason. The probes now reveal the planets as they really are; and photographs of the Earth from space put our planet in its context among the others. Lander probes are now exposing the intimate details of the surfaces of Venus and Mars, too, exploring these distant worlds in very much the same way as geologists have investigated the Earth.

Spacecraft flying past a planet see it not only in close-up; they can also feel its magnetic field, and sniff out the types of atoms in the outermost atmosphere. With the exploration of the planets by manned missions still a science fiction scenario for the future, our long-range sense organs on these willing, automatic workhorses are nevertheless unravelling at least some of the planets' secrets, even now. But, as always in astronomy, as some mysteries are solved, others seem inevitably to emerge to take their place.

KING OF THE PLANETS

Jupiter reigns supreme as the undisputed head of the planetary family. It is 11 times the Earth in diameter, which means that a hollow Jupiter could hold over a thousand Earths inside. In a balance, Jupiter would outweigh all the other planets together. The solar system is almost just the partnership of the Sun and Jupiter, with the other planets as incidental pieces. But even so, the partnership is unequal. It would take a thousand Jupiters to make up one Sun, although this is a reflection on our star's bulk rather than on Jupiter's status as the giant of the planets.

Oddly enough, though Jupiter does not have an active core like a star, it produces a small amount of heat. Every planet absorbs light and heat from the Sun, and radiates it out into space again. To stay at the same temperature a planet must radiate away exactly as much heat as it receives; paying out as much as the Sun pays in, so that its heat account won't either increase or decrease over the aeons that the solar system has existed. But Jupiter's heat budget does not balance: it pays out twice as much heat as it receives, so it can evidently produce its own heat internally.

Several theories have been proposed to explain this discrepancy. One is that the gases which make up Jupiter are changing their atomic arrangements near its heart, and liberating heat in the process; another that it has not yet finished contracting. Its appearance suggests that it has reached its end-point—from being a diffuse cloud of gas the planet seems to have settled down to being a stable globe—but if it is still shrinking slowly, the contraction would supply a continual source of heat. To balance its heat books, Jupiter need only be shrinking at a rate of one millimetre per year. Needless to say there is no way to test such a minute shrinkage directly on a planet 142,800 kilometres (88,730 miles) across, which is a mushy mixture of ices and gases without any real surface.

The bulk of Jupiter, in fact, is made up of a mixture of hydrogen and helium, as an intermingled liquid. Only in the topmost thousand kilometres, thin as the peel on an orange, are these elements in their 'normal' state as gases, and in this atmosphere float layers of coloured clouds. Jupiter turns once in 9 hours 55 minutes, and with its colossal size,

Three of the giant planets are known to have systems of rings—but all are different. Jupiter's ring consists of tiny dark grains of rock, forming a continuous sheet extending down to the planet's cloud-tops. The main rings of Saturn are split into over a thousand closely-spaced ringlets. Each is composed of billions of icy blocks, a few metres in size, whose dazzling-white surfaces make Saturn's rings far brighter than those of Jupiter and Uranus, and easily visible from Earth. Uranus, too, has narrow ringlets, but only a few, and these are widely separated. They are made of solid chunks a few metres across, and are dark rocks rather than ice. Uranus's dark rings are not directly visible from Earth, but can be detected when they pass in front of a star and block off its light.

The two Voyager spaceprobes sent back detailed views of Jupiter's cloud structures, including the Great Red Spot (*lower left*), the adjacent white ovals, and elongated white plumes (*centre left*). At centre right is the reddish moon Io.

the regions at its equator feel a powerful tug of centrifugal force as they are whirled round, pulling the gases here outwards. As a result, Jupiter's equator is 6 percent wider than its pole-to-pole diameter—it is flattened by its own rotation into a tangerine shape so pronounced that it can be made out even by a small telescope.

And even a small telescope will show the different coloured bands across Jupiter's face. Through a larger one, these are revealed as complex cloud patterns, and coloured spots.

The lighter bands are regions where gases are rising and producing white clouds in the topmost layers of the atmosphere. In the browner bands, we are looking down into the lower layers, where the clouds are tinged with unknown coloured compounds. The many hues of orange and brown represented suggest that sulphur is Jupiter's colouring matter, with its many different compounds and combinations producing the varied tones.

The clouds seethe with ceaseless activity. White cloud-heads erupt as a new upcurrent breaks through; spots continually turn round like huge bathplug whirlpools, while the streaks and banners around them move the other way as if to prevent them from winding up too much. The equatorial gases of Jupiter are moving faster than the planet as a whole turns, coming round it in only 9 hours 50 minutes and so gradually lapping the gases further from the equator. After a hundred turns of the planet, the equatorial gases have caught up a whole revolution on the rest of the planet.

Between the fast-moving equatorial gas and the rest of Jupiter's atmosphere there is a region where gases are moving past each other. It is here that the spots are found, and it seems they are eddies, turning between the two streams of gas like ball bearings. Detailed analysis shows, in fact, that it is the rotation of the spots which drives the gas streams at different speeds, rather than the other way round. The energy to turn the spots seems to be Jupiter's internally-generated heat making its way upwards through the planet. The spots are generally white in colour, but there are a few small red ones as well as the most famous blemish, the Great Red Spot. It is not certain why the spots should come in these two colours, but the white seems to be simply cloud over the spot; the red is the more tantalizing puzzle, and the favourite—though not yet proven—theory is that its colour is due to red phosphorus, the element in match-heads that ignites when a match is struck. Among the poisonous gases in Jupiter's atmosphere is phosphine, a compound of phosphorus and hydrogen. Phosphine may be brought up from the depths in the whirl of a spot, and broken down to red phosphorus when it is exposed to the Sun's glare above Jupiter's cloud layers.

The Voyager probes sent back enough information from Jupiter to keep the mission scientists occupied for years. But, more important, the existence of the 32,000 Voyager photographs means that we can at last treat the outer regions of the solar system in the same way as the inner planets: a place where we can deal with hard facts, figures and measurements, rather than disputable and hazy glimpses from our remote station on Earth. And perhaps the most far-reaching consequence of all is that, as the broad outlines of this alien planet are becoming known, scientists and astronomers are developing a 'feel' for it; in the realm of the least Earth-like planet, man is beginning to feel at home.

SATURN

In mythology, Saturn was Jupiter's father but as a planet it is more like a smaller brother. Saturn's claim to fame is its beautiful set of rings, which make it the most magnificent object in the solar system through a moderate telescope. But these are incidental to its real nature. Saturn is a gas giant like Jupiter. It contains only a third as much matter, but it is not an exact scaled-down version of its larger brother. It is slightly larger than we would expect a one-third mass replica to be, so its matter is evidently not as compacted—probably because, with less matter in the planet, its centre is not as crushed by the weight of the overlying layers. Saturn's matter is so light, in fact, that the planet would float on water if there were an ocean big enough to hold it!

The Voyager spacecraft have unveiled many of Saturn's secrets, as they have Jupiter's. Voyager 1 passed it in November 1980, and Voyager 2 in August 1981; their major achievement was a set of fantastically-detailed photographs of the rings and moons, but they also took a close look at the planet itself. Like Jupiter, Saturn radiates more heat to space than it receives from the Sun, but it is more puzzling here, for Saturn's shrinking could provide only one-third the observed heat. Many astronomers favour the new view that hydrogen and helium gases are separating within the planet, with helium settling towards the core. But if this is happening, the atmosphere should contain less helium than Jupiter's—and first results from Voyager indicated that it does not.

Saturn's internal heat source—whatever it is—drives whirling eddies, which in turn produce strong winds running in zones parallel to its equator. The powerful internal heat gives the eddies 10 times as much energy as Jupiter's spots, and the equatorial winds whip round the planet five times faster than Jupiter's equatorial belt—at 1600 kilometres per hour (1000 miles per hour). The eddies show up on Voyager photographs as faint spots, hardly different in colour and tone from the rest of the planet's atmosphere. They are apparently hidden by a thick layer of overlying haze, deeper on Saturn than on Jupiter because its greater distance from the Sun gives the former's atmosphere a different run of temperature with height.

Oddly enough, what spots are seen on Saturn do not occur at boundaries between wind streams of different speeds: the ball-bearing analogy does not apply exactly, and the rotating spots must have some more subtle way of driving the high-speed winds. Meteorologists are studying closely the differences between the atmospheres of the two planets. Quite apart from its intrinsic interest, one of the practical benefits of the Voyager missions will be a greater understanding of atmospheres in general, and so an improved ability to predict weather on Earth.

URANUS

Slow-moving Saturn is the most distant planet known since ancient times, its dull golden speck of light taking almost 30 years to travel once around the sky. No one expected that the Sun's family contained more planets until William Herschel stumbled across the next one in 1781. Herschel had left the Prussian army and settled as a musician in Bath, England, but his abiding passion was astronomy. In his spare time he built reflecting telescopes which were capable of more detail than the refracting (lens) telescopes used by professional astronomers at the time. These latter were better suited to measuring positions in the sky accurately, the main concern of the professionals, but during Herschel's systematic searches of the sky his ability to discern detail paid off.

On the night of 13 March 1781, he was looking through the faint stars in the constellation Gemini when he noticed that one appeared, not as a point of light, but as a tiny disc, and it moved from night to night. It was evidently not a star, and Herschel at first thought he had found a comet. But when its orbit was calculated, it proved to be a new planet, larger than Earth, and circling the Sun beyond the orbit of Saturn.

Uranus is just visible to the eye without a telescope if you have dark skies and know exactly where to look; binoculars show it as a point of light, crawling so slowly that it travels round the sky in 84 years. Today's telescopes do not show many more details on Uranus than did Herschel's. Although the planet is four times larger than Earth, it is so distant that it appears no larger than a pinhead at 50 paces. It is undoubtedly a gas giant planet, but its make-up is different from Jupiter or Saturn. Its average density is higher than either, yet, as we saw in the case of Saturn, a smaller hydrogen-helium planet should be *less* dense, since its centre is less compressed. So Uranus must have a heavy core of rock and, despite its smaller size, this core must be roughly the

Space around Jupiter is far from empty. Beyond the orbiting dust grains making its ring, fast electrons whirl in a cloud, trapped by the planet's magnetic field. They produce radio waves that can be detected from Earth, as well as posing a hazard to the electronics on spaceprobes passing through.

Farther out, the volcanic moon Io ejects atoms of sodium, sulphur and hydrogen into a cloud which extends right round Io's orbit. Io is linked to Jupiter by a 'flux tube' through the magnetic field, which carries an electric current of five million amps between Io and Jupiter's magnetic poles. As this current penetrates Jupiter's atmosphere, it produces intense aurorae.

same size as Jupiter's. Uranus is also slimmer than its companions because it has far less in the way of overlying liquid layers of hydrogen-helium fluid.

The oddest feature of Herschel's planet is that it rotates backwards; and not just backwards but tipped so much that its poles lie almost in the same plane as its orbit. Uranus spins virtually on its side as it travels around the Sun. Since a planet's axis always points in the same direction in space—as Earth's points towards the Pole Star—a Uranian citizen would experience strange seasons during the course of his long 'year'. In 1966 the Sun was shining directly down onto its equatorial region, but by 1987, when it is a quarter the way round its orbit, it will be pole-on to the Sun. Twenty-one years later the Sun will be over the equator again; and the other pole will be exposed to the Sun's rays after a 'winter' lasting 42 Earth-years—a winter of perpetual darkness.

The warmest place on Uranus, incidentally, is not the equator for, during its 'summer', one of the poles becomes warmer than any other part of the planet can ever be. The Sun is constantly overhead then, neither rising nor setting, and so provides a constant source of heat. At the times when the Sun is over the equator, a Uranian would see it rise and set as the planet spins, and the warmth of the day would be tempered by the coldness of night. Lest this should sound like a travel advertisement for the poles of Uranus, it is worth pointing out that the planet's average temperature is −210°C (−350°F)!

NEPTUNE

Although it was Herschel's reflecting telescope which made him the first astronomer to see Uranus as a disc, and so realize it was not a star, the next planet's discovery was due to the professionals' smaller refracting telescopes. They could measure positions extremely accurately, and as a result they could follow Uranus in its orbit and confirm that it was a planet. But over the succeeding decades, Uranus did not move quite as it should have done. Its path was being perturbed by an invisible influence, and the only logical culprit was the gravitational tug of another planet beyond.

Two astronomers set out to calculate where this planet should lie, neither knowing of the other's work. John Couch Adams at Cambridge University completed his calculations first, and the director of the Cambridge Observatory began, after some delay, to measure the positions of some thousands of stars in that part of the sky, so that he could go back later and find out if any had moved. But Urbain Leverrier in France was luckier. He sent his calculated position for the unknown planet to Berlin Observatory, which just happened to have a chart of the stars in that part of the sky. The interloper was quickly spotted. On 23 September 1846 another new member of the Sun's family became known to man. Leverrier named it Neptune, after the god of the sea.

As a planet, Neptune is very much Uranus's twin, but its position in the solar system means that it is more difficult to study from Earth. It does not share its twin's amiable eccentricity of spinning the wrong way round but, even more strangely, its large satellite Triton circles the planet backwards.

Although astronomers know which way their axes point, the time it takes for both Uranus and Neptune to turn is still unknown. Older books quote what seem to be very accurate measurements of 10 hours 49 minutes and 15 hours 48 minutes as the 'day' lengths on Uranus and Neptune respectively, but in all the present uncertainties the one fact that seems to be generally acknowledged is that these figures are wrong. The discs of these planets do not show any spots or other features that can be watched to see them turn, so measurements must be more indirect. According to different results, Uranus could spin as quickly as once in 12 hours, or take as long as 24 hours over one rotation. Neptune's 'day' is probably similar, somewhere around 20 to 24 hours in length. If astronomers do not reconcile their conflicting results before then, the Voyager 2 spaceprobe should answer the question—as well as many others about these little-known worlds—if it survives to carry out its reconnaissance of Uranus in 1986 and Neptune three years later.

THE RINGS AROUND THE GIANTS

Saturn has long been known as 'the ringed planet': its wide flat rings are so much a part of its personality that it is difficult to imagine it without them. For centuries astronomers have wondered why only Saturn had rings, and why they were so flat and extensive—they cannot be seen from Earth when presented edge-on, so they must be less than a couple of kilometres thick, although they are as wide as 20 Earths. Detailed Earth-based investigations and spaceprobe pictures have now revealed that both Jupiter and Uranus also have rings—but that Saturn's rings in close-up are very different from what had been expected.

All rings systems consist of small fragments of rock or ices, their numbers running into billions, which circle their planet as miniature satellites. But despite the fact that we now have detailed studies of three systems to help us, planetary rings seem as mysterious as ever. Voyager 1 discovered Jupiter's ring as it

Saturn's braided F-ring (*above*) consists of two ringlets twisted about each other: the twisting may be caused by Saturn's magnetic field or by the gravity of nearby satellites.

passed in 1979; Voyager 2 sent back detailed pictures a few months later. The ring is quite tenuous, and made of dark matter, so it does not reflect enough light to be visible from Earth. It extends from a sharp outer edge some 60,000 kilometres (36,000 miles) above the clouds inwards as a continuous sheet right down to the cloud tops. From the way it scatters light, astronomers calculate the ring is made up of very tiny rock particles no larger than the particles in wood smoke, perhaps dust particles swept up from a comet which disintegrated near Jupiter; or it may indeed be a smoke ring, composed of smoke particles erupted from the volcanoes of Jupiter's satellite Io.

The rings of Uranus were also an unexpected discovery. Two years before Jupiter's ring was found, astronomers watching Uranus passing slowly in front of a star were astonished to see the star flicker out several times when the planet was well away from it, both before and after the planet itself blocked off the star's light. The only reasonable explanation was that Uranus is girdled by several rings whose particles obscured the star's light as they moved in front of it.

Uranus has at least nine rings, and because they block off starlight so effectively, they cannot be made up of small particles like the constituents of Jupiter's ring, but must be solid chunks of rock, probably a few metres across. We cannot see the rings from Earth, even using sensitive electronic light detectors, so they must be dark rocks, darker even than coal. What has haunted astronomers since the discovery is their extreme narrowness. Apart from the outermost, the rings are each only 10 kilometres (6 miles) wide, although they circle some 50,000 kilometres (30,000 miles) from the planet. Why should chunks of rock orbit at only these particular distances?

A close-up of Saturn's rings from Voyager 1 (*right*) reveals that they are split into hundreds of ringlets. Faint ringlets are even found in the wide gaps between the major rings, such as the Cassini division (running across rings in foreground).

Astronomers agree that they must be 'herded' into these orbits by the gravitational tug of small, unseen satellites; but there are two, apparently contradictory, theories. The first invokes two satellites for each ring, one just inside and the other outside. In this case, Uranus must have at least 18 small moons, as yet undetected from Earth, which are arranged in nine pairs so that the orbits of each pair sandwich a ring between them. Alternatively, each ring may follow the orbit of a single moon, and consist of the debris lost from that satellite's surface. This theory invokes only nine unknown moons, one in each ring. That is as far as observations of Uranus's rings can take us at present; but the answers to their problems may well come instead from the staggering pictures of Saturn's rings sent back by the Voyager spacecraft.

The rings of Saturn

From Earth, Saturn's rings look like a continuous sheet, broken into two major parts by an empty gap called the Cassini division. The ring outside this division is the A-ring; the B-ring lies inside. Both are undoubtedly made up mainly of icy fragments, a metre or so in diameter, whose snowy surfaces reflect light

so well that they are far more brilliant than Jupiter's or Uranus's rocky rings. A fainter ring lies within the B-ring; this C-ring, or crêpe ring, contains rocky debris, too, and reflects less light than the outer rings. Earth-based astronomers have also reported even fainter inner and outer rings, but the most interesting new one was discovered by the first spaceprobe to visit Saturn. A year before the Voyager 1 flyby, the simpler cameras on Pioneer 11 revealed a ring just outside the A-ring. This F-ring was unlike the others: it was so narrow that it resembled far more the rings of Uranus.

Even so, scientists were completely unprepared for the revelations from the Voyagers. The rings of Saturn, seen in close-up, are not continuous sheets as they appear from Earth. Each is made up of literally hundreds of narrow ringlets, closely spaced, but completely distinct from one another. Almost all of them are accurately circular but, to compound the enigma, there are a few which are slightly oval.

The observations have to some extent solved one riddle, because Saturn's rings do not seem to be different in kind from Uranus's—they are not broader, but simply far more numerous—so whatever process confines one probably operates in both. Whichever of the two mechanisms is responsible, Saturn's rings must contain hundreds, or even thousands, of tiny satellites which herd up the smaller particles into ringlets. These satellites need be only a few kilometres across, and may simply be the largest of the multitude of particles making up the rings.

The Voyager pictures certainly show that satellites *can* herd ring particles. Voyager 2 photographs of Jupiter revealed a previously-known satellite, only 30 kilometres (20 miles) in diameter, orbiting right at the outer edge of the ring, while Voyager 1 found a small satellite just outside Saturn's bright A-ring. These moons undoubtedly keep the particles from straying outwards, and so define the outer edge of the ring by their presence. Voyager 1 also discovered small satellites just within and beyond Saturn's narrow F-ring, confirming exactly the theory that a pair of satellites can confine a narrow ring.

A close-up photograph revealed just how odd this F-ring is: it is composed of three narrow ringlets, two of which seem to be twisted around each other like a braided rope.

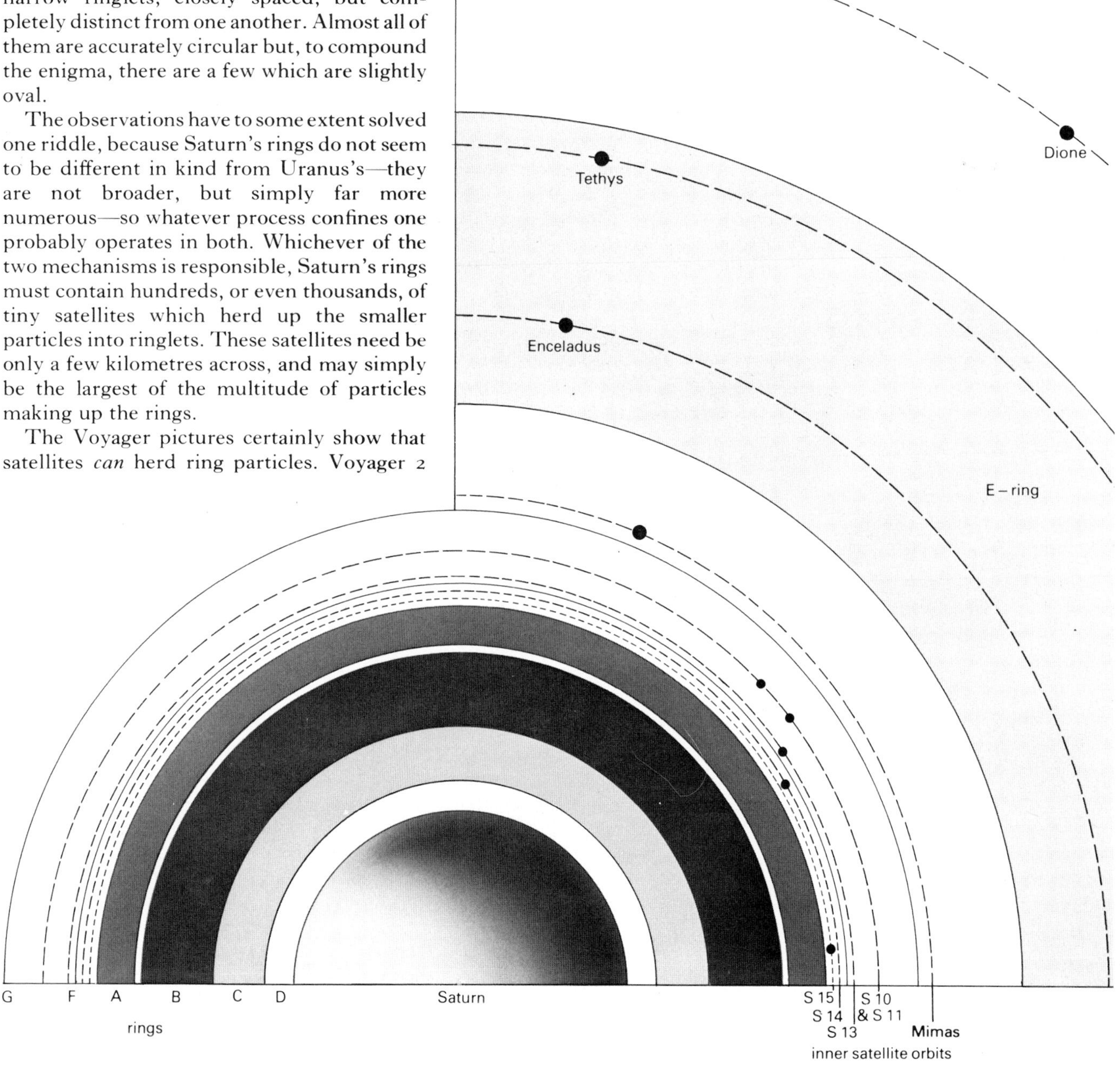

Observations from Earth and from Voyager 1 have revealed a complex system of satellites and rings surrounding Saturn. The named rings (A to G) vary from the very bright, wide B-ring to the extremely faint D- and E-rings, and to the narrow F- and G-rings. The small innermost satellites control the adjacent rings: satellite 15 causes the sharp outer edge to the A-ring, while satellites 13 and 14 'shepherd' the particles making up the F-ring between them. Satellites 10 and 11 share virtually the same orbit, and are probably two halves of a disrupted moon. The small moon Dione B (satellite 12) follows the orbit of the large satellite Dione; gravitational effects keep it at the third corner of an equilateral triangle with Dione and Saturn.

First press reports announced: 'It defies the laws of orbital motion'—and even ... 'defies the known laws of physics!' But the explanation is not likely to require rewriting science. Already two theories have been advanced. One is that the gravitational pull of the controlling moons makes the ring particles follow non-circular, intersecting orbits; the other that one ringlet is circular and a well-behaved collection of icy and rocky chunks, while the other is the oddity, consisting of very fine, dust-like particles, which can become electrically charged. They are then affected, not just by gravitational forces, but also by Saturn's magnetic field, and this can twist them away from their normal circular orbit, even to the extent of twining them around the normal ringlet.

Such magnetic influence on small particles has been seen in Jupiter's rings. And it provides a plausible explanation for another phenomenon in Saturn's rings. Voyager 1 photographed darker 'spokes' running outwards across the bright A and B rings, and turning with the planet rather than keeping pace with the motion of the ring particles. The spokes may well be regions where finer particles have been lifted out of the plane of the rings, above and below, where because they are electrically charged they have become 'tied' to Saturn's magnetic field. As a result, they rotate with the planet where the magnetic field is rooted.

The startling Voyager pictures are still being analyzed. The true explanation for the mysterious spokes, the braided F-ring and the hundreds of ringlets may well differ from the early ideas described here. And the most fundamental mystery remains: Why does Saturn have so many bright, icy ringlets, while those of Jupiter and Uranus are dark and dusty or rocky? Were the ice blocks left over from Saturn's birth? Or are they the shattered remains of a disintegrated satellite? Although we are beginning to understand the structure of planetary rings, only further investigation will reveal their origin.

PLANETS MASQUERADING AS MOONS

Where can you find the smoothest, the most cratered, the most active and the least active worlds in the solar system—and those with the youngest and oldest surfaces? A space traveller would not have to expend much effort to travel between these worlds, for the Voyager spaceprobes revealed in 1979 that Jupiter's four largest satellites fill all these roles among them. Until then, 'satellite' had been a synonym for dullness and inactivity. Astronomers thought that Earth's Moon, now well explored, was typical. Mars's two tiny moons are just hunks of rock, large pitted boulders. What more could be expected from the satellites of the other planets?

As it turned out, plenty. The outer planets' satellites are a mixed bunch: the smallest, probably just captured asteroids; the larger, regularly spaced moons produced by the planets' formation. Although they are technically 'satellites' because they orbit a more massive body, some of these worlds ought to be thought of as planets themselves. Saturn's Titan, Neptune's Triton and Jupiter's two largest moons are bigger than the planets Mercury and Pluto, with Jupiter's other two major satellites not far behind. And the Voyager spaceprobes have shown that they are just as exciting—perhaps more so than the planets themselves.

Not only are Jupiter's four major moons different from our Moon, but the Voyager spacecraft found that there is little family resemblance among them. The outermost, Callisto, is the most cratered body known in the solar system, with craters standing shoulder-to-shoulder, overlapping one another until there is not a portion of surface which is smooth.

Next-in Ganymede seems to have the most complicated 'geology', so alien that astronomers aided by geologists have yet to decipher it. Much of its surface is also crater-scarred, though here the craters are diluted by smooth, lava-filled plains. These moon-like regions are separated from one another by lighter-coloured bands some hundred kilometres wide, unlike anything seen on any planet. The light bands cover Ganymede like a net, with the spaces of the net filled by the moonlike crater and lava plain terrain. They are composed of swarms of narrow parallel cracks, each perhaps like a geological fault on Earth, where blocks of land have moved up, down or sideways. It seems that the cratered sections of the surface have shifted during its lifetime, producing cracks and rifts between them. Geologists can compare these movements with continental drift on Earth, where enormous portions of the Earth's crust are continuously in motion, carrying the continents around. Ganymede is too small for its internal heat to melt rock, but it does not need to. Out in these regions of the solar system, ice was abundant when the planets formed and Ganymede is probably a mixture, half rock, half ice. The solid crust, therefore, could be floating on a rock-ice slush below. That is one possible explanation; but on such an alien world the answer may be even more fantastic.

Jupiter's next satellite is also a network of lines; but these are long and thin—and darker than the rest of the surface. Europa is also the smoothest body in the solar system. Although it is a moderately large world—a quarter of Earth's diameter—its tallest hills are no higher

Artist's impression of a volcano on Jupiter's satellite Io. The cutaway shows the underground reservoir of molten sulphur under Io's crust of solidified sulphur, with molten rocks below. In Io's low gravity, the volcanic plumes rise to almost 300 kilometres (200 miles).

than a sand dune. Scientists have explained both its smoothness and its flatness by assuming that it has an icy surface, with some rock 'dirt' mixed in. At some time in the past, the entire surface melted then gradually refroze into large ice-floes. Floating on a liquid layer covering the entire planet, the floe tops came to be at the same level. As Europa froze again, its surface floes welded together as a smooth, mountainless surface. And the cracks show up dark, just as satellite photographs of the Earth show dark lines between polar ice floes welded together as they have refrozen. Again, it is a convincing theory, but perhaps the truth is stranger.

With Io, we come to perhaps the most bizarre world in the solar system. Scientists were astonished by Voyager 1's pictures of this satellite, a world slightly smaller than our Moon, but with more geological activity than any planet. Io's is the only surface in the solar system which bears no craters. Its surface must be continually wiped clean, replaced by new layers and refurbished completely over a time of roughly a million years—a small fraction of the solar system's 4600 million-year age.

The reason became obvious even as Voyager 1 was completing its highly successful flyby. An astronomer studying positions of stars near Io's edge turned up the brightness on a Voyager picture—and saw a faint mushroom cloud at the planet's limb, standing out palely against the black sky. It was the plume from a volcanic eruption. Searching through the other pictures, mission scientists found seven other active volcanoes—each larger than any on Earth, on a world one-quarter the size. They were erupting jets of sulphur dioxide gas, and their flanks were stained with streams of molten sulphur, dark brown near the

volcanic vent and paler orange farther off.

Voyager 2, passing the satellite at a slightly greater distance some months later, discovered that one of the previously active volcanoes was now dormant. Io is constantly seething with activity. Volcanoes appear, spew sulphur over surrounding districts, and fade away again. The gases ejected into space follow Io's orbit around Jupiter, in a glowing cloud which had been discovered from Earth even before the Voyagers arrived, although its origin was then unknown. And it may be volcanic smoke from Io which forms Jupiter's ring of dust-sized particles.

Why should a small world like Io have reserves enough of heat to melt sulphur, and probably the rocks it is composed of, too? An answer was proposed even before the volcanoes were discovered. Stanton Peale and his colleagues at the University of California pointed out that Io is undoubtedly stretched by the tidal forces of Jupiter's gravity. As a result, it must be a slightly egg-shaped world, and it tries to keep the long axis of the egg pointing towards its planet. But the gravitational tug of the next two satellites, Europa and Ganymede, keep tugging it back and forth. The resulting strain in the rocks heats them up and melts Io throughout, except for a thin solid crust. And when the competing gravitational pulls flex Io enough, the crust cracks and a volcano bursts into life.

Other scientists have estimated that Io has spewed out so much sulphur from its volcanoes that its surface must be coated with it. Its outermost layers would then be oceans of molten sulphur capped by a thin skin of frozen sulphur, and volcanoes powered by escapes of sulphur dioxide gas through cracks in the congealed skin. Liquid sulphur—brimstone—and volcanoes must make Io a close approach to the Bible's account of Hell.

The moons of Saturn

A year later, Voyager 1 revealed the moons of Saturn to be as interesting, if not as varied, as those of Jupiter. Two small satellites, provisionally called S10 and S11, follow nearly the same orbit around the planet, almost certainly fragments of a single satellite that has been broken in two by the impact of a huge meteorite. The satellite Mimas, 390 kilometres (240 miles) in diameter, has only by a narrow margin escaped such a fate; it bears a crater one-quarter as large as itself, the testimony of an impact almost powerful enough to break it apart.

Saturn's largest moon, Titan, is one of the largest satellites in the solar system although Voyager pictures now reveal it to be slightly smaller than Jupiter's Ganymede. Being farther from the Sun, though, Titan has managed to hold onto an atmosphere—it is at least 50 percent thicker than Earth's—the only satellite in the solar system known to have done so. Before the Voyager encounter, some astronomers had hoped that Titan's atmosphere might trap enough heat to make it a possible abode of life, but Voyager 1 discovered that it is mainly nitrogen, a poor blanket, so the surface temperature is around −180°C (−290°F). If there are any 'oceans' on Titan, they are of liquid nitrogen, not water. Voyager could not, however, photograph the surface itself: only a thick layer of reddish cloud could be seen enveloping the satellite and obscuring what lies beneath. Scientists are keenly interested in knowing

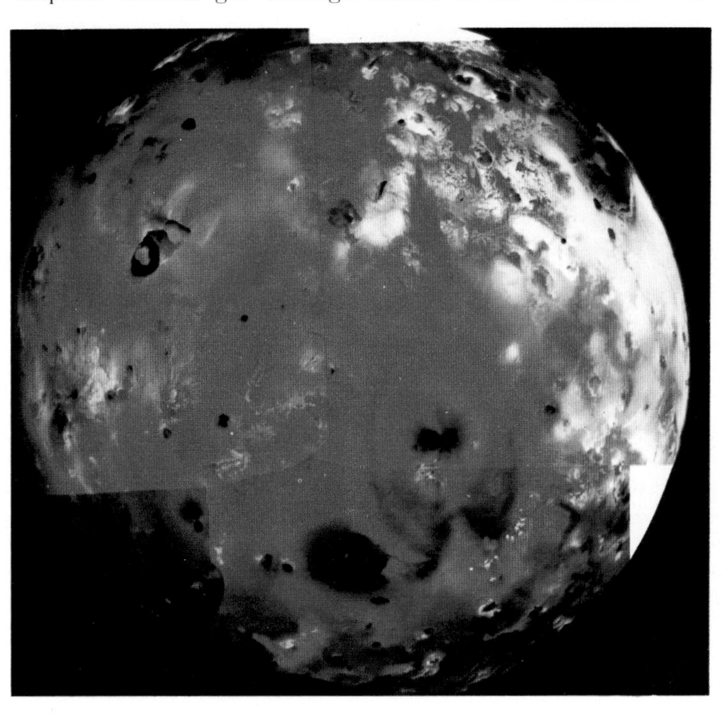

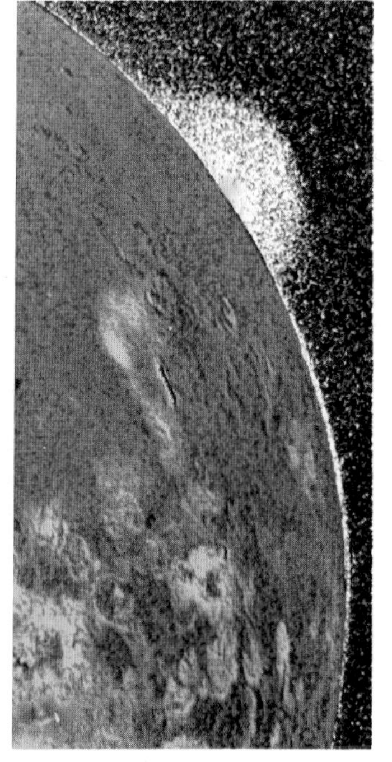

Voyager 1 photographs of Io. The volcanic eruptions (*far right*) are more powerful than any on Earth, and sulphur dioxide gas gives them a blue tinge. Sulphur from the eruptions coats Io with reddish sulphur deposits (*right*), covering any craters that may lie below. The inverted heart-shaped swelling just below Io's centre is one of its active volcanoes; dark markings are streams of molten sulphur pouring from the satellite's volcanic vents.

what is below the unbroken cloud, for this is the only known world which resembles Earth in having a nitrogen-rich atmosphere. And Titan's clouds are made of simple organic compounds, like methane, acetylene and hydrogen cyanide, molecules that scientists believe were synthesized in Earth's early atmosphere, and from which living cells formed. Even though Titan is too cold for life, it is possible to view it as an early Earth in deep-freeze since the birth of the solar system.

The two Voyager probes have shown that 'moons' of the major planets are by no means as dull as Earth's drab satellite, but worlds with individual characteristics, many as fascinating as their parent planets, and all bearing clues to the planets' birth. If man ever visits the outer solar system, it will be the satellites' solid surfaces which will provide his bases rather than the deep gas-liquid-slushes that make up the giant planets themselves.

THE SMALLEST PLANET

The outermost member of our planetary system left its post in January 1979. At the moment, Pluto is not the farthest planet from the Sun, for its oval path has brought it within Neptune's almost circular orbit. Neptune is guardian of the system's outer edge until March 1999, when Pluto will reclaim its position.

Pluto's elongated orbit, which takes it within Neptune's for 20 years of its 248-year round trip about the Sun, is only one of its oddities. It is a tantalizingly peculiar place.

Pluto's moon Charon is shown by a special technique called speckle interferometry (*top right*). Pluto's image is the lower semicircular area, bisected by the dark border—it appears noticeably larger than the image of a point-like star imaged by the same technique (*middle right*). Charon is the almost-separate image to the top left of Pluto. These photographs from the Canada–France–Hawaii Telescope show that Charon is half Pluto's size.

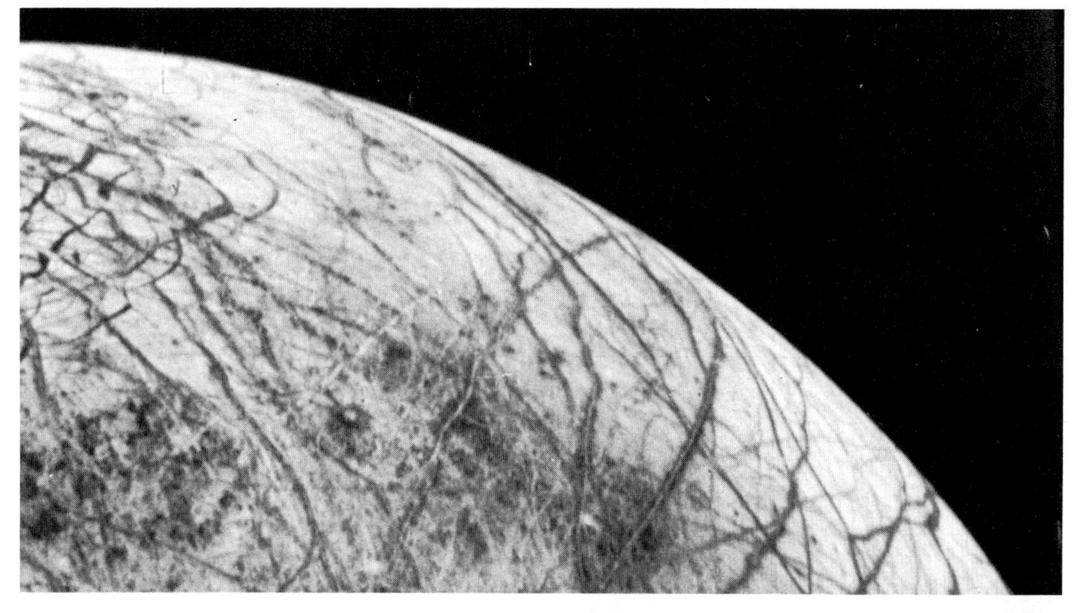

The frozen surface of Jupiter's satellite Europa (*left*) bears a network of dark lines, probably muddy water from below that has welded the surface 'ice floes' together.

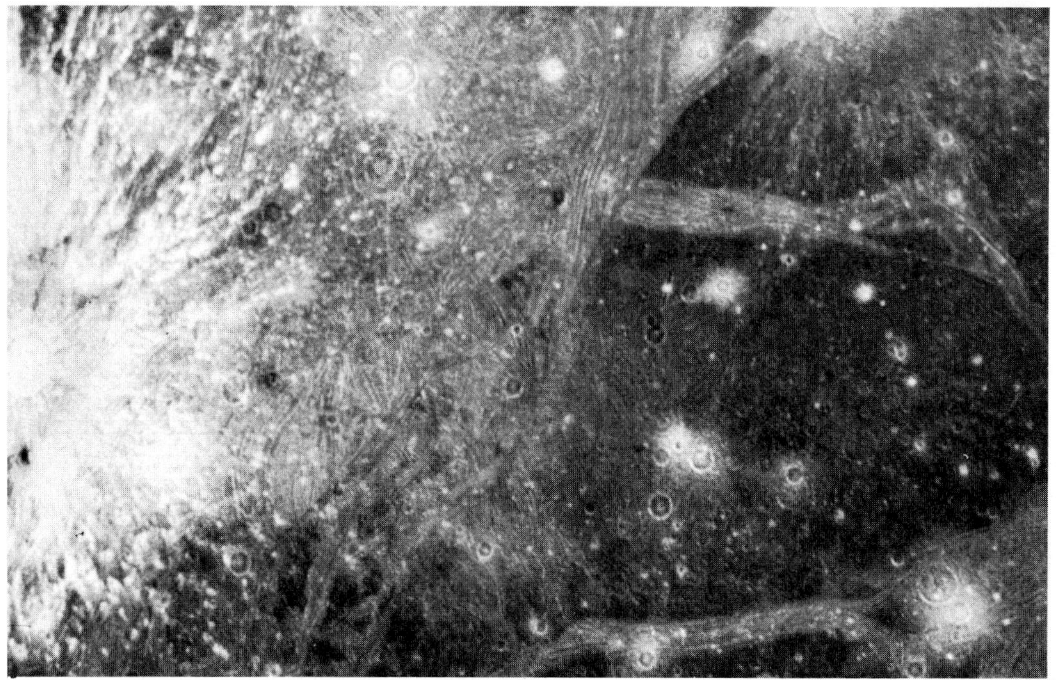

A close-up photograph of Jupiter's largest moon, Ganymede, shows swarms of parallel cracks separating regions which are heavily cratered.

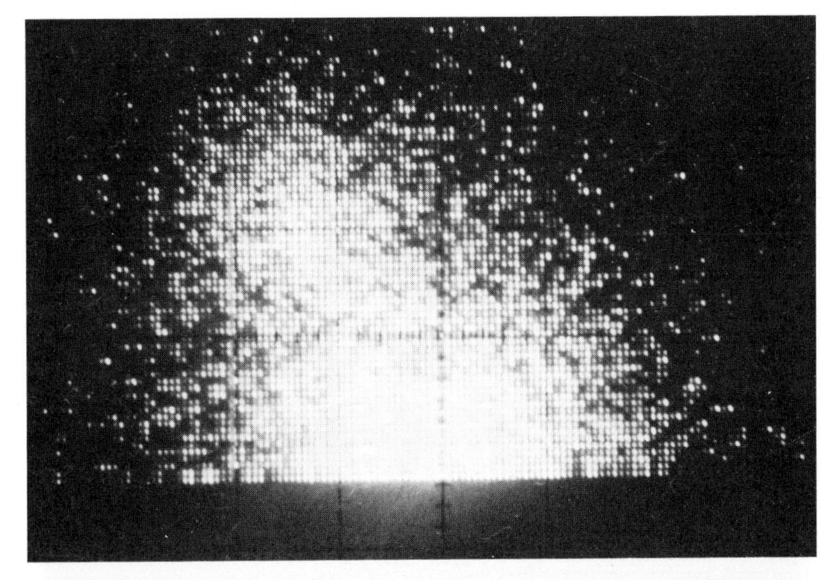

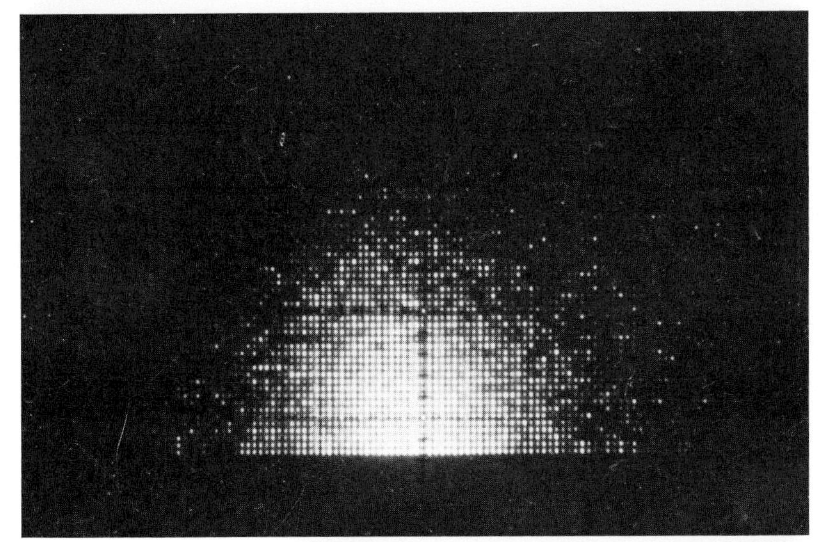

Although it lies in the realm of the gas giants it is far too small to be one of them, and because it lies so far from us, it appears only as a faint point of light, making its mysteries even more difficult to unravel.

Pluto's story, like that of Neptune, begins even before its discovery. In the closing years of the last century, astronomers found that Uranus was once again straying from its expected path. Neptune's gravitational tug on the planet was now known and allowed for, and the new departures were much smaller than those that had led to Neptune's discovery. They seemed to tell of a more distant planet. Astronomers and mathematicians now repeated the calculations made by Adams and Leverrier, in the search for a further planet.

Percival Lowell was one man bitten by this bug. A wealthy Boston businessman, he had built a private observatory in the clear air of Arizona, mainly to study Mars. But the lure of the unknown planet occupied him more and more, and his observatory staff continued to search the region of sky where it should have been, even after Lowell's death in 1916. Since it had not turned up in the first eager surveys, the planet was evidently faint, probably no more than a dim point of light that could only be distinguished from the myriad of faint stars by its slow motion from night to night.

Through sheer perseverance the planet was eventually tracked down. In early 1930, Clyde Tombaugh, who had inherited Lowell's mantle in the trans-Neptunian planet search, found a faint 'star' which had moved slightly between photographs taken on 23 and 29 January. A new planet had been added to the solar system, and it is no coincidence that its

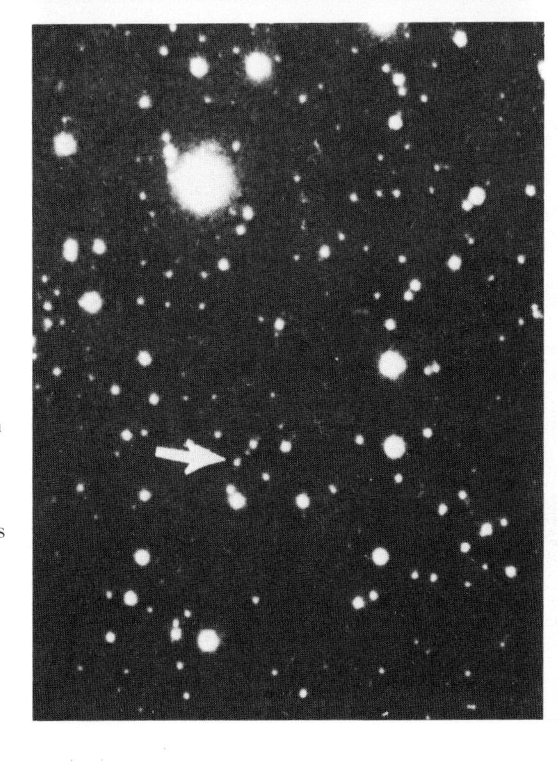

The photographs which led to Pluto's discovery in 1930 (*right*). On ordinary photographs like these, Pluto appears merely as a point of light, no different from star images. Its planetary nature was revealed only by its motion from night to night.

name of Pluto, ruler of the underworld, starts with the initials of Percival Lowell.

Pluto's orbit marked it out at once as unusual. As well as coming within Neptune's orbit, its path is tilted away from the plane in which all the other planets revolve about the Sun. Its faintness meant it could not be a gas giant; indeed it must be smaller than Earth. Its gravitational pull on Uranus and Neptune, however, meant it must be heavier than Earth—and putting these figures together, it turned out that Pluto's matter would have to be extremely compacted, as dense as lead! In the 1950s, astronomers had to ponder the idea of a small, exceptionally dense planet out in that part of the solar system where only lightweight gas, ice and ice-rock worlds should exist.

Pluto *has* turned out to be odd, according to recent research, but not as odd as this. It is now known that it reflects sunlight as if it were covered by snow—not snow as we know it on Earth, but a layer of frozen methane. Like ordinary snow, though, methane snow is dazzlingly bright. For a Pluto with a dazzling white surface to appear as faint as we see it, it must have only a small area of surface to reflect light. If it is entirely covered by snow, it must be not only smaller than Earth, but also smaller than the Moon.

Methane is a common gas in the outer solar system; Uranus and Neptune contain a great deal of it. At Pluto's chilly distance from the Sun, a methane atmosphere would freeze out and lie as snow on its surface. So it seems to be akin to the outer planets' satellites, formed from a similar mix of gases and rocks at the solar system's birth. And in that case, it must be a very lightweight world, about the size of the Moon. Composed mainly of methane ice, it would weigh only a third as much as the Moon—one two hundred and fiftieth of the Earth's weight.

This indirect weight estimate was confirmed in 1978. James Christy of the US Naval Observatory was looking at a recent photograph of Pluto when he realized that the image was not circular. Although it appears scarcely more than a point of light, the Earth's unsteady atmosphere smears its light out into a blurred circle on a long-exposure photograph. Christy saw that the image was not round, but pear-shaped. He searched through older photographs of the planet and found that sometimes the image appeared round and sometimes the mysterious pear-shape. The only reasonable explanation is that the 'neck' of the pear must be the blurred image of a fainter point of light, overlapping with Pluto's image and, from the way it comes and goes, this second point of light must be orbiting around the planet every 6 days 9 hours—a moon of Pluto. Christy has called the satellite Charon, the mythical boatman who ferried the departed across the river Styx to the land of Hades.

Charon is a strange moon. It is over a third of Pluto's size, judging by its relative brightness, making Pluto more a double planet than a planet plus satellite, and its orbit is tipped up, even relative to Pluto's own tipped orbit about the Sun. From Pluto's regular changes in brightness, astronomers have long known that the planet turns on its own axis once in 6 days 9 hours. This now turns out to be the same time as Charon takes to circle Pluto. Since Pluto is turning under Charon exactly in time with the satellite's motion around it, Charon appears from the planet's surface to hang motionless in the sky, like an artificial communications satellite in synchronous orbit above the Earth.

Charon is more than just another exotic beast in the solar system menagerie. From the size of its orbit and orbital period, Pluto's gravitational pull can be easily calculated—and since this depends on the amount of its matter, Pluto can thus be accurately weighed. Charon's orbit confirms that Pluto is indeed a lightweight—perhaps featherweight—world, only one-four hundredth as heavy as the Earth. Astronomers are now convinced that Pluto is no more than an overgrown snowball, similar to the satellites of the outer planets.

Indeed, it may actually be an escaped satellite. One early explanation for its odd orbit is that Pluto once orbited Neptune as a similar moon to Triton, which at that time orbited Neptune the correct way around. But at one stage the two moons approached too close. The gravitational pull of each on the other swung Triton back on itself into a reversed orbit, while Pluto was thrown out to orbit the Sun as a 'planet'.

This theory now seems less likely, but a modern version can perhaps save it: that another planet slightly larger than Earth passed through Neptune's satellite family, its gravity disrupting Triton's orbit and expelling Pluto. The intruding planet itself would have been flung into an even larger orbit, twice as far out as Pluto. According to most astronomers, however, it is more likely that Pluto formed where it is now. Perhaps it is the largest of a new group of asteroids at the fringes of the solar system. Closer in to the Sun lies the asteroid Chiron (not to be confused with Pluto's moon), orbiting the Sun in an elongated orbit between Saturn and Uranus, while a strange comet, called Schwassmann-Wachmann 1 follows an almost circular path just outside Jupiter's.

Chiron will eventually be thrown out of the solar system by the gravitational pull of one of the giant planets, but Pluto is luckier. Although it crosses Neptune's path, it is never close enough to that planet to be affected by its gravity. Pluto circles the Sun twice for every three orbits by Neptune, and the two planets always maintain a respectable distance from each other. Pluto could possibly be the survivor of a small family of eccentrically moving planets and asteroids. It just happened to lie in a 'safe' orbit while its less fortunate brethren were flung to the wilds of outer space as they encountered one or other of the giant planets.

With its diminutive size, the ninth planet is the runt of the solar system, minute compared to the next smallest planet, Mercury, and smaller even than many of the satellites circling the other planets—including our own Moon. Yet Tombaugh discovered Pluto by following calculations based on the fact that the unknown planet was tugging at the giant planets, Uranus and Neptune, perturbing their paths about the Sun. The new measurements of Pluto's mass show that it could not, in fact, have any discernible effect on either giant planet. Mathematicians have therefore looked again at the calculations, and shown that the suspected discrepancies in Uranus's orbit are probably not real anyway, but the result either of tiny errors in measurements, many made at the telescope eyepiece before photography was possible, or the way in which the effects of the other giant planets were calculated.

The final irony of Pluto, then, is that its discovery was luck after all. Only by chance did it happen to lie in the part of the sky which Lowell's calculations had indicated. It was really Tombaugh's painstaking search, guided by what has proved to be merely a fortunate guess, that led to its discovery. Although it was hailed at the time as a supreme success of mathematical astronomy, the tracking-down of Pluto was really one of those coincidences that sometimes occur in science, as in everyday life. In the last resort, it was due to the steady perseverance of astronomers actually scanning hundreds of photographic plates over the years to find a faint, slow-moving point of light patrolling the borders of the solar system.

Measuring the distances to the planets

The planets follow orbits under the influence of the Sun's gravitational pull, and the law of gravitation ensures that there is an exact relation between a planet's average position from the Sun and the time it takes to complete one orbit. Kepler's Third Law states that the cube of the planet's distance is proportional to the square of its period. If we measure the distance (D) in astronomical units (the Earth–Sun distance) and the period (P) in Earth-years, then

$$D^3 = P^2$$

The revolution periods of planets can be measured extremely accurately, so Kepler's law tells us their distances from the Sun very precisely, in terms of the astronomical unit. This yields a scale model of the solar system.

To find the distances in kilometres (or miles) we must know the size of the astronomical unit. Historically, many methods have been used, but today one method is so precise that it has displaced all others: radar distance ranging. A radio transmitter sends out pulses; these bounce off a planet and a radio telescope 'listens' for the echo. Because the speed of the radio waves is accurately known, the delay time tells us the distance to the planet. Once one planetary distance has been determined, the length of the astronomical unit can be calculated, and hence all the distances in the solar system deduced.

Radar astronomers use Venus as their target, because it approaches Earth more closely than any other planet. When Venus is closest to Earth, the time taken for a radar 'echo' to return is

276.034,38 seconds.

The one-way trip must be half this, namely

138.017,19 seconds.

We get the distance by multiplying by the speed of radio waves (equal to the speed of light):

$$138.017,19 \text{ seconds} \times 299,792.458 \text{ km/second} = 41,376,512 \text{ km.}$$

This is the distance between the *surfaces* of the two planets: to calculate the distance between the centres of Venus and the Earth, we must add the radii of the two planets:

$$41,376,512 \text{ km} + 6052 \text{ km (Venus)} + 6378 \text{ km (Earth)} = 41,388,942 \text{ km.}$$

From Kepler's Third Law, we know the Earth–Venus distance is 0.276,668 astronomical units (a.u.). Thus

$$0.276,668 \text{ a.u.} = 41,388,942 \text{ km}$$

Hence

$$1 \text{ a.u} = \frac{41,388,942}{0.276,668} \text{ km} = 149,597,860 \text{ km.}$$

(Since one mile is now defined to be 1.609,344 km, the a.u. is 92,955,800 miles.)

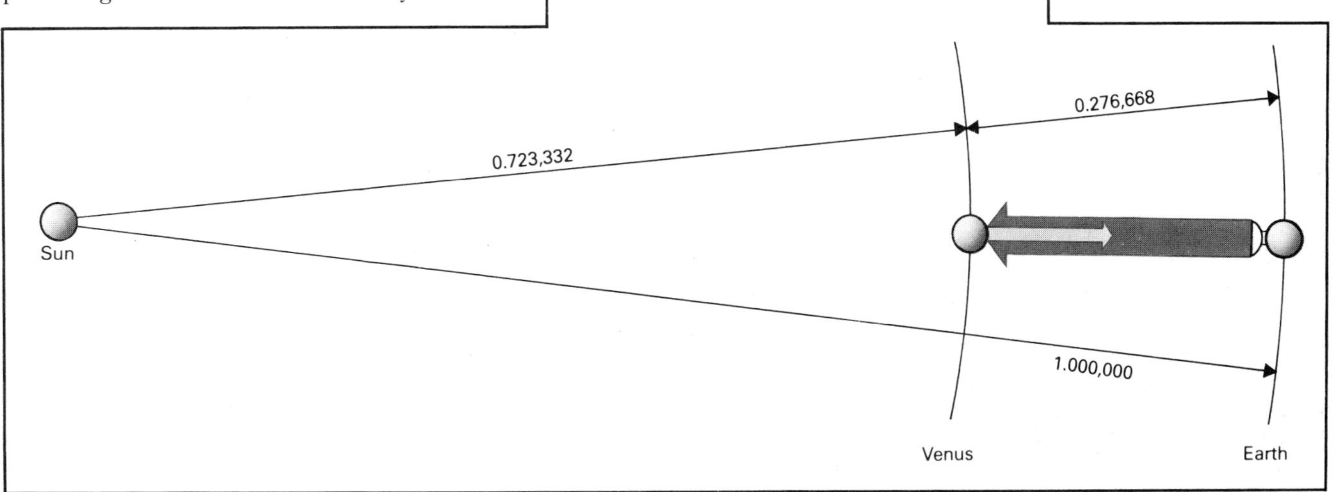

CHAPTER 3

THE INNER PLANETS

Small planets mark both of the limits of our solar system. While Pluto guards the frozen outer edges, the second-smallest planet patrols the torrid regions closest to the Sun. Mercury is roughly one-third the size of Earth, and even although it is practically a next-door neighbour in comparison to the outer planets, little was known about it until the Mariner 10 probe dropped in to visit it in the mid-1970s.

MERCURY

Because it circles the Sun so closely, Mercury never strays far from the Sun as seen from Earth. Occasionally, it can be seen twinkling low down just after sunset or before sunrise, but the twilight glow usually masks it effectively. Through a telescope the planet appears as a small pink globe bearing vague, darkish markings. Astronomers thought for a long time that these markings were in the same position, meaning that Mercury kept the same face turned towards the Sun on its 88-day orbit. But in 1965 radar astronomers bouncing radio waves off Mercury gave them a rude awakening: the faint echoes they received not only gave Mercury's distance, as in ordinary aircraft radar ranging, but confirmed that the planet was not rotating as expected. It is now known to turn once every 59 days.

This is the time Mercury takes to turn relative to the stars. But it does not correspond to our usual idea of Mercury's 'day', that is, the time from sunrise to sunrise. After 59 days Mercury has turned once on its axis as seen from outside, but it has also travelled two-thirds of the way round its orbit in this time. Since Mercury at that point is on the far side of the Sun, parts of its surface originally facing the Sun are now facing away from it—the night-side of the planet. Mercury's day from sunrise to sunrise is actually 176 Earth-days—twice the length of its 'year'.

Mercury has an elongated orbit, too, second only to Pluto in its departure from a circle. As a result, its speed around the Sun changes constantly; and a Mercurian would see the Sun speed across his sky change during the 'day'. When the planet is at its closest, the Sun slows to a standstill and backtracks for a while before resuming its slow progress from east to west. The region of Mercury where the Sun is overhead during this manoeuvre naturally gets more heated than the rest of its surface, and because of the peculiar relationship between 'day' and 'year' there are only two places on the planet at which this can happen. These regions, diametrically opposite each other and both lying on Mercury's equator, become the hottest places on a roasted planet, where the average noon temperature would melt lead.

All we know of Mercury's surface is due to the efforts of one sturdy spaceprobe. After passing and photographing Venus in February 1974, Mariner 10 went into an elongated orbit about the Sun which enabled it to rendezvous with Mercury every 176 days, two Mercurian 'years'. Its photographs showed that Mercury looks remarkably like our Moon. Its surface is splattered by craters of all sizes, with here and there flat plains where molten lava has leaked out and solidified. There is one huge crater, the Caloris Basin (so named because it happens to lie at one of the hot points on the equator) which takes up a tenth of the planet's surface. Here a huge rock from space, as large as Cyprus and as thick as it was wide, must have collided with Mercury to gouge out the hollow.

Mercury also has a patch of jumbled mountains, valleys and cracks. Astronomers in puzzlement labelled it 'weird terrain' and then noticed that it lies exactly opposite the Caloris Basin. Earthquake waves from the titanic impact must have spread out across the planet's surface and converged again at the opposite side. Here the waves running together must have produced one of the largest 'earthquakes' in the history of the solar system. Some mountains were shattered; others were raised in cataclysmic upthrusts; the crust was cracked up; the weird terrain was born.

And Mercury's surface is wrinkled, like an old apple. The 'wrinkles' are long, meandering ridges, crossing mountains, plains and craters alike, and are often 3000 metres (10,000 feet) high. The planet seems, in fact, to have shrunk, a fact which has been related to its denseness (Mercury rivals Earth as the solar system's densest planet). Since it is also a small world, whose central parts cannot be compressed much by the weight of the overlying layers, the only explanation for its density is that it has a huge, heavy core of iron-rich metal. Most of the interior of the planet must be taken up by this core, with only a thin crust

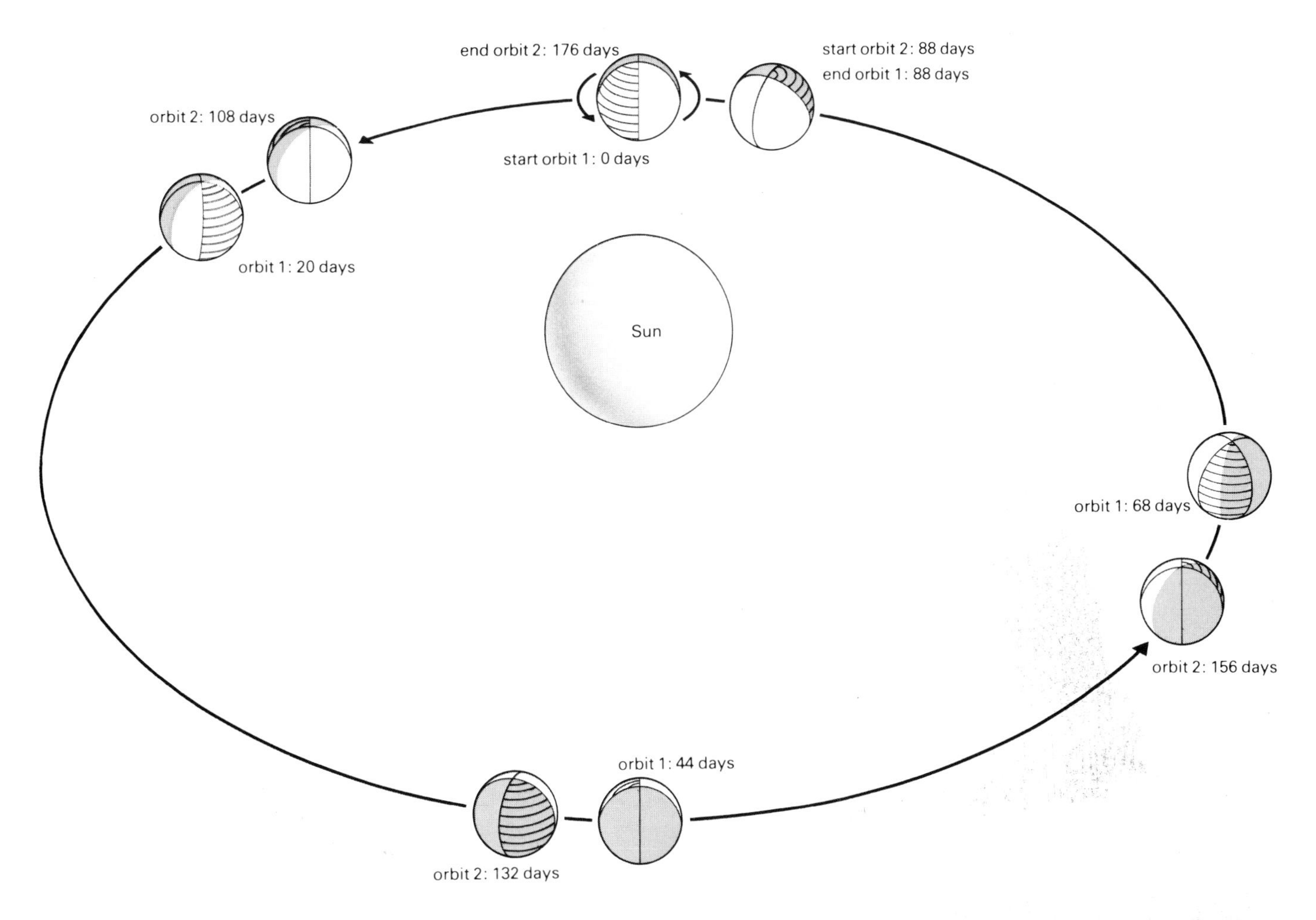

of rock on top. Since the planet's formation, the core has cooled and contracted in the process which, given the proportion of the planet it occupies, has had a drastic effect on the surface. Mercury's crust, therefore, no longer fits its shrunken centre.

All these interpretations are based on the Mariner 10 results. Superb as they are, however, the probe's pictures cover only half the planet, and less than this is shown in detail. When probes first went to Mars they sent back photographs of parts of the surface which turned out to be fairly unrepresentative of the planet as a whole. Perhaps the same will happen with Mercury. Astronomers do not expect the unknown half to be radically different from that shown by Mariner 10—but until a spaceprobe has been to see it, they will be wary of claiming that these are the final answers.

VENUS

The hottest planet in the solar system is not Mercury: Venus's surface is hotter even than the plains of Mercury subjected to the scorching Sun at the near point of that planet's orbit—yet Venus is twice as far from the Sun, where the heating effect should be far less.

Two consecutive orbits of Mercury (*above*) mark two of the planet's 'years' but only one 'day'. The striped segment indicates the planet's rotation. In the period shown, Mercury rotates three times as seen from outside; but its changing position relative to the Sun means the 'day'—is longer. Note that the striped portion has 'daytime' during the whole of the first orbit, and 'nighttime' for the second orbit.

The Mariner 10 photograph of Mercury (*right*) shows its cratered moonlike surface. Half of the huge walled plain called the Caloris Basin is visible (*centre left*), and sinuous ridges (*top left*) testify to the planet's shrinking.

This is just one of the paradoxes of Venus. Earthlike in size, and probably in its internal make-up, its surface and atmosphere are as alien as any in the solar system. It has for long been *the* mystery planet, for it is covered by perpetual, unbroken clouds. When astronomers had to rely only on the view from Earth, it seemed we could never know anything of Venus as a world, and guesses ranged from a barren desert to a lush tropical paradise.

Only in the 1960s did enlightenment come. As in the case of Mercury, it was astronomers bouncing radio waves off the planet who discovered the first hidden secret. Under her heavy veil, Venus turns the opposite way to all the other planets, Uranus and Pluto excepted. Unlike them, however, Venus is not tilted at a remarkable angle; its equator is inclined very little to its orbit.

And Venus rotates very slowly, turning once in 243 days, unlike the other planets which nearly all rotate within periods between a few hours and a day. The Sun's gravitation could have slowed its rotation, as it has undoubtedly done for Mercury, but that does not make the problem any easier to solve—it is even more difficult to explain if Venus was spinning *rapidly* backwards in the past. Could it perhaps be coincidence that all the other planets bar Uranus turn the same way? It depends on which theory of the solar system's birth you back; the conventional picture certainly finds Venus's spin a difficult and embarrassing problem, for all the planets should turn the same way as they move about the Sun.

The 243-day period is significant, too. It

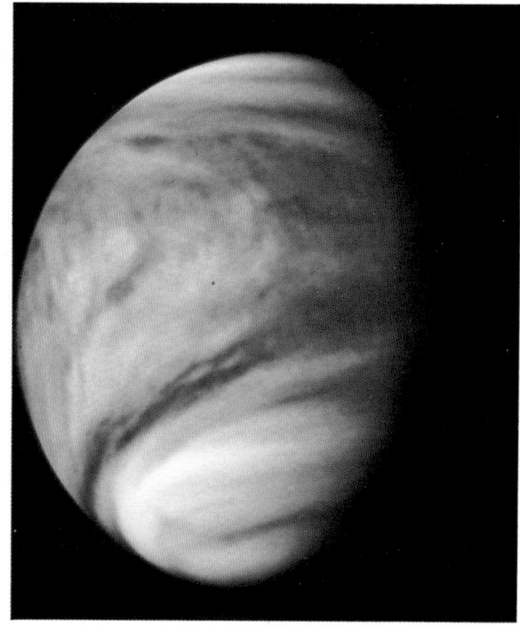

The unbroken sulphuric acid clouds of Venus are tinged yellow by sulphur, as seen in this Pioneer Venus Orbiter photograph (*left*).

Radar equipment on the Pioneer Venus Orbiter mapped the planet's surface using radio waves which penetrate the clouds. The contour map (*below*) is coloured at successive steps of 500 metres (1640 feet) in height. Most of the surface is low-lying and relatively smooth (purple, blue and green), but there are two major raised areas, Ishtar Terra and Aphrodite Terra. Aphrodite is the larger, about half the size of Africa, although the Mercator projection makes it appear smaller than Ishtar on the map.

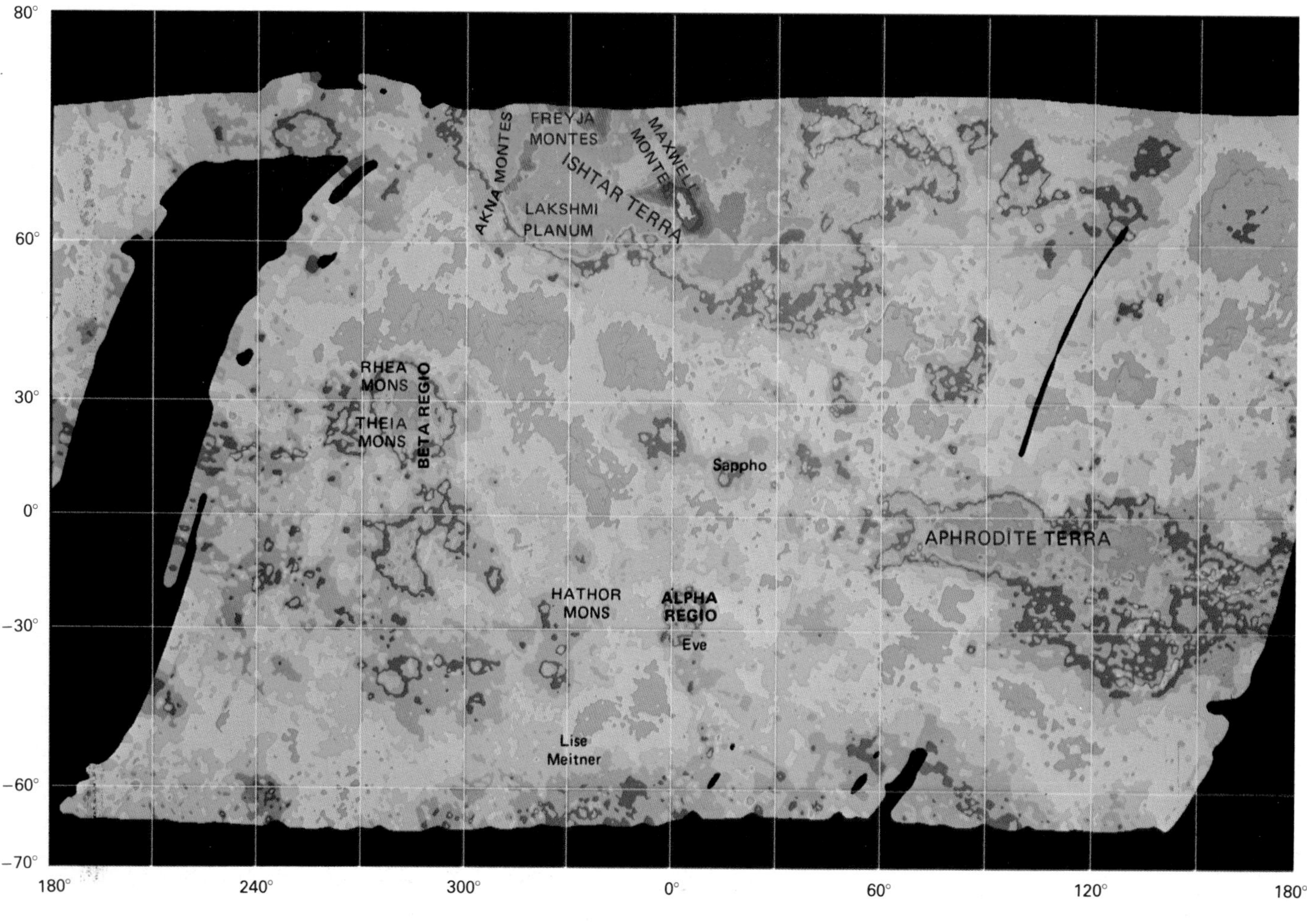

The radar data show Venus's surface more dramatically when converted to a topographic map (*below*). (Black areas are regions not mapped by Pioneer.) The mountain ranges ringing Ishtar Terra are clearly seen, including Maxwell Montes, the highest point on Venus. The isolated peaks of Beta Regio are probably extinct volcanoes. Venus does not have any evidence for 'continental drift': it lacks the long split ridges and the deep curved troughs that mark the edges of crustal plates on Earth—compare with the topographic map of Earth on page 63.

means that whenever Venus and Earth are closest together, the same side of Venus is turned towards us. During the time between one close approach and the next, 19 months, Venus has gone round its orbit just over $2\frac{1}{2}$ times, while it has rotated just under $2\frac{1}{2}$ times the other way; the two fractions add up to exactly one, meaning that precisely the same point on Venus appears at the centre of its disc as seen from Earth. Unless this is just chance, it is difficult to explain. If Venus were slightly egg-shaped, then Earth's gravity could have regimented its turning over the aeons of the solar system's history, but spaceprobes have measured Venus's shape and found that it is not nearly elongated enough. And the latest radar results show that the rule is not quite precise. When Earth and Venus are at their closest, Venus has not turned quite far enough—it is just a few hours slow. Over 19 months that does not mean much, but it gradually builds up. This makes the problem even more complicated. Why is Venus's rotation period so very close to that needed for Earth-resonance, but isn't quite? There are now two questions instead of one.

Venus is currently being surveyed by radar sets both on Earth and in the Pioneer probe put into orbit in December 1978. Their radio waves can penetrate the obscuring veil of clouds and show Venus's geography, its mountains and valleys. And there are spectacular examples of both. One mountain, named Beta, is as high as Mount Everest and 700 kilometres (420 miles) across, with tongues of rough ground extending as far again from it—it is probably a huge volcano surrounded by solidified lava flows that have overwhelmed its neighbourhood. There is an enormous plâteau, bounded on three sides by mountain chains, each a rival to the Himalayas. One contains Venus's highest mountain, named Maxwell after the physicist who predicted radio waves. Large circular rings mark the craters where the planet has been scarred by huge rocks from space, like all the other bodies of the solar system.

In addition to the mountains and craters, Venus's surface is split by huge valleys. These 'Grand Canyons' are far larger than Arizona's: they resemble the enormous cracks in the Earth's crust that make up rift valleys like the Red Sea. Although the surface rocks have cracked into slabs reminiscent of the continents of Earth, these slabs have not moved around the planet's surface as the continents have 'drifted' here. Since Venus is so similar in size to our planet, astronomers are puzzled by the fact that its geology lacks the most important feature of Earth's.

Man's only views of the land beneath the clouds were sent back by the Russian Venera 9 and 10 landing probes in 1975. They survived for about an hour on the surface of the planet and their television cameras revealed a rocky desert. One probe landed amid rounded rocks, where it seems that erosion has worn them smooth, and its sister-craft in a region of jagged rocks, perhaps thrown out in a recent volcanic eruption. But these two probes have looked at only a few metres around their landing sites, on a planet almost Earth-sized, and there is no guarantee that the other regions of the planet are similar—after all, two probes landing on Earth at random could both end up in the sea, and send back pictures suggesting that Earth is entirely covered by water.

The greenhouse effect

But Venus is certainly *not* the watery tropical paradise once thought. Its surface rocks are 465°C (870°F), so hot that rocks on the night side of the planet must glow dull-red in the dark. And as spaceprobes have come down through the atmosphere they have also discovered that even Venus's 'air' is mainly composed of unbreathable carbon dioxide gas. But the surprise has been that it is so thick: at surface level the atmosphere is 90 times denser than the Earth's, and the weight of the thick overlying atmosphere presses down with a force 90 times as hard as Earth's atmospheric pressure.

It is no surprise that none of the specially designed armoured landing probes have survived for more than an hour on such baked and oppressive soil. It is perplexing that two planets, roughly equal in size and distance from the Sun, have turned out to be so different. Venus's thick atmosphere is not even made of the same gases as Earth's; rather than oxygen and nitrogen, it consists almost entirely of carbon dioxide, a gas that forms only a small fraction of Earth's air. The oddity here is actually Earth. In the early days of the solar system, the inner planets were probably stripped bare of gases by the high-speed 'wind' from the Sun. Their present atmospheres must have arisen later. They may be the volatile matter from comets which hit the young planets, or they may be volcanic gases, puffed out by active volcanoes and exhaled from cracks early in their lives. In either case, the result should be a poisonous mixture of gases, including carbon dioxide and water vapour, with smaller traces of the noxious-

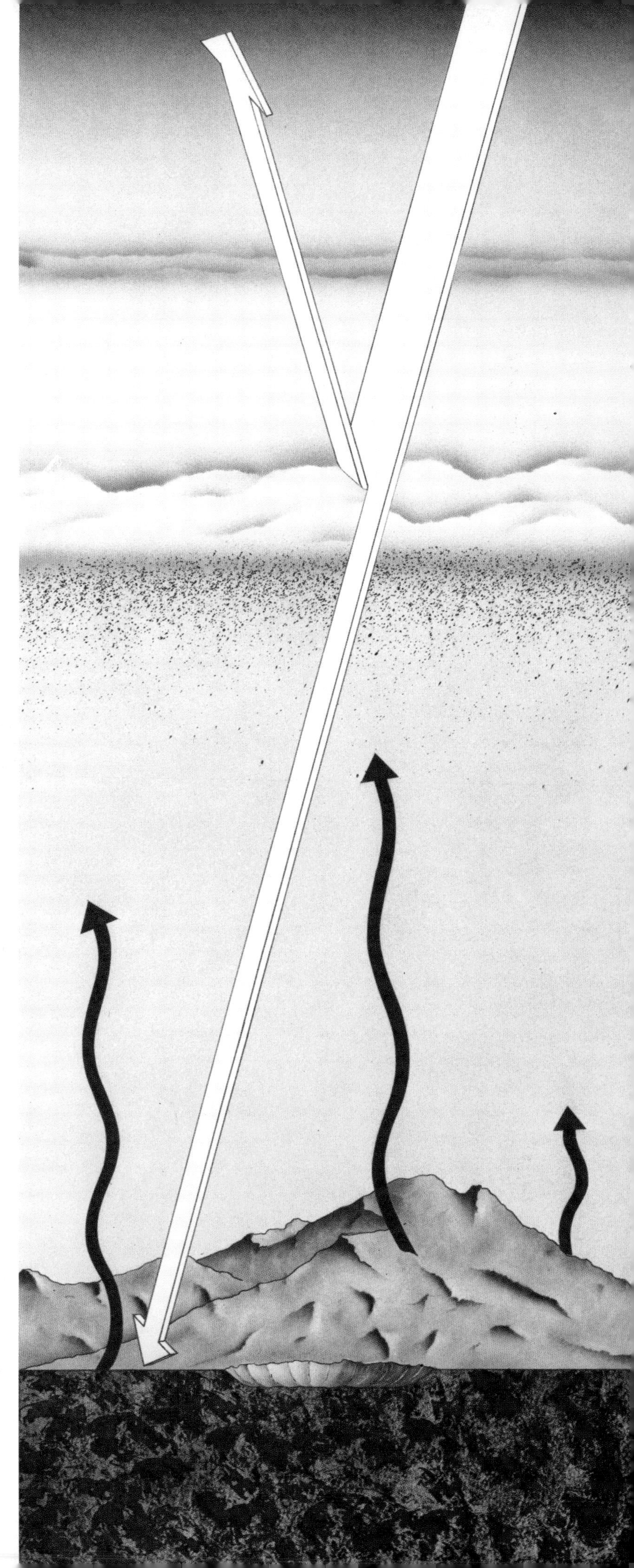

Venus's atmosphere creates a 'greenhouse' effect. Radiation from the Sun which penetrates the high clouds can heat up the surface rocks and the resulting infrared emission is absorbed by the thick carbon dioxide atmosphere, raising its temperature to 465°C.

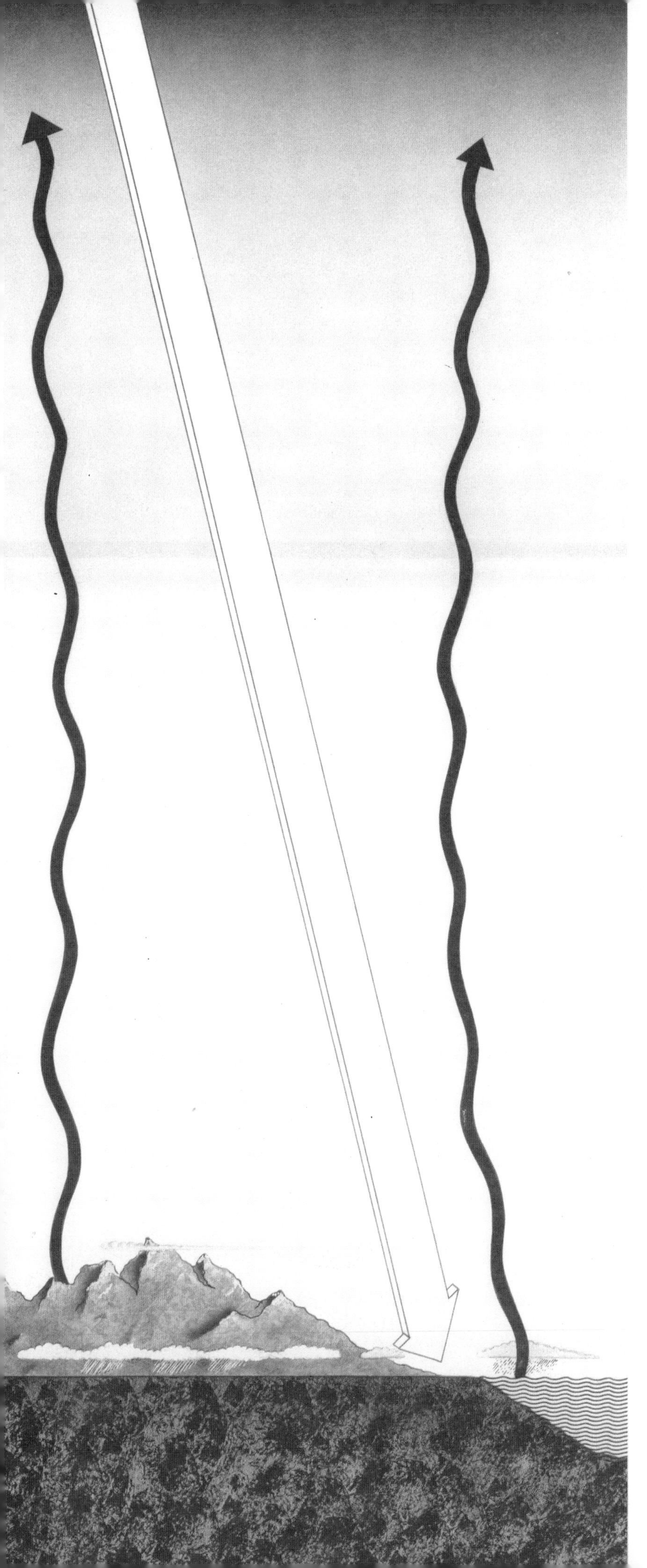

smelling sulphurous gases and the poisons carbon monoxide and hydrogen cyanide. So a vile, choking, unbreathable atmosphere is what we would expect—and why Earth does not have such an atmosphere we shall take up in the next chapter.

But Venus's atmospheric oddity is that it has hardly any water vapour. Its surface is so hot that any oceans would boil, and so the atmosphere should be thick with steam—but it is not. Only sensitive measurements by the most recent probes have found any at all.

However Venus acquired its present atmosphere, all that carbon dioxide should be matched by a huge quantity of water. No one knows where the water has gone. The most plausible theory is that water vapour at the top of the atmosphere is broken up by sunlight, and the fragments of the molecules speed up enough to escape into space. Water vapour from lower down takes its place, and is in turn ejected into space. Venus's potential oceans seem to have been jettisoned molecule by broken molecule into interplanetary space.

Venus's clouds are not like ours, then. They are not made of water drops at all; and we are still uncertain about the composition of this yellow-tinged veil. It is most likely to be droplets of concentrated sulphuric acid for, bizarre as this idea seems, this is the only substance yet investigated that reflects light as the clouds do, and the yellow tinge would simply be due to sulphur in the drops.

As the Pioneer probes descended, they found five layers of cloud, all much higher than Earth's. On Earth, clouds are mostly below the level of the highest mountain peaks, whereas on Venus the lowest layer lies at three times the height of Mount Everest. Hundreds of lightning strokes flash between adjacent clouds every second; so many, in fact, that the clouds must appear to be continuously lit from within. And the cloud tops whirl around the planet, completing a circuit in only four days while the planet below turns in a leisurely 243 days. The circulation of gases and clouds in the atmosphere is completely different from Earth's—and the reasons why are just as interesting to meteorologists trying to understand Earth's weather as to the astronomer with his eye on Venus.

It is Venus's thick atmosphere that keeps its surface hot. Although the clouds reflect most of the sunlight reaching Venus straight back into space, a small proportion filters down to the surface rocks and heats them. On a planet

In contrast to Venus, the infrared radiation from Earth's surface can escape largely unhindered into space, keeping our planet reasonably cool. Earth's watery clouds are at a much lower level than the sulphuric acid 'clouds' of Venus.

The solid bodies of the solar system differ greatly in their 'roughness'; below is a comparison of the range of heights on some of these bodies. In each case, the heights (and depths) are relative to the size of the body itself, and have been exaggerated a hundred times. Mars's volcano Olympus Mons is huge compared to the planet itself, while Jupiter's moon Europa is completely smooth.

like Earth, heat radiation from the rocks then escapes back into space, and the balance between the Sun's heating and this cooling keeps our planet at a reasonably comfortable temperature. But heat (infrared) radiation cannot easily penetrate carbon dioxide. Venus's heat is thus trapped at its surface, and the rocks have become blisteringly hot. Because a greenhouse works in a similar way (though on a minor scale!) as its glass traps heat radiation within, this is known as the 'greenhouse effect'. It happens to some extent on all planets with atmospheres, but Venus has taken it to extremes.

Carbon dioxide alone could not keep Venus *quite* as hot as it is. The Pioneer probes found some traces of water vapour, and although the amount is small, the water molecules in the atmosphere help to plug the leak of heat radiation. Ironically enough, it is water that helps to keep Venus too hot for life to survive there. But just suppose that man could, at least briefly, stand on its dimly-lit surface. Perched on red-hot rocks, crushed by the heavy atmosphere of unbreathable gases, and exposed to a slow drizzle of sulphuric acid, the surroundings would be macabre indeed—competing only with Io for the title of Hell in the solar system.

MARS

It is a relief to turn from Venus to the most Earth-like of our planetary neighbours. Through a telescope Mars looks inviting: We see the planet unhidden by cloud, with dark markings and white polar caps. And it turns in almost the same time as the Earth, $24\frac{1}{2}$ hours.

A century ago astronomers were misled by Mars's similarities into thinking it was indeed a second Earth, peopled by intelligent beings. The dark patches change with the seasons during Mars's long year, almost two Earth years in length, and this seemed to show that there were patches of vegetation on an otherwise barren desert of a world. The close approach of Mars to Earth in 1877 really started the Martian craze. Italian astronomer Giovanni Schiaparelli saw narrow dark lines linking some of the dark patches, and he called them 'channels': in Italian 'canali'. Inevitably, this was mis-translated as the 'canals of Mars', and since the word 'canal' implied they were artificial, they were thus believed to have been built by intelligent Martians!

The doyen of Martian canals was Percival Lowell, who founded the observatory where Pluto was later discovered. He saw literally dozens of canals stretching across the Martian deserts. His explanation, too, was more than plausible. Mars harbours an intelligent race, whose world is drying out; they have constructed irrigation channels to bring water from the icy polar caps to irrigate the equatorial deserts, and the 'canals' visible from Earth are waterlogged fields beside the channels.

Even at the time, other astronomers were doubtful. Probably the most keen-sighted observer of Lowell's time was E. E. Barnard, and he saw no canals. Mars has many small dark patches just at the limit of visibility for an Earth-based telescope, and he thought that an eye straining at its limits could connect these to give the appearance of straight lines. With the benefit of hindsight—and interplanetary probes—we know that Lowell was indeed wrong: there are no straight canals on Mars. The dark patches are not vegetation, but exposed rocks, the colour of which alters

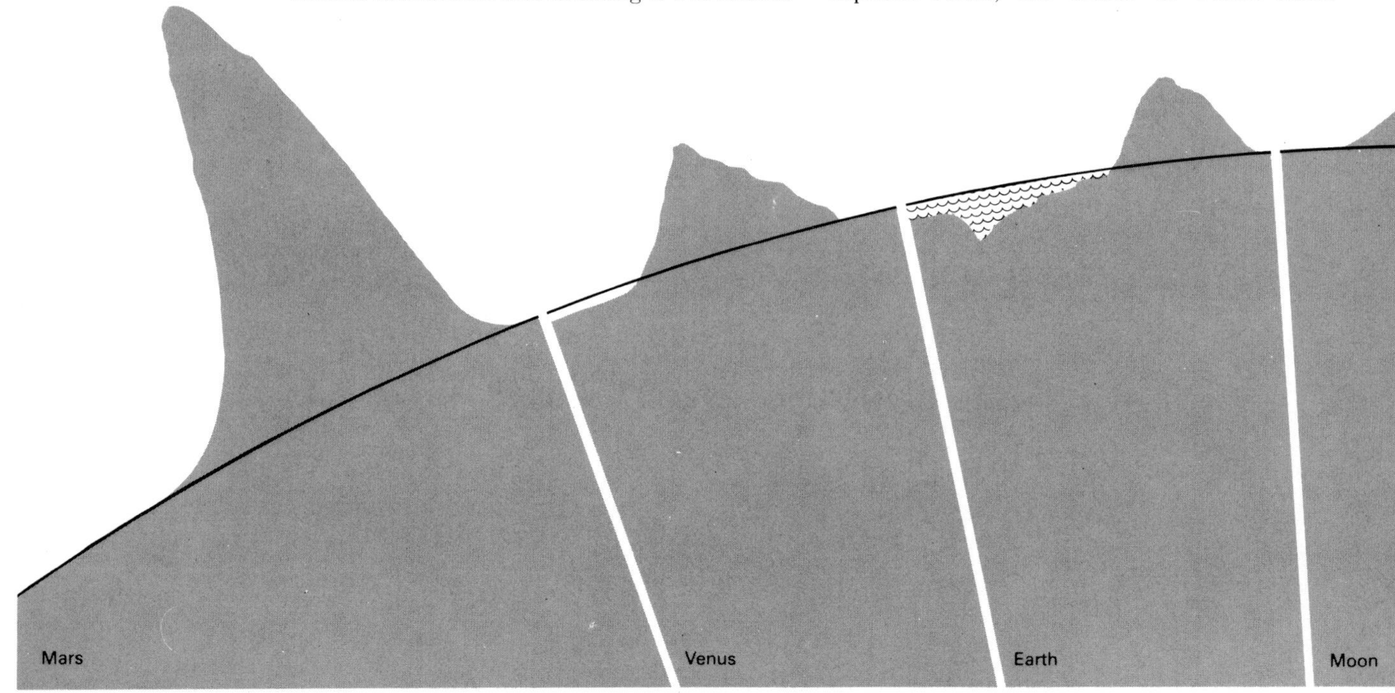

throughout the Martian year as seasonal winds sweep the brighter desert sands on to them and off again.

But the controversy is not quite over. The American Mariner 9 and Viking 1 and 2 probes have photographed the entire surface of the planet in great detail, and these pictures show a myriad of craters, huge volcanoes and gaping canyons. Very few of these correspond to Lowell's canals, though. A huge equatorial valley, now called Valles Marineris (after the Mariner probes), is Lowell's canal Agathodaemon; a couple of canals have turned out to be ridges; and another two are actually chains of craters. These seem to bear out Barnard's idea—but they are only a small fraction of the canals mapped by Lowell. If Lowell and Schiaparelli had been the only astronomers to see Martian canals, it would be easy to believe that they were just poor observers with vivid imaginations. That would be unfair, however, for many other astronomers since have seen some, and though few claim to have spotted as many as Lowell, these canals do seem to coincide exactly with Lowell's darkest and most prominent ones. Is this just psychological suggestion, a case of astronomers seeing someone else's Mars maps and unintentionally reproducing them when straining to see detail at the telescope? Or could there be darker regions, forming roughly straight lines without Lowell's precision or narrowness, that could be seen at a distance by the human eye (a very good detector of faint contrasts), but don't show to the television cameras skimming Mars's surface? Even if they exist, the Martian canals are not telling us anything interesting about Mars as a world, as the probes have revealed in their detailed pictures; but for all that, it is a controversy that refuses to die altogether.

And while on historical controversies, the 1877 approach of Mars raised another, more light-hearted puzzle. Asaph Hall, testing a large new telescope at the US Naval Observatory, spotted two points of light near Mars, tiny moons circling the planet. As companions for Mars, god of war, they were named Phobos (fear) and Deimos (terror). But in one sense they were not new. Jonathan Swift, the satirical writer, had included an account of advanced observations made by a fictional civilization on the flying island of Laputa, one of the lands visited by Gulliver in the course of his travels. And in this book, published in 1726, his fictional astronomers discovered two moons circling Mars. In this account they are small and orbiting very close to the planet—remarkably similar to Phobos and Deimos.

In recent years Phobos has generated more excitement. Accurate measurements show it is very gradually spiralling in towards Mars, and will hit the planet in around 100 million years time—to gouge out another crater on its already pockmarked surface. Phobos is feeling the drag of Mars's tenuous outermost layers of atmosphere, like the ill-fated space station Skylab which plummeted to a fiery end above Australia in July 1979. Before the first spaceprobes reached Mars, some scientists thought the analogy could be exact. In the 1950s the leading Russian astronomer Iosef Shlovskii calculated that Mars's outer atmosphere could not affect Phobos unless the satellite were very lightweight, and that would mean the 20 kilometre (12 mile) diameter moon must be hollow. It obviously could not be natural, and Shlovskii thought it might be a vast space station put into orbit by now-extinct Martians.

Spaceprobes have since shown that Phobos is no more than a battered, cratered chunk of rock; Mars's atmosphere simply has more effect than expected. In close-up, the world shows strange, parallel grooves extending over its surface, possibly because the tidal forces raised by Mars's gravity are stretching it. Perhaps Phobos is so fragile that it will disintegrate into a set of rings instead of crashing on to Mars, adding to the solar system's tally of ringed planets. Phobos and

Europa
Ganymede
Io

Deimos may be the final fragments left over from Mars's formation, soon to join their brethren already incorporated into the planet; or they may be asteroids which have been captured fairly recently—perhaps Deimos orbited Phobos as its satellite while they were both in the asteroid belt.

As well as Phobos and Deimos, Mars now has three artificial satellites: Mariner 9 and Vikings 1 and 2. Even before these ever-circling probes were sent to the planet, the first 'flyby' spacecraft revealed Mars to be cratered. Remarkably enough, E. E. Barnard saw some of the larger craters on Mars at the turn of the century, but never published the fact. By unhappy coincidence, the first three probes to fly past Mars only photographed some of the least interesting parts, moonlike regions populated only by craters and flat plains. The orbiting probes have found Mars to be a much more interesting world. Like Venus, it has mountains and valleys larger than any on Earth—yet Mars is only half Earth's size.

The largest volcano is Olympus Mons, named after the abode of the Greek gods. It towers over the surrounding plains to a height three times that of Everest, and it is broad enough at its base to cover a country the size of Spain. Like Mars's dozen other giant volcanoes it is now inactive. Near the equator is the huge canyon Valles Marineris, a large rift valley where the Martian surface has cracked open; it is large enough and deep enough to hide the entire Alps.

Mars probably does not suffer 'continental drift' in the same way as Earth. How then were its mountains built, its canyons cracked open? It must have interior pockets where rocks have become molten, and erupted as volcanoes. Next to the volcanoes of the Olympus Mons region, there is a huge swelling, where lava has apparently pushed the surface up, and the vast canyon of Valles Marineris seems to be split on the side where the swelling has overstretched the surface rocks.

The volcanoes are no longer active. Scientists are in dispute over how long it has been since erupting volcanoes lit up Mars's red surface. Most of the eruptions probably took place thousands of millions of years ago, halfway through Mars's life as a planet. Olympus Mons was probably active around 100 million years ago (comparatively recently on the solar system's timescale) and some astronomers think activity may have continued up to a million years ago. In this case Mars may not be volcanically dead, but just dormant at the moment. It is more likely, though, that the planet has by now cooled too much for molten rock ever again to belch out through its oversized volcanoes.

Another surprise awaited the circling probes. Mars has winding, branched channels, looking for all the world like dried-up river beds—and after trying other explanations, astronomers have been forced to conclude that that is exactly what they are. (They are no relation to the 'canals', incidentally, being far too small to be visible from Earth.) But Mars is

Percival Lowell mapped dozens of straight dark 'canals' on Mars, (*below*). None of them exist, but a few correspond to real Martian features.

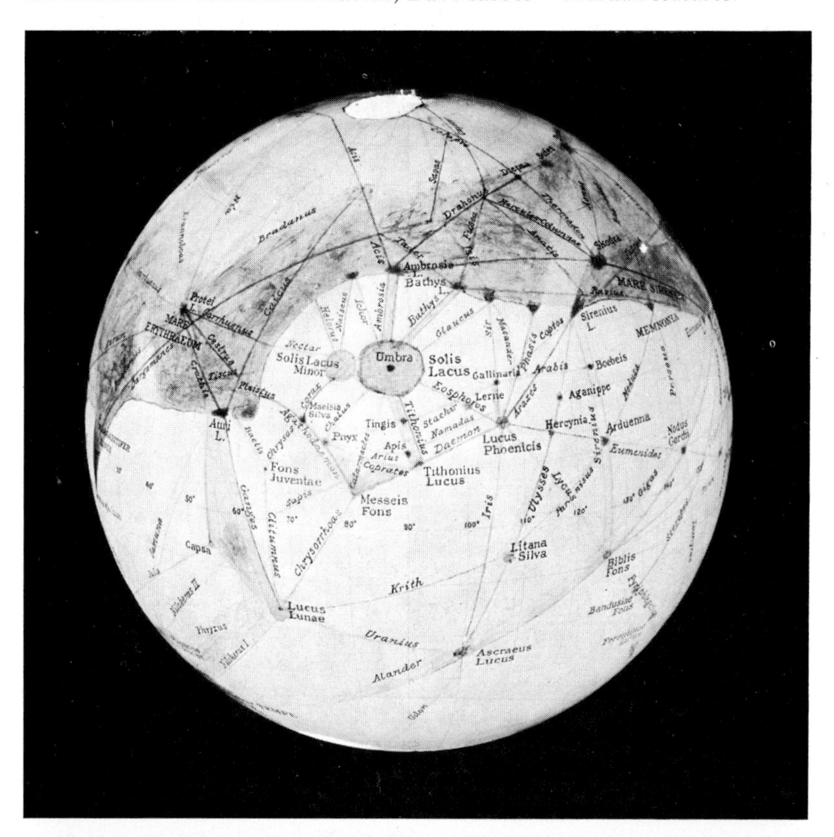

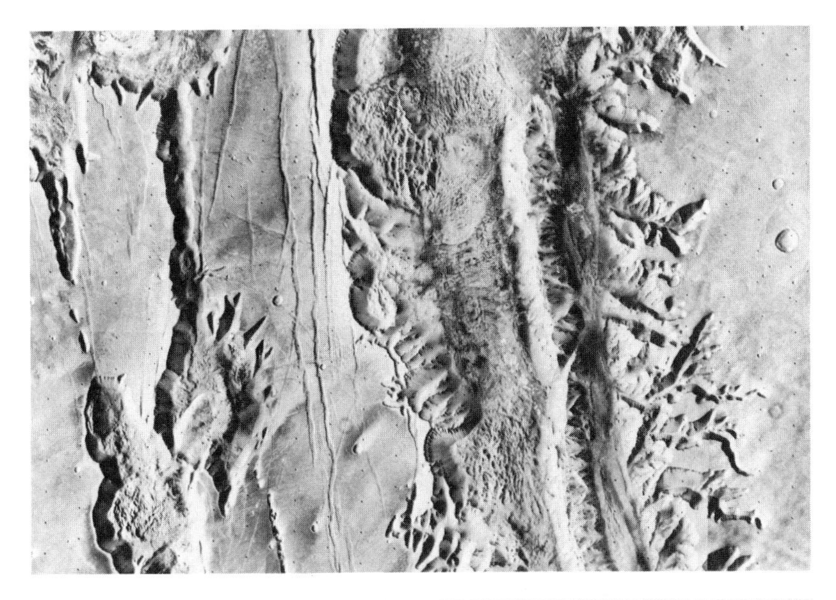

Valles Marineris (*above*) is a complex rift valley. Lowell saw the Valles Marineris, and drew it as two canals—Agathodaemon and Coprates.

The extinct Martian volcano Olympus Mons (*right*) is a wide cone, of the type known as a 'shield volcano'. It is far larger than any volcano on Earth, rising to 26,000 metres (86,000 feet), and stretching over 600 kilometres (400 miles) at the base.

Topographic view of Mars (*left*) highlights the extensive terracing around the polar cap regions. Some of these scarps are 500 metres (1600 feet) high. The effect is probably due to the wind eroding loose dust particles from the surface.

too cold to have running water—its average temperature is a frigid −40°C (−40°F), way below the ice point. On the other hand, astronomers think that Mars's soil has ice frozen into it, like the similar frozen wastes of Siberia and Northern Canada. The top layers are therefore a soil-ice mixture: permafrost. And each polar cap contains thick layers of ice at its base; even in summer, temperatures here are too low to melt the ice, while in winter they plummet so low that carbon dioxide gas freezes out of the atmosphere to create a cap of 'dry ice' that dissipates again in the summer.

There is enough water on Mars then to carve out the channels—if it could be melted. Another problem has to be faced: Mars's atmosphere is a hundred times thinner than the Earth's, which means there is so little atmospheric pressure that liquid water would instantly boil away as water vapour, even at low temperatures. If rivers ran for any length of time in Mars's past, the planet must have had a thicker atmosphere than it does now.

Some of the rivers were probably only transitory. Where molten rocks reached the surface, or stopped just below, their heat could have melted the ice in the permafrost layers and produced a sudden torrent of water, which would flow for a while before all the liquid evaporated into the atmosphere. A meteorite hitting the frozen surface could also melt the ice locally, or open a new channel to water below the surface. Some channels on Mars do, in fact, look very like the valleys excavated by flash floods on Earth caused, for example, by a dam bursting.

But others seem to be telling of a radical change in climate. These patterns occur mainly in the oldest, unchanged parts of Mars. When the planet first formed, and the volcanoes were cloaking it with gas, the atmosphere was probably considerably thicker than it is today, and it looks as though rivers could flow then. As the volcanoes have one by one cooled off, gas escaping from the weak gravity into space has not been replaced. And as its atmosphere has leaked away and thinned out, Mars has frozen.

The layers of dust near its poles show a more complicated history than this. These successive layers of different thickness indicate that Mars, like Earth, has incessantly been coming and going through series of Ice Ages. And on Mars the fluctuations in climate were probably more extreme, for its tilt (one of the factors that affects a planet's average temperature) changes more than the Earth's over hundreds of millions of years. So even while Mars was more hospitable, it must have had periods when the ice advanced and froze its surface. Conversely, is its present cold barren surface only temporary, just an Ice Age affecting the planet? Perhaps Mars, despite its thin atmosphere, will one day warm up again. And some astronomers believe that the amount of carbon dioxide and water vapour that would fill out Mars's atmosphere then would create a 'greenhouse effect' which would keep its surface warm.

Attractive as this idea is, the planet's volcanoes may well have halted the procession of climatic swings. The huge swelling near Olympus Mons has affected the amount that Mars's tilt can change, and as a result it may never again fully recover from its present frigid state.

For all its frozen austerity, its cooled-off volcanoes and its heritage of ancient craters, Mars is the only planet which seems remotely like home. The outer solar system, of huge liquid planets and their frozen ice-ball moons, is alien to man; Pluto still more deep frozen. Mercury is barren and parched, Venus a Hell mercifully hidden by its off-white clouds. Beyond our Moon, Mars is man's next target in space. Among its mountains, deserts and dried-up valleys, not so different from our own deserts, man would not feel too homesick for Earth.

CHAPTER 4

THE DOUBLE PLANET: EARTH AND MOON

Between the orbits of Venus and Mars lies the strangest planet of our solar system. It is, in fact, a double planet. Earth and Moon circle each other like a pair of dissimilar twins as they together orbit about the Sun.

Although it has been traditional to call the smaller of the two the 'satellite' or 'moon' of the larger, the Moon is fully one-quarter Earth's diameter, far larger in proportion to its controlling planet than any other satellite in the solar system (with the exception of Pluto's moon, Charon). The Moon is not much short of Mercury in size, and would make a respectable planet if it were orbiting the Sun in its own right. And it is not strictly fair to say that the Moon orbits the Earth: the Moon has an appreciable gravitational tug on the Earth, too. Both bodies are orbiting around a balance point that is not at the centre of the Earth but lies along the Earth–Moon line just below Earth's surface. This balance point traces out the 'Earth's orbit' around the Sun, while both Earth and Moon swing around it once a month.

The Moon is just the kind of small, rocky, cratered world we would expect to find at this position in the solar system. Its companion is not. Even leaving aside its double-planet status, Earth is odd. Its exposed rocky surface bears hardly any sign at all of craters, and its atmosphere is unique in the solar system in consisting of nitrogen and oxygen gases, with barely a trace of carbon dioxide. Even a casual glance at Earth from a distance reveals that it is different from any other planet. The others range in colour from yellow pale, through shades of orange to ochre-red; Earth, seen at midnight from above Venus's clouds, or as an 'evening star' in Mars's pink skies, is a vivid blue. Just as man has always known Mars as 'the red planet', so others would know our home as 'the blue planet'.

THE LIFELESS MOON

The Moon's closeness to Earth has been an exceptional stroke of luck for astronomers studying the planets. It has meant, for instance, that men have been able to travel to and explore another planet-like world. Well before Neil Armstrong and Buzz Aldrin set foot on the Moon in 1969, astronomers knew that it was a dead, barren place. It has no atmosphere, for its weak gravity cannot hold

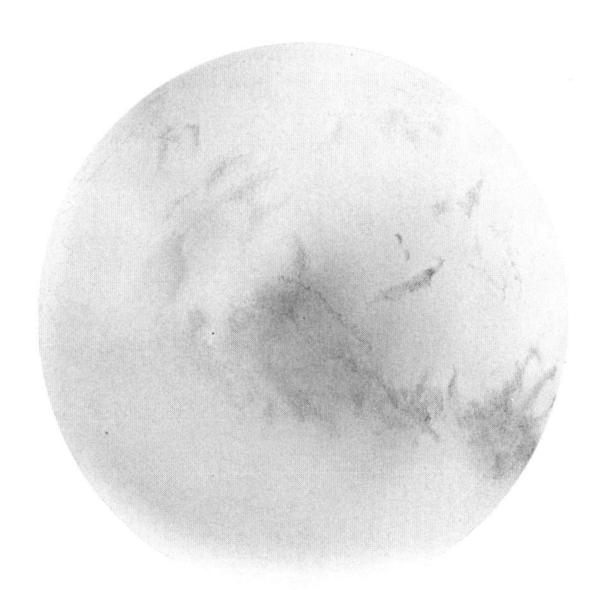

Earth and Moon (*left*) are effectively a double planet, with the 'satellite' world fully one-quarter as large as the major body. In contrast, the moons of Mars (*centre*) and Jupiter (*right*) are miniscule when compared to the planet itself.

onto gas. As a result, it is exposed to all the radiation from space, and extremes of heat and cold, far greater than ever encountered on Earth.

The Moon is now better known in some ways than Earth. Its main features have long been familiar: through a telescope, the dark patches (lunar 'seas') making up the face of the 'Man in the Moon' turn out to be large flat plains of solidified lava, while the lighter patches between are higher-standing 'highlands'. Ring-shaped rock walls, the craters of the Moon, are strewn over both plains and highlands, peppering the latter so thickly that there is scarcely space between them. Spacecraft circling the Moon in low orbits have now photographed and mapped it as completely as they have studied the Earth.

Astronauts in the six successful Apollo landings left instruments which have measured the heat flowing out of the Moon. They also returned with rocks from half a dozen different locations. Unmanned Russian probes have added to these samples by bringing back to Earth small amounts of soil from three other sites. And although this may seem a small number upon which to assess the geology of a whole world, compared, say, to the huge number of outcrops and borings that would have been studied for a similar exercise on Earth, scientists have found that most types of rock are present in all nine sites.

Placing together the evidence, it is possible to come up with a convincing picture of the Moon's early days. It most likely formed about 4600 million years ago at the birth of the solar system, and its outer crust of rocks, still exposed in the highland areas, must also date from about this time. (These rocks are unusual, in that they contain a great deal of the elements aluminium and titanium—which could make the lunar highlands a prime site for future mining operations, since titanium is a useful and expensive metal.) They probably separated out of molten rock of a more ordinary composition, and floated to the top like scum.

The original lava, covered by this solid scum, quickly solidified. From then on, the infalling planetesimals making up the Moon hit a solid surface and gouged out craters. Some were the craters we see today in the highlands, others were larger— up to 1000 kilometres (620 miles) across. The walls of these craters still stand as the towering curved mountain chains surrounding the lava plains, but at this time they contained no lava.

For its first 700 million years the Moon was pounded by planetesimals, in its final sweeping up of the rock-infested space around. Its surface rocks were pummelled and strewn over the rest of the surface by the impacts; they were broken into fragments, then welded together again as conglomerates containing bits of rock of widely different ages and provenances (a type known as *breccia*).

A second chapter began as the rain of planetesimals petered out. The Moon was now solid throughout, but at this time the layers of rock immediately below the surface began to get hot. Heat from radioactive atoms, like uranium, melted pockets of rock. The lava flowed up through the Moon's cracked surface layers, and spread out on the lowest-lying parts of the surface, the huge craters left by the biggest impacts, to produce the lava plains. Although the dark plains are so large they can be seen from Earth without a telescope, they are so thin that they constitute less than one-thousandth of the Moon's material.

This second chapter also lasted 700 million years. Since the end of that time, 3200 million years ago, the Moon's surface has hardly changed. Small meteorites have peppered the once-smooth lava plains with a myriad of small craters, while the occasional large one has blasted out a few bigger ones—those known as Copernicus and Tycho are less than 1000 million years old.

Moonquake recorders left by the astronauts have confirmed that the Moon is virtually dead. Some quakes do occur, but their sources are deep, halfway to the centre—probably around small still-molten rock pockets. The total energy released by them in a year is less than that involved in firing an artillery shell, whereas quakes on Earth total some 250 million Hiroshima-type atomic bombs over the course of a year.

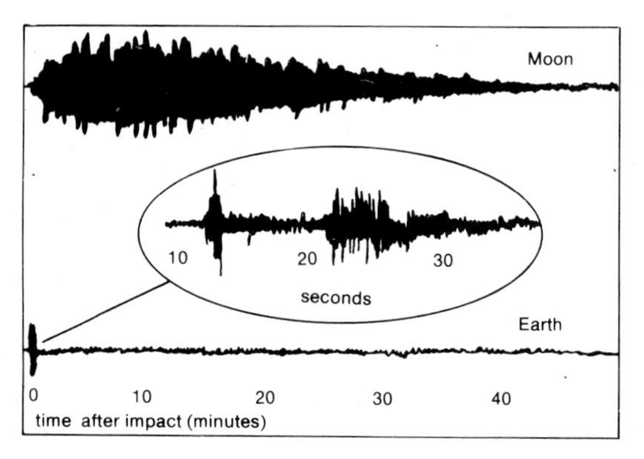

The major enigmas of the Moon's early life now seem to have been resolved, but this does not mean that it has been stripped of all its mysteries yet. For a start, there is the strange preponderance of titanium in the highland rocks. And mixed in with these ancient rocks is another odd type, which has been named KREEP. The name certainly fits, but it has a prosaic explanation: the rocks contain an unusually large amount of potassium (chemical symbol K), 'rare earth' elements (REE) and phosphorus (P). Their distribution shows that KREEP basalts must have been molten after the crust had solidified, but well before the dark molten lavas flooded the lowlands to make the lunar seas. Because it is so ancient, KREEP rock has been thoroughly broken up and mixed in with the crust rock by infalling meteorites, and that makes its origin even more difficult to interpret. It does, however, show that the first chapter of the Moon's history was not quite as simple as the rest of the evidence suggests—and that there must be some other episodes in its past which remain to be unveiled.

TIDE AND TIME

According to astrologers, all the planets have an important effect on human life and personality, but none more so than the Moon. Its changing shape, as the Sun illuminates it from different angles, speaks, they say, of changeability and moodiness; the 'Moon characteristics'. Astronomers, of course, have no truck with astrology. It is inconceivable that remote planets such as Saturn or Neptune could affect either humans, or even the Earth itself. But the Moon is the one body in the sky (apart from the Sun) that does undoubtedly influence us. Sailors, dockers, fishermen—all those connected with the sea for their livelihood—have their lives governed by the rise and fall of the tides; and the ocean tides are raised by the Moon's gravitational pull.

It is easy to see that the Moon can pull up the water on the side of the Earth nearest to it, but there is also high water on the opposite side: the oceans are stretched out into a slight oval shape. There is no mystery. The Earth is swinging around the balance-point of the

The Moon's surface (*above*) is scarred by meteorite impacts dating from soon after its formation. Seismometers on the Moon show that it vibrates for much longer than Earth when hit by a meteorite (*top left*). Although this came as a surprise at first, geologists now explain it as a consequence of the Moon's drier rocks. Manned expeditions have returned selected rock samples which have revealed details of the Moon's past history. Apollo 17 astronaut and geologist Harrison Schmitt investigates a large boulder in the Taurus mountains (*right*).

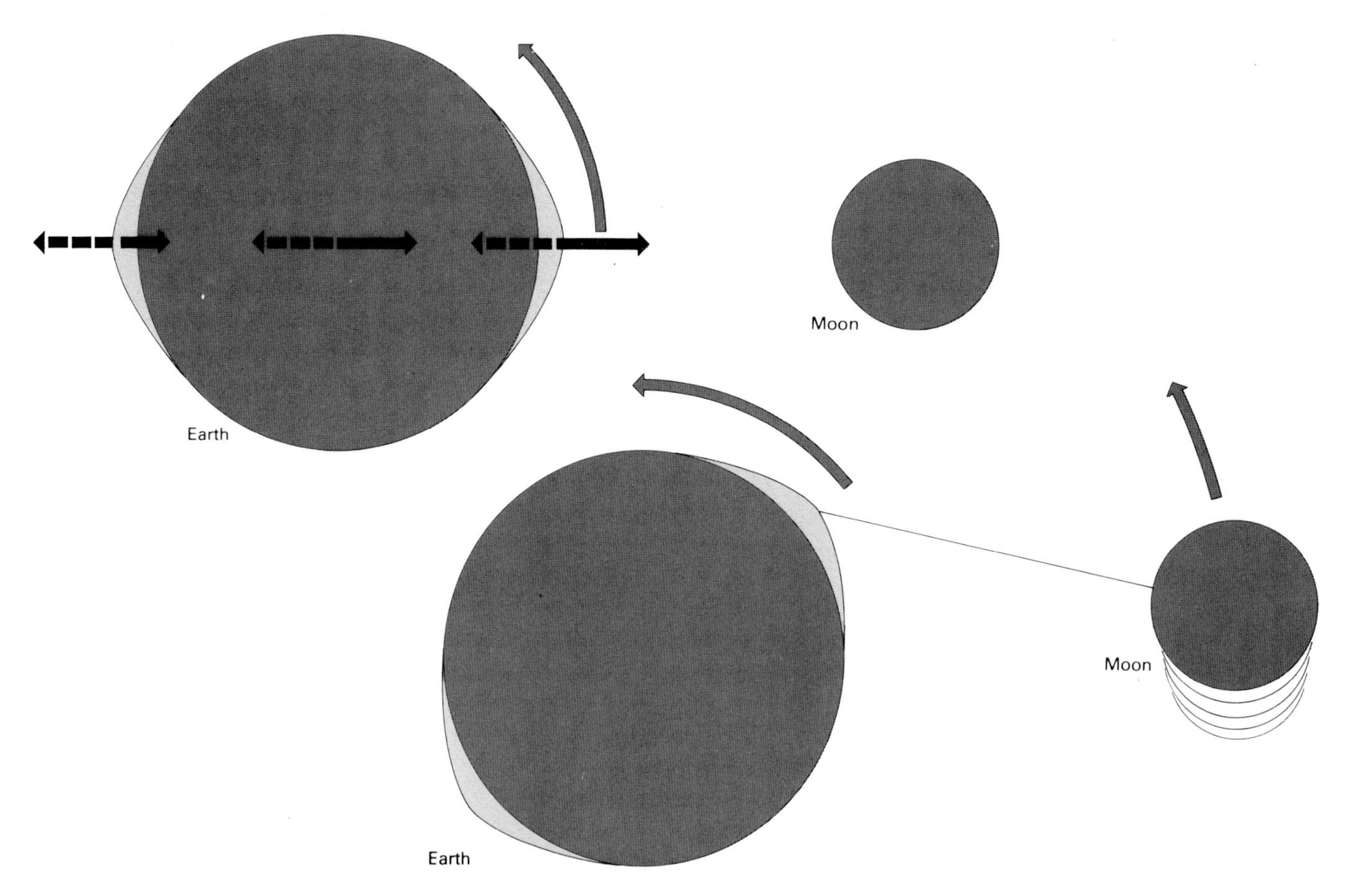

The Moon raises two tidal bulges in the Earth's oceans. The 'centrifugal force' due to Earth's motion in the double-planet system is the same everywhere on Earth (dashed arrows); the Moon's gravitational pull (solid arrows) differs. On the side facing the Moon, the gravitational pull is stronger; on the other side, centrifugal force is the more powerful. In both cases, the result is an outward force, raising the sea level. Friction between the tidal bulges and the ocean floor slows down the Earth's rotation and pulls the bulges ahead of the Earth–Moon line (lower diagram). The displaced bulges tend to pull the Moon ahead in its orbit, gradually increasing its size.

Earth–Moon double planet system once a month, and the centrifugal force on the far side as it wheels around the balance-point pulls the ocean water outwards here as well.

So the Earth's oceans bulge out both towards the Moon and away from it. As we turn once a day, every seaport on the planet is carried through regions of high water twice a day, with low tides in between. High tide occurs some 25 minutes later each day, in fact, because the Moon has moved part of the way round in its orbit during the intervening 24 hours. So the tides do not keep pace with the rhythm of day and night, and the life pattern of those who work immediately with the sea is correspondingly complicated.

The tides influence the Earth itself. As water ebbs and flows in the shallow seas, it drags against the sea floor. The ocean bulges lying on either side of the Earth act as a pair of huge brake-shoes on the rotating solid planet beneath. As a result, the Earth is slowing down: the day is getting longer.

The most accurate way to measure day-length is to time exactly when stars pass overhead, and astronomers armed with atomic clocks to give accurate time-keeping have indeed found that they reach the overhead point slightly later each day. The slow-down *is* very gradual—during the course of a century the length of a day increases by one-thousandth of a second—but, since there are 365 days in a year, the discrepancy does mount up to become significant. An atomic clock running at the rate appropriate to the day-length of a century ago would now be gaining on a clock set to Earth's present rotation period by about one-third of a second per year. These odd fractions pile up relentlessly, year after year, and in a world where time must often be quoted to fractions of a second it is important that everyday time, based on the passing of days, is kept in line with the uniform flow of time revealed by atomic clocks. This is achieved by inserting leap-seconds into the atomic count of time, whenever the discrepancy amounts to more than 0.7 seconds, to bring it back into line with the Earth's faulty time-keeping. Each time we hear the extra 'pip' of a leap second in a time signal, it is a reminder that Earth is not alone in space.

As Earth rotates, friction between the sea beds and the water in them not only slows it down, but also pulls the tidal bulges slightly *ahead* of the Earth–Moon line. As a result, the Moon feels a slight extra tug in a forwards direction from the gravitation of the displaced tidal bulges, which results in its orbit gradually increasing in size; it is very slowly spiralling away from us. The rate is again small, about 4 centimetres ($1\frac{1}{2}$ inches) per year, and that in an orbit 384,000 kilometres (239,000 miles) in size.

Astronomers rely on a fascinating variety of

indirect methods for measuring the slow-down of the Earth, or the related increase in the Moon's orbit. Amateurs equipped with no more than a small telescope and stopwatch can time when the Moon hides a star; and thousands of such measurements have been analyzed to chart its motion accurately, and to show that it is indeed pulling gradually ahead. Another approach is through history. If an ancient author has recorded that a total solar eclipse was seen at Babylon on 31 July 1063 BC, we know that Babylon, the centre of the Moon and the centre of the Sun were precisely in line at that time—and calculations show that only if the Earth is slowing, and the Moon's orbit expanding, would the eclipse have been seen there at that time, rather than somewhere farther east.

Finally, geologists also have their say. Coral organisms change their growth rate during the day, and from season to season during the year. As a result, it is possible to count the number of days in a year from fossilized coral growths, and since the length of the year does not change, the day-length can be calculated. Fossil corals from 400 million years ago bear silent witness that a 'day' then was only 22 of our present hours long.

BIRTH OF THE MOON

Since tides on the Earth are gradually forcing the Moon away from us, the Moon must have been much closer in the past. When the lava flows flooded out to make the dark plains, around 3500 million years ago, it was at only one-sixth its present distance from Earth. At that time it would have loomed huge and luminous in the sky, six times its present size, and over 30 times brighter. The hot lava flows would have painted the now-dark plains a fluorescent dull-red on the Moon's unlit side.

But what of its earlier relationship with the Earth? And its birth? Before the Apollo flights, astronomers had put forward three possibilities for the Moon's origin: it was a part of Earth that had broken off, and its distance had increased continuously since that traumatic parting; it formed near Earth; or it came into being somewhere else in the solar system, as an independent planet that was later 'captured' by Earth. One of the aims of the expensive and complex Apollo Moon missions was to settle this question. Yet despite all the efforts of the past 10 years, the riddle remains unsolved.

The first theory, that the Moon broke away from the Earth, has been around for at least a century. If the early Earth had been spinning rapidly enough, centrifugal force could perhaps have split it in two. Modern versions of this idea have the original planet splitting into Earth and Mars, the latter flying off to take up its present orbit about the Sun. The Moon was a small droplet from the break-up that remained in orbit about the Earth. William McCrea has taken this theory to its logical conclusion: in his bizarre picture, all the planets are the remains of split-up early proto-planets.

Most astronomers expected that the rocks brought back by the Apollo astronauts would rule out this theory. Its original appeal had rested largely on the fact that the Moon's overall density had looked very similar to the rocky outer layers of Earth, suggesting it could simply have been a broken-off piece. The Moon's rocks have, indeed, turned out to be very different from Earth's in their composition, as sceptics had predicted. But for all that, the theory won't completely lie down. Its supporters point out that lunar rocks differ considerably among themselves, and that the surface may thus be an unreliable guide to the overall composition of the Moon. But most scientists interpret the evidence as showing that the Moon *is* significantly different on the whole from the Earth, with more aluminium, calcium and uranium in its rocks, and less gold, bismuth and potassium. The majority, therefore, think that our companion came into being as a world in its own right.

But did it form close to Earth, or in distant parts of the solar system? Opinion is divided. It might have formed from a ring of rocky particles orbiting the early Earth, rather like Saturn's rings today. If this is the case, though, it is difficult to see why Venus and Mars do not have large satellites, too.

Earth's early ring, on this theory, should have been made of much the same kind of rocks that went into the Earth, and again we have the problem that Moon rocks are so different. It looks as though the ring must have formed into the Moon at much the same time as Earth itself was aggregating at the centre; and in this tumult of whirling gases and rocks, some of the elements must have been concentrated at different distances from the centre. Scientists are now investigating whether the elements could actually have been distributed as they are, in fact, between Earth and Moon. In support of the theory, lunar rocks bear one indication, in the ratio of different kinds of oxygen atoms, that the Moon formed at very much the same distance from the Sun as the Earth. If this is taken at its face value, it is proof that the Earth and Moon were born together. Lunar scientists would dearly love to study the oxygen in rocks from other planets, and see just how faithful an indicator it is of their birthplace in the solar system.

As a serious contender to this theory is the idea that the Moon formed elsewhere in the solar system and was 'captured' by the Earth's gravity. This would naturally explain why

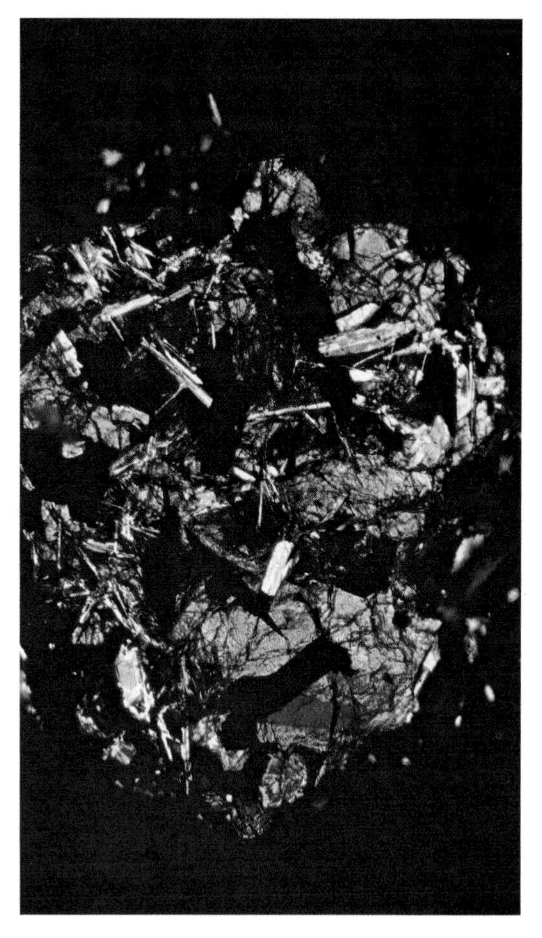

When photographed with a polarizing microscope, a Moon rock appears as a jumble of different types of crystals. Meteorite impacts have flung small fragments thousands of kilometres from their original site, and much of the lunar surface is composed of dissimilar rock crystals from different sites welded together as *breccia*.

Moon-rocks are different from the Earth's, for the Moon's composition is much what scientists expect for a planet born in a slightly hotter part of the original gas disc, closer to the forming Sun.

But the main problem with this idea is simply that it is extremely unlikely that the Moon could have been in an orbit that brought it so very close to the Earth. Such a 'close encounter' must have brought the two planets to within 50,000 kilometres (30,000 miles) of each other—a large figure in everyday terms but, to put it in context, only three-thousandths of Earth's distance from the Sun. Still, if the Earth did acquire its Moon through such an unlikely accident, it would not then be surprising that none of the other inner planets have a moon of comparable size. And it would also explain why the Moon's orbit is at a tilt to the Earth's equator, while the satellites of the other planets orbit exactly in the equatorial plane of their rulers.

The shape of the Moon lends some support to this latter theory. Because it turns in such a way that it always keeps the same side facing Earth, the far side can never be seen from here. Only space flights have revealed its hidden hemisphere, and allowed us to measure its exact shape. The far side is not very different from the near side in appearance, except that it has hardly any dark lava plains. In contrast to the near side, which bears the face of the 'Man in the Moon', the back of his head is bald, pitted and featureless.

The overall shape of the Moon was fixed when it first melted and the crust separated, shortly after its birth. If the Moon has always been associated with the Earth, it must have been at its closest then, and the tides raised by the Earth in the Moon's liquid rocks should have been at their strongest, deforming it into a slight oval. When the lava plains flooded, some thousand million years later, the Moon would have receded somewhat. The tides would be less, and the shape marked by the surface of the plains should be closer to an exact sphere. Detailed studies have now revealed both the Moon's overall shape, and the shape defined by the surface of the plains. And in fact the opposite is true. It looks as though its closest approach to Earth occurred at the time the lava plains were flooding; before that it was more distant, presumably a separate planet in its own orbit. According to this evidence, the Earth and Moon joined company as an indissolubly wedded double planet after some thousand million years of bachelordom, about 3700 million years ago.

THE ORIGIN OF TEKTITES

Whatever the Moon's origin, it has moved a long way from us now. Over the past few thousand million years, the Earth and Moon have evolved in their own separate ways, apparently linked only by the invisible threads of gravity. But the Moon may be the key to understanding the most extraordinary rocks on Earth—the tektites. Tektites are glassy-looking pebbles, the largest a few centimetres across, and tinged in colours ranging from light green to black. For 200 years they have defied explanation.

The first intriguing fact about them is that they occur in only a few groups, each found in a different part of the world. Although each tektite is small in itself, the huge number in each group adds up to a total weight of over a million tonnes. All the tektites in a particular group are made of the same kind of glass, and are the same age and same colour, but the different groups vary tremendously. The Czechoslovakian Moldavites are greenish in colour, while the most recently-discovered group, the Russian Irghizites, are black. The oldest known tektites, found in North America, date back 35 million years while those found in Australasia are only three-quarters of a million years old.

Some groups are spread over vast areas of the Earth. The North American group is distributed over the entire south and east coast of the United States, from New Mexico to Cape Cod, and ocean bed sediments show that

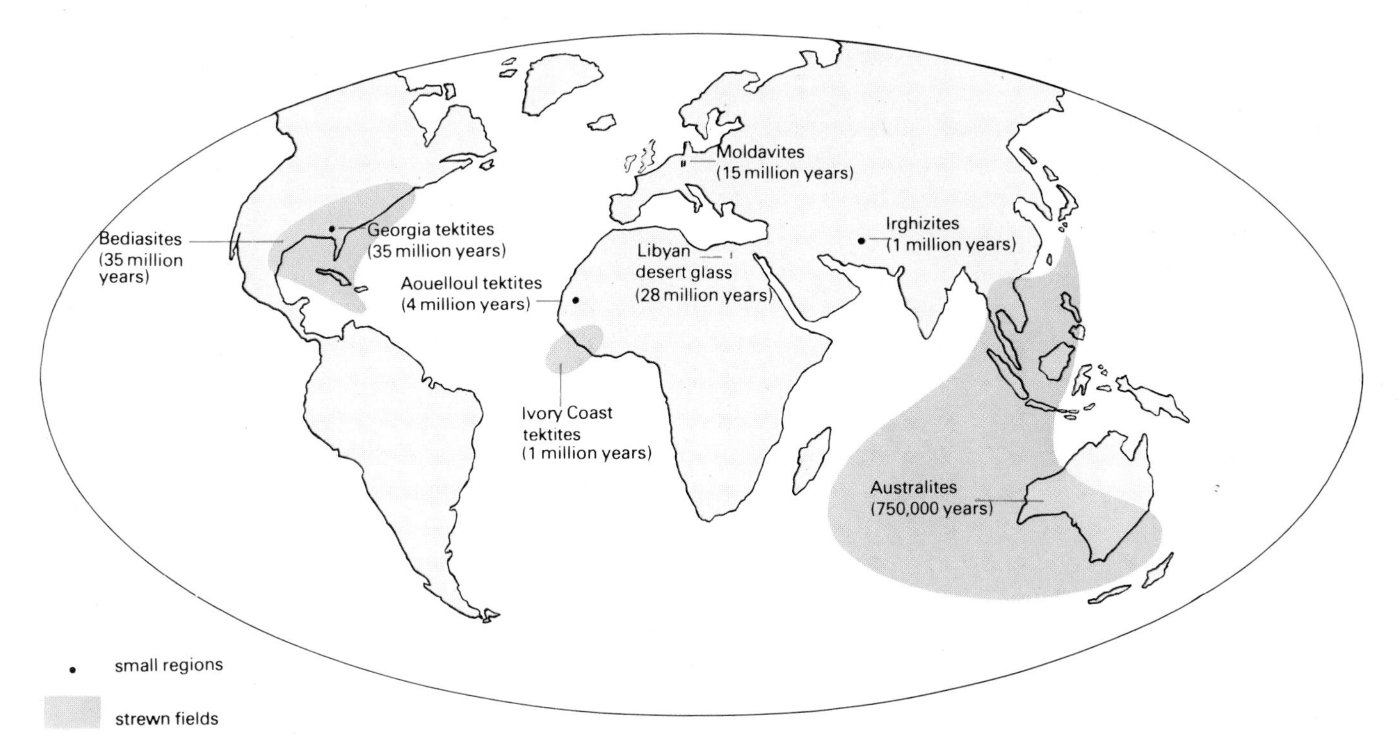

The glassy meteorites called tektites are found only in certain locations, called 'fields'. Each field contains tektites of one particular type and age.

it extends out to sea over most of the Caribbean. The Australasian tektite 'field' is even larger, being strewn over the southern and western parts of Australia, north through the East Indies as far as the sea off Japan, and almost as far west as Malagasy.

Early theories, such as that the Czechoslovakian tektites were primitive man-made glass, or that the Australasian ones were pebbles worn smooth in emu's gizzards, have now been discarded. Many of the smoother tektites are basically spherical, but with a flange around, as if they have travelled through the air at high speed while molten; wind tunnel tests on molten glass blobs have produced just such shapes artificially. Some kind of huge explosion must have thrown out molten rock, which subsequently solidified to glass and rained down over a wide area.

But tektites are not simply meteorites from interplanetary space. A stream of meteoritic bodies travelling together through space would spread out into a stream wider than the Earth, and an encounter with such a stream would shower half the Earth, not just isolated fields. So tektites must have originated in explosions on Earth, or possibly on the Moon.

Such explosions could have been either volcanoes, or the impact of large meteorites. Volcanoes on Earth can be ruled out: there are none near most of the fields; they would not project rocks fast enough to cover thousands of kilometres around; and, anyway, tektites are very different in composition from volcanic glasses (obsidian). The composition of tektites also rules out their being splashed off the Moon's surface by a meteorite impact. Although the lunar soil does contain a myriad of small glassy beads from meteorite impacts, they are composed of a very different mix of elements.

Tektites most closely resemble sandstone, particularly the type called subgreywackes. Most astronomers and geologists believe that each group was splashed out as molten globules when a meteorite hit a sandstone region of the Earth. Tektites contain small bubbles of gas, at a very low pressure, which seems to show that they congealed into glassy drops high in the atmosphere, after the impact had flung them some 40 kilometres (25 miles) high. In some fields, there are indeed nearby craters of the same age as the strewn tektites: the Ries crater in Germany, for example, is 15 million years old, contemporary with the Czech Moldavite tektites.

Some astronomers, however, disagree. John O'Keefe of NASA is one leading opponent. He points out that the impacts that are invoked to create the North American and Australasian fields would have blasted craters some 300 kilometres (200 miles) across, and fairly recently on the geological timescale. Where are these fresh craters, each the size of Ireland? He also points out that sandstones contain traces of water and elements like lead—yet the tektites show no sign of contamination by these trace substances. And, finally, O'Keefe says that the wind-tunnel experiment on high-speed drops of molten glass show that the glass blobs must have been moving at a rate of 40,000 kilometres per hour (25,000 miles per hour). He has confirmed this speed by calculations which are based on the formula that NASA uses for the heatshields that protect astronauts returning from space. Such a

Tektites from different fields differ in composition, age, colour and shape. All show evidence of having travelled at high speed through the air while still molten. This flight has caused the characteristic raised rim on the Australites.

speed is far too high for meteorite splash, but is exactly what we would expect for something falling in from beyond the Earth.

Since meteor-like particles orbiting the Sun are ruled out, and so are impacts splashing glass off the Moon, O'Keefe concludes that tektites must have been ejected by volcanoes on the Moon. According to his analysis, the interior of the Moon may be similar in composition to sandstones, even though the surface is so different. Since water is scarce, volcanic eruptions would be accompanied by outbursts of hydrogen rather than steam and, as a result, the fragments could be thrown out with sufficient speed to escape from the Moon and fall to the Earth. The craters excavated near some tektite fields may have been blasted out by single super-sized tektites.

O'Keefe is criticized by other astronomers. The main problem with his theory is that there is no sign of recently active volcanoes on the Moon: its surface seems to have remained unchanged since the last lava plains solidified some 3200 million years ago. Professional astronomers spend very little time observing the Moon, but it is a favourite target for amateurs, and for many years they have noticed 'transient lunar phenomena', occasional small patches of obscuration that seem to recur when it is closest to the Earth, and mainly around the edges of the lava plains. These are probably clouds of dust raised as old lines of weakness around the plains boundaries settle slightly, rather than signs of volcanic activity. Yet O'Keefe would have at least half a dozen different lunar volcanoes erupting over a very short and recent period. It is not impossible that the Moon is very inactive between periods of tektite eruption, but it does seem unlikely to most astronomers.

Tektites indeed present astronomers with an intriguing puzzle. One can raise objections to the idea that either meteorite splashes on Earth or lunar volcanoes account for their presence, but these, apparently, are the only possibilities left, for irrefutable arguments can be marshalled against any other explanation of their origin. Only a more detailed study of the Earth, the Moon and the tektites themselves can unravel the secrets of these strange glass fragments.

THE ODD PLANET

The larger world in this double system is perhaps the most peculiar planet in the solar system. The very fact that ours is the only hospitable world (for life like us) demonstrates its oddity: in many ways, Earth is *the* mystery planet.

Much of our world's unique character springs from the fact that it is at just the right distance from the Sun for water to exist as a liquid on its surface. A little closer, and it would have boiled away; a little farther from the Sun and it would have become locked in the icy grip that Mars suffers. The blue planet owes its colour to the intense cloak of water that shrouds two-thirds of its surface from extraterrestrial view. And many of its other oddities are indirect results of that water. Life probably came into being in Earth's seas, and the evolution of living organisms has influenced our home planet immensely.

But water has had another, inanimate, role. The constant fall of rain, and the resulting flow of rivers, is all the time carving up the rocks on Earth's surface. Mountain ranges are worn down to their roots, razed to flat soggy plains, as rivers and glaciers cut through their rocks and carry their substance away as fine sand and silt to the sea. On the ocean floor they build up into thick layers of sand and mud. The lowest layers are crushed so much that water is squeezed out, and over vast stretches of geological time the tiny particles bind together as solid sedimentary rocks, like sandstone and shale.

Yet erosion and deposition cannot be the whole story. If they were, the Earth's surface would long ago have been worn flat, and its shallower seas silted up. Land must move up and down, too. Mountains are raised out of the rocks that once lay on the sea bed, leaving marine fossils on top of mountain ranges, such as the Alps. Other areas must sink, to create new shallow seas where sediment can accumulate.

Until recently, geologists could not easily account for this factor. The now-accepted theory first raised its head as 'continental drift'. The edges of continents on either side of the oceans look as though they are adjacent pieces of a jigsaw, taken apart by some gigantic hand. This is obviously the case with the west coast of Africa and the east coast of South America, on either side of the South Atlantic Ocean. German meteorologist Alfred Wegener proposed the theory in the 1920s, but it was quashed by geologists of the time on the grounds that there was no known way a continent could break up and have its pieces forced thousands of kilometres apart to leave an ocean in between. But by the 1960s, geologists were faced with a variety of mysteries in studying the Earth—all of which would be solved neatly by the simple assumption that the continents could indeed move.

The fact that rocks are similar on the facing edges of continents separated by wide expanses of ocean, and that similar types of animal and plant life are often separated by oceans, too, had long been recognized. A worse conundrum arose when geologists constructed the climate of early Earth and discovered that, between 250 and 350 million years ago, huge ice sheets covered central Africa and parts of South America (regions now near to the equator), while now-polar regions such as Greenland were hot deserts.

Earth's surface rocks also record our planet's magnetism, and another problem was revealed when the position of the Earth's north and south magnetic poles throughout the ages was calculated. As we go back in time, the rocks show that these poles have apparently moved around—but, worse than this, different continents show them to have moved in different ways.

All these facts can be explained if the continents did indeed 'drift'. Their climates would have altered as they moved from pole to equator and back, and the apparent motions of the magnetic poles are, in reality, due to the continents' motions. The different paths of the magnetic poles recorded in each continent are due to the fact that the continents have moved relative to one another. Relative motions calculated from the apparent pole movements, in fact, tie in well with other continental drift indicators, like the fit of eastern South America to western Africa.

The rocks of the ocean floor provided the final and clinching evidence. If Africa and South America *are* drifting apart, then a new ocean bed must be forming between them. The central line of the Atlantic Ocean *is* marked by a ridge, paralleling the two coasts, and it is natural to think that fluid rock here is welling up to fill the gap left as the continents separate. The ocean floor should thus be youngest at the centre, and increasingly older towards the ocean margins. As the rocks solidify at the ridge, they become magnetized by the Earth's field, and retain that magnetism as they move away on either side.

The ocean floor is like magnetic tape, spreading out on each side of the ridge, and carrying with it the magnetic imprint of the Earth's field over the past 120 million years. If our magnetic field were constant, the magnetism of the ocean floor would tell us nothing new. But the Earth's field does change. It switches over, erratically, so that while the magnetic pole in the Arctic is what physicists call 'South', it has at some periods in the past been 'North'. We should expect, then, that the ocean floor rocks should be a magnetic tape displaying alternate periods of 'normal' and 'reversed' magnetic field—and this is exactly what has been found, a pattern of magnetic stripes matching precisely across the mid-ocean ridge. It is proof positive that the ocean floors are spreading, and thus that the continents on either side must be moving.

Continental drift has come back in from the cold, as the only possible explanation. In its modern reincarnation, 'plate tectonics', it sees the continents and the attached portions of ocean floor as rigid 'plates' moving over the Earth's surface. And although the theory is basically about horizontal movements, it also neatly explains the mystery of the elevated mountain chains.

Some ocean floors are growing, so it seems that others must be shrinking. The rocks of these ocean floors must go somewhere, and their fate is to end up back inside the Earth. South America, for example, is travelling

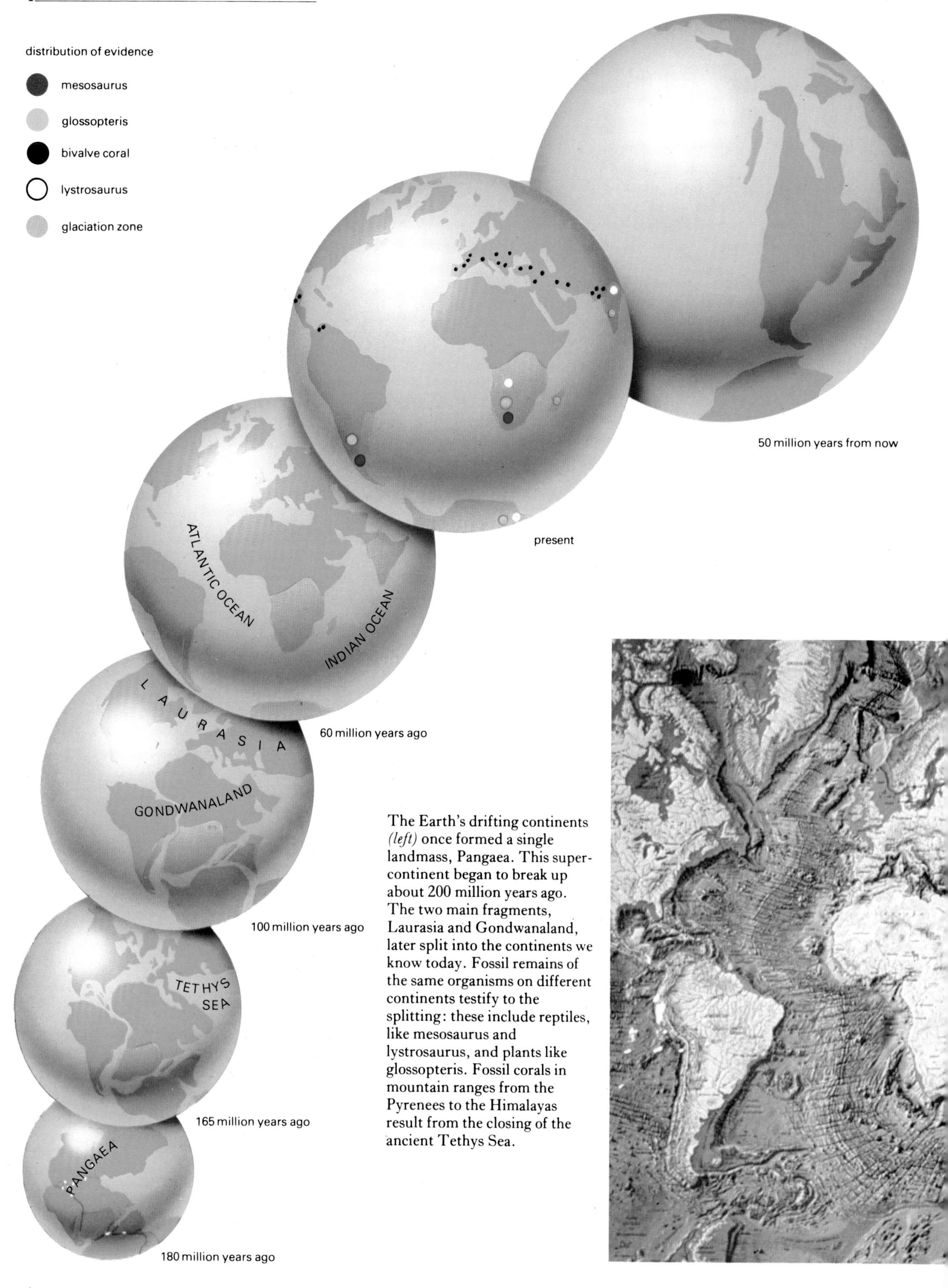

The Earth's drifting continents *(left)* once formed a single landmass, Pangaea. This supercontinent began to break up about 200 million years ago. The two main fragments, Laurasia and Gondwanaland, later split into the continents we know today. Fossil remains of the same organisms on different continents testify to the splitting: these include reptiles, like mesosaurus and lystrosaurus, and plants like glossopteris. Fossil corals in mountain ranges from the Pyrenees to the Himalayas result from the closing of the ancient Tethys Sea.

The continents are carried on the backs of huge 'plates' of crust, which move over the softer rocks beneath. At the plate boundaries, whole sections of crust are colliding or separating, and the result is intense geological activity. A map of volcanoes and earthquake locations *(below)* shows up the plate boundaries in detail. It also shows activity in East Africa, indicating that the African plate is currently splitting in two. A topographic map of the Earth, with the oceans removed, also reveals the plate boundaries *(bottom)*. Where plates collide, one is forced under the other and there are deep trenches, like those around the Philippines. Where plates are drawing apart, molten rock from below oozes up to make the mid-ocean ridges which meander across the globe.

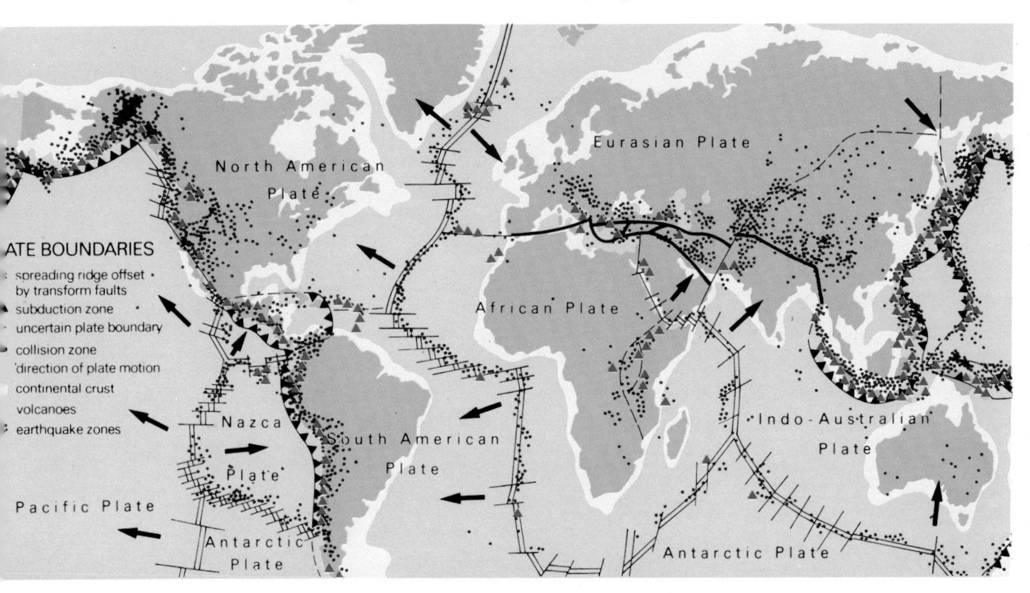

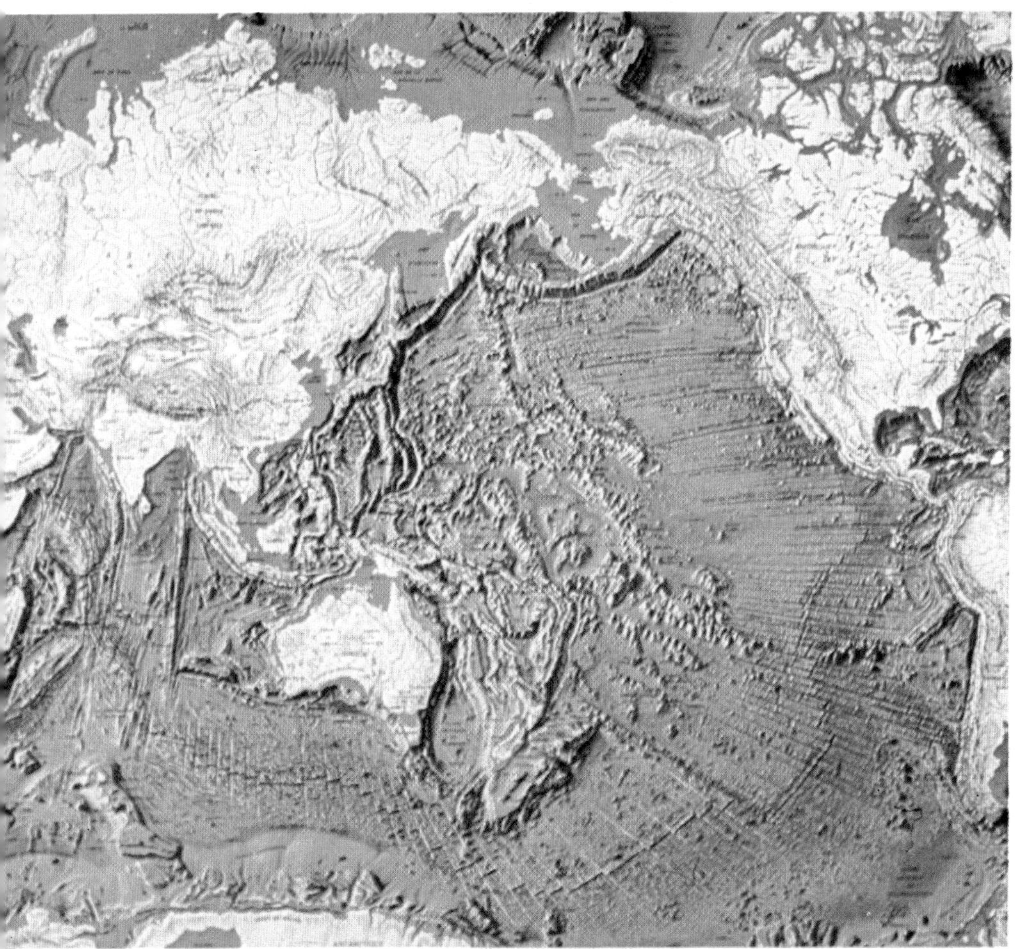

The scars of continental drift show up on land as well as on the sea-bed. The San Andreas fault in California *(top)* is the boundary between land attached to the Pacific plate *(top of picture)* and the North American plate. As the Pacific plate moves north-westwards *(above right)*, the stream bed is moving with it. The Red Sea *(above centre)* is an embryonic ocean, gradually widening as Africa and Arabia move apart.
The island of Iceland lies right on the mid-Atlantic ridge. Its jagged rift valley *(above)* marks the division between the American and Eurasian plates, and Iceland is gradually growing as the plates separate.

westwards at a rate of 2.5 centimetres (1 inch) per year, and its western edge is riding up over the floor of the Pacific, forcing the ocean bed rocks downwards below it. The rocks do not accept this without a struggle. As they are forced down, they crack apart to cause earthquakes above, and some of the rock melts. The molten rock forces its way up to make the long chain of volcanoes that we know as the Andes. The same thing is happening all around the Pacific, and as a result the Ocean is surrounded by an almost continuous chain of earthquake-prone volcanic regions, including Japan; a chain known to geologists as the Ring of Fire.

Although a continent can over-ride the ocean floor, it cannot push its way over another continent. When two continents collide, their edges crumple together. In the pile-up, the once low-lying rocks of the continental margin, and the sedimentary rocks in the shallow seas off their coasts, are forced up into towering mountain ranges. The Himalayas are the mangled edges of Asia and the sub-continent of India, crumpled up as India has charged across the Indian Ocean from a position off East Africa to plough into the larger land mass. And the Alps are a shallow sea-bed pushed up as Europe and Africa have drifted together.

Now that continental drift is accepted, scientists are trying to solve the riddle of what powers its motion: it takes a lot of energy to move a continent around at the rate of a few centimetres a year. Motions of viscous rock deep in the Earth seems to be the most likely answer. But Warren Carey from Tasmania has a more unorthodox view, at first sight appealing in its simplicity. He believes that the ocean floors grow because the Earth is gradually getting larger, expanding as new matter appears spontaneously at its core and cracking the surface in ever-growing rifts. He estimates that the Earth was only four-fifths its present size 200 million years ago. The rate of growth must have been slower before then, however, because calculating backwards at the present rate would mean it was zero size only 1000 million years ago!

There are good reasons for thinking Carey is wrong, however. The Ring of Fire can only be easily explained if the ocean floors are being over-ridden by the continents, and although it is difficult to work out the exact rate at which rocks are consumed, the best estimates make it very similar to the rate at which the mid-ocean ridges produce new ocean floor. The total surface area of the Earth must therefore be staying the same, contrary to Carey's claim. Just as telling is the fact that the other planets of the inner solar system show no signs of expanding: Mercury, indeed, appears to have shrunk slightly. And if matter is created at the Earth's core, where temperatures and pressures are high, even more should come into existence in the centres of stars—and the consequences of this would make the Universe very different from the way we observe it.

It seems that the force driving the continents must be motions of rocks below them, within the Earth, just as ice floes in the Arctic oceans are moved and broken up by water currents below. The light rocks of the Earth's surface crust extend about 30 kilometres (20 miles) down; they are welded on to the top layers of the 'mantle', denser rocks that make up most of the planet. The Earth's interior is hot, warmed by the energy from radioactive atoms, and under pressure from the weight of the overlying rock. In the mantle, it behaves as something between a solid and a liquid: pitch and 'potty putty' are similar substances, shattering like a solid when hit with a hammer, but flowing slowly like a viscous liquid when left alone. Mantle rocks must move about very slowly, rising in some places and sinking in others. At the top of the mantle, the rocks move horizontally between the rising and falling regions, carrying the surface plates with them.

THE EARTH'S CORE

Earthquakes are one of the worst scourges inflicted upon us, but to geologists they are a blessing. Although they occur just below the Earth's surface, the waves of disturbance pass right through the planet, effectively 'X-raying' it. One type of wave can travel only through solids (including 'solid-liquid' mantle rocks), and if the Earth were entirely solid such waves could be picked up by sensitive seismometers all over the planet. But in fact seismometers exactly opposite an earthquake never pick up this type of wave: the Earth must therefore have a liquid core that absorbs the waves and produces a shadow-zone on the far side.

Our liquid core is almost certainly not made of rock, but of molten metal, an iron-nickel mixture. The planet's average density is much higher than that of rock, showing that somewhere there must be denser matter. Meteorites indicate the solution to this, for some of them are not made of stone, but are pure metal, obviously solidified from a liquid iron-nickel melt. As some meteorites are asteroid fragments, some of the asteroids must contain metal cores—and this in itself suggests that rocky planets like Earth should have a metallic centre. Iron is twice as dense as rock, and a metal core of the size indicated by the shadow-zone would nicely account for our high average density. In fact, it would make Earth slightly *too* dense, and geologists have yet to settle the controversy over what other element is mixed into the core to lighten it slightly.

A metal core accounts very neatly for

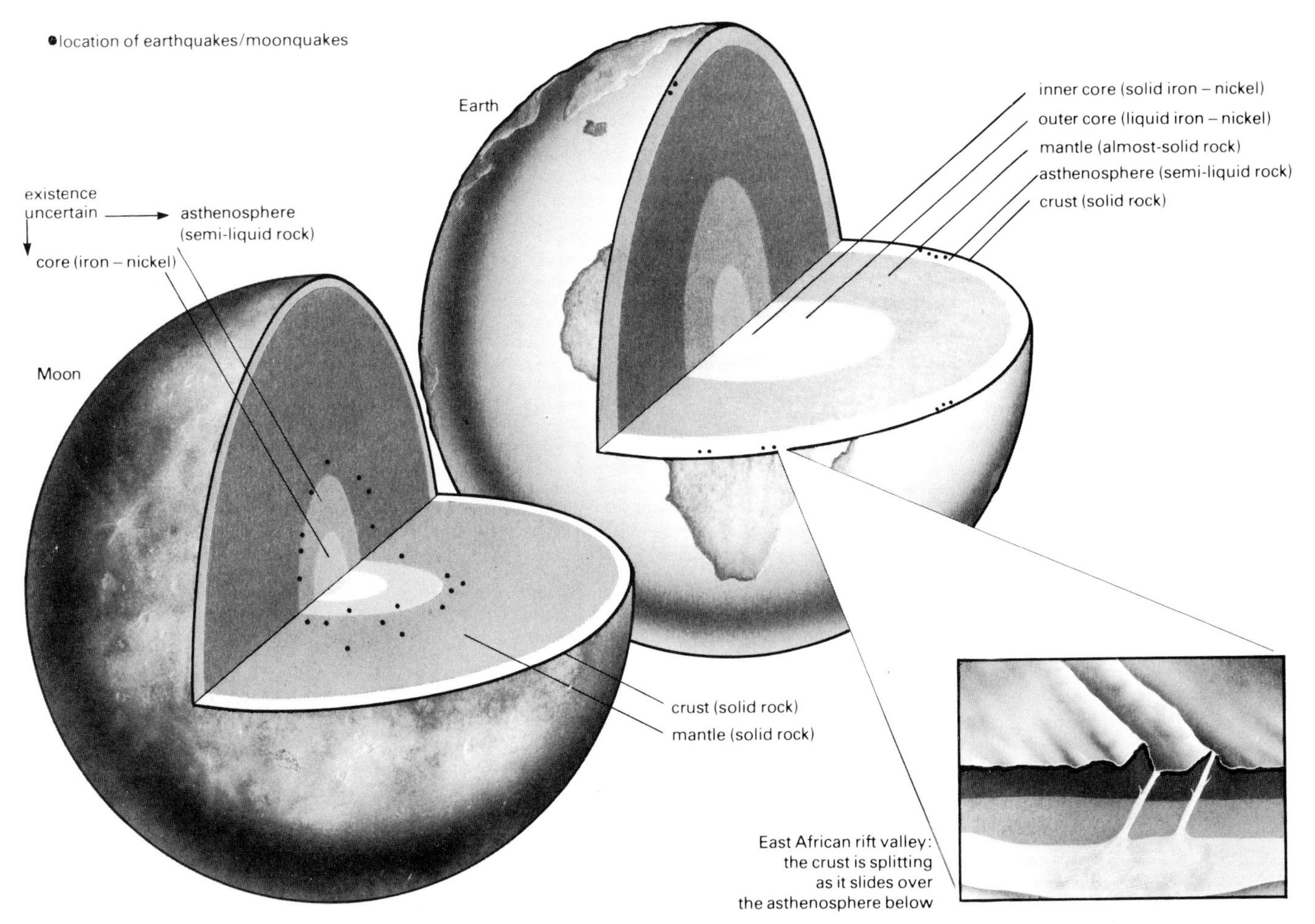

another feature: magnetic field. The Earth behaves as though it has a simple straight bar magnet at its centre, with the 'South' pole below Canada and the 'North' pole opposite, not quite coincident with the geographical poles. (A compass needle's 'North' pole points northwards; because 'unlike' poles attract, the Earth's magnetic pole in the Arctic must be the opposite type 'South'.) But Earth's magnetism varies. The field's exact shape and strength change from decade to decade, while over a period of half a million years or so it reverses completely. Whatever produces the magnetism is less like a permanent magnet than an electromagnet, where changing the electric current fed to it alters the magnetic field strength.

In our core, liquid metal is sloshing about in the magnetic field, and this can produce electric current. This current can then produce more magnetic field. But there is no reason why the field should be one particular way round, and the changes in the Earth's field show that the electric currents in its core can reverse spontaneously. What causes the metal in the core to move about is, however, still unknown. At first sight, the core should turn at the same rate as the rest of the Earth and, if this were the case, there would be no distinct streams of moving metal there, and hence no electric currents and no magnetic field generated. It is possible that heat from radioactive atoms causes distinct rising and falling metal streams, and that these motions combined with the Earth's turning generate the magnetism—rather in the way that up and down currents of air in the atmosphere are twirled into spiral patterns by the Earth's spinning. Other scientists suggest that it is the Moon's gravitational tug on the Earth that causes the core to slosh about inside the semi-solid mantle rocks.

Spaceprobe measurements of other planets' magnetism may help to solve this riddle. Mars is too small to have a liquid core at present, and—as predicted—it has no magnetic field. At the other extreme, Venus should have a molten core, but its slow rate of turning means its magnetic field ought to be negligible—which it is.

But there are two exceptions which are difficult to fit into this neat scheme. The Moon has no magnetic field now, but its surface rocks bear signs that they solidified in a magnetic field of roughly the Earth's strength. The Moon is too small for it ever to have melted at the centre, according to conventional ideas, so perhaps the material from which the planets formed contained short-lived radioactive elements whose energy heated the Moon's

The interior structure of the Moon and Earth are quite different, because the smaller Moon has cooled down much more. Both worlds must once have been largely molten, but the Moon has now solidified and become inert. Only near its centre are some pockets of rock still molten, and it is here that moonquakes can occur. Earth is still mainly molten or semi-molten. It has a solid crust which can slide about in a few rigid 'plates'. Earthquakes occur at plate boundaries, and lie only a short distance below the surface. Earth has an iron-nickel alloy core, mainly liquid but with a solid centre.

centre only during its first few million years of life. Some astronomers believe this substance was an unstable form of aluminium, for which there is evidence in some meteorites.

The planet Mercury is the other magnetic oddity. Although it is so small that its core should have cooled and frozen solid by now, and it rotates too slowly to churn up any liquid core, it does have an appreciable magnetic field (albeit only one-hundredth of the Earth's strength). The spaceprobe Mariner 10's discovery of Mercury's field came as a shock to planetary scientists, and it still ranks as a major unsolved mystery of the solar system.

The giant planets, as befits their status, have the strongest magnetic fields in the solar system. Jupiter outranks Earth 20,000 times, while Saturn's magnetism is a thousand times that of Earth. Fast moving electrons trapped in Jupiter's magnetic field broadcast radio waves that make it, even at Earth's distance, one of the strongest radio sources in the sky. The giant planets are far too light in density to have heavy iron-nickel cores: they must be mainly hydrogen throughout. But when hydrogen is compressed enough, it changes from its normal compressed liquid form to become a metallic liquid. If we could see Jupiter's centre, the mix would look more like the mercury metal in a thermometer than a traditional liquid like water. Electric currents in these huge sloshing cores must generate the powerful magnetism of the giant planets.

THE IMPACT OF LIFE

The Earth's surface is sculpted by the joint action of the moving continents, propelled from below, and the fall of water from above, as rain. As well as smoothing the solid crust through the erosive action of rivers, water has had more far-reaching effects, which can explain most of the remaining mysteries of our world. The key is life.

Life probably began in the shallow seas of the early Earth, as tiny single-celled creatures like algae. From these, a whole complexity of

Magnetic fields

	Strength of magnetic field (relative to Earth)	Tilt of magnetic axis relative to rotation axis	Orientation of magnetic field at present; pole lying in northern hemisphere is	Switches over in period of
Sun: average over globe large sunspot	3 10,000	0° —	magnetic S —	11 years
Mercury	0.01	15°	magnetic S	?
Venus	0	—	—	
Earth	1	11°	magnetic S	500,000 years (approx)
Mars	0	—	—	—
Jupiter	20,000	10°	magnetic N	?
Saturn	1000	0°	magnetic S	?

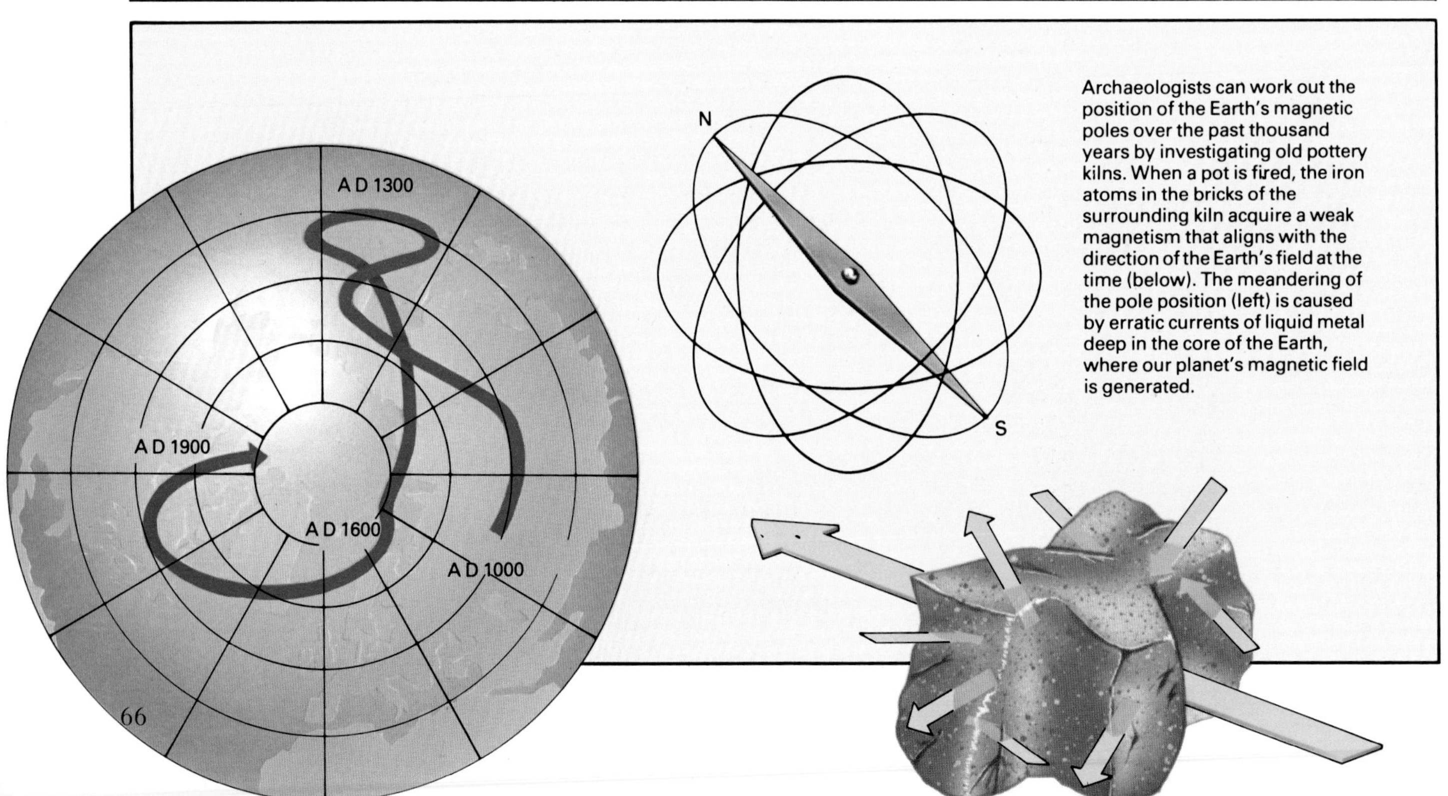

Archaeologists can work out the position of the Earth's magnetic poles over the past thousand years by investigating old pottery kilns. When a pot is fired, the iron atoms in the bricks of the surrounding kiln acquire a weak magnetism that aligns with the direction of the Earth's field at the time (below). The meandering of the pole position (left) is caused by erratic currents of liquid metal deep in the core of the Earth, where our planet's magnetic field is generated.

life has evolved. Even though many plants and animals no longer need to live in the sea, the cells making up the bodies of all living things are three-quarters water (the rest being the actual chemicals of life: proteins, fats and the hereditary chemicals such as DNA). Only within such a wet environment can the chemical reactions of life take place; today's life still needs liquid-water conditions to survive.

In its evolution, life has progressed from single cells to organisms as large as the towering redwood tree and as complex as the body of man, with its intricate nervous system and brain. In the process, it has altered the face of Earth. Most obvious from space are the huge expanses of green, where plants have colonized all of the available land space apart from deserts and frozen polar lands.

Earth's unique atmosphere is another legacy of life. All living matter is built up in a framework of carbon atoms; and life has proliferated because plants have been able to incorporate carbon atoms from carbon dioxide gas in the atmosphere. Earth's first 'air', in fact, must have been the noxious breath of volcanoes—like Venus's and Mars's today. Plants have breathed in this gas, extracted the carbon, and regurgitated the spare oxygen back into the atmosphere. Over the 3500 million years since life first appeared on Earth, they have reduced the carbon dioxide to a minute proportion, leaving oxygen as the major constituent of the atmosphere (along with the inert gas, nitrogen).

The unique properties of Earth can thus be explained, in broad terms at least. Until spaceprobes reached other planets, scientists had no idea our Earth was so unusual; now, with comparisons to guide them, they can solve several more mysteries: we live at the right distance from the Sun for water to be liquid, and that is the major clue. Continental drift reshapes the Earth's surface, but virtually all the other features of our planet are due, directly or indirectly, to liquid water and its progeny, life.

Earth (*above*) is the 'blue planet', the only world where liquid water can exist. Apart from its purely geological effect in causing erosion, water has led to the appearance of life on Earth.

The invisible influence of Earth's magnetic field stretches out into space around (*right*). It shields us from the wind of energetic particles streaming from the Sun. Some of these particles are rerouted into the Van Allen radiation belts (depicted in orange) that girdle the Earth.

CHAPTER 5

COSMIC CATASTROPHES

We know more about our own planet than any other. By studying its rocks, geologists interpret its history as well as its present state. They find the Earth has changed remarkably little in many ways—and, ironically enough, it is as difficult to explain why it is so constant as it is the changes that have occurred. And to explain some of these changes, we have had to look to influences from beyond Earth itself: our planet has suffered cosmic catastrophes.

Geologists can interpret the history of Earth back to some 3800 million years ago—only 800 million years after its formation—by studying the successive layers of rocks that make up its surface. There is an almost complete record of development in sedimentary rocks, laid down in succession in the shallow seas, and igneous (volcanic) rocks that erupted at various times onto the surface. This evidence shows that our planet has remained very similar in its broad properties for the past few thousand million years.

The oldest known rocks, outcroppings near Gödthab in Greenland, are sedimentary (although much altered since by volcanic intrusions). They indicate that Earth had oceans of liquid water 3800 million years ago, in which rocks were laid down just as they are today. Rather less ancient rocks from around the world provide evidence that continental drift has been going on for at least 2700 million years.

As seen from space, then, Earth would always have looked much as it does now. A snapshot taken at any time during the past 4000 million years would have shown continents covering a minority of the surface (although in different positions from today's arrangement), oceans covering the remainder apart from ice caps at the poles, and cloud patterns hanging in the atmosphere. The major difference would be a lack of greenery on the continents. Until 400 million years ago, when plant life first emerged from the seas, the continents were as red and bare as the lifeless deserts of Mars.

This changelessness perhaps comes as no great surprise, for we are used to such a view. And there is, indeed, little reason why geological processes such as continental drift and volcanic activity should change much with time. But there has been one powerful, continuous trend in the record of the rocks: the evolution of life. Fossilized fragments of plant and animal life are mute witnesses to the gradual development of life from simple to more complex organisms. Indeed, life's increasing contribution to our planet has meant that, at some relatively recent periods, the sediments on the sea floors have been made up primarily of fossil remains. Hence there are bands of new types of rock: the chalk layers, thick beds of 100 million year old shells, and the coal reserves derived from the Carboniferous forests of 300 million years ago. Plants like the giant Carboniferous ferns have

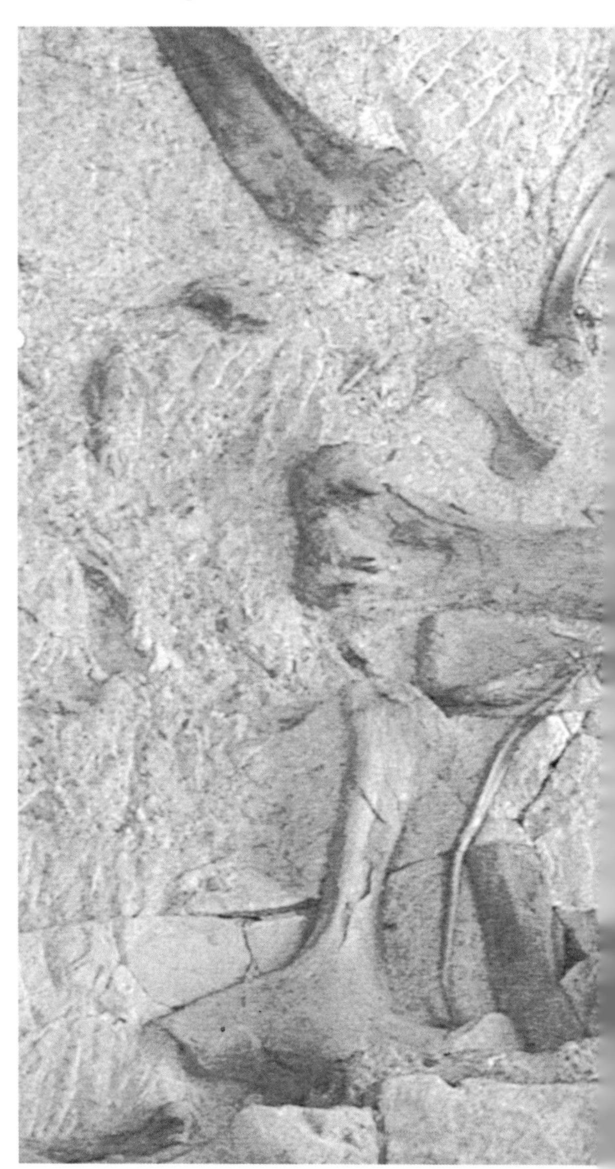

extracted carbon dioxide from Earth's primitive atmosphere of carbon dioxide, and replaced it with the oxygen-rich atmosphere we have today.

Originally, carbon dioxide probably made up 98 percent of Earth's atmosphere. Much of this was dissolved in the seas, and some was incorporated into rocks. Plants have converted what remained, until carbon dioxide is now only a very minor constituent of our air—making up only one molecule in 3000.

The changing carbon dioxide content, indeed, should have made a profound change in our climate. This gas traps heat from the Sun, in the 'greenhouse' effect, which on Venus has run to extremes, with its thick atmosphere raising the surface temperature to 465°C (870°F). The early Earth, farther from the Sun, never became this hot. The primeval oceans were lukewarm, ideal conditions for the first living cells to form. But as the carbon dioxide was depleted by plants, the temperature should have dropped. By now, it should be below 0°C (32°F) all over the planet: we should be a frozen icy desert, like Mars.

What has saved us is a quite unconnected, compensating factor: the Sun has gradually become hotter. The Sun's energy comes from nuclear reactions at its core, in which hydrogen changes to helium. The rate of the reaction depends on a balance between two factors. One is the gravitational force in its heart, which squeezes the nuclei of the hydrogen atoms together; and the other is an electrical repulsion between the nuclei. Over billions of years, helium 'ash' has accumulated in the Sun's core, and its presence has increased the gravitational effect. The hydrogen atoms react slightly more quickly now, with a resulting greater amount of heat and light from the Sun.

Since its birth, the Sun's heat output has increased by 40 percent. As far as Earth is concerned, this gradually increasing warmth has exactly compensated for the decreasing carbon dioxide blanket in our atmosphere, and has kept the surface temperature constant.

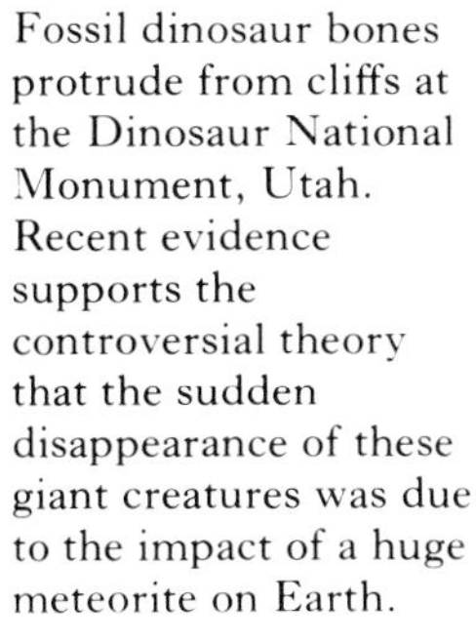

Fossil dinosaur bones protrude from cliffs at the Dinosaur National Monument, Utah. Recent evidence supports the controversial theory that the sudden disappearance of these giant creatures was due to the impact of a huge meteorite on Earth.

This temperature balance is quite remarkable, particularly as other minor factors affect both sides of the balance. The layer of ozone 50 kilometres (30 miles) high absorbs some of the solar radiation and turns it to heat. Ozone comes from oxygen, which green plants have produced from carbon dioxide; and this extra absorption of heat has, to a small extent, offset the decreasing greenhouse effect lower in the atmosphere. Earth's snowy polar caps and brilliant white clouds reflect solar heat back into space, and constitute a factor which can swing the balance either way. If the polar caps are covered by dark ash from volcanic eruptions, they will reflect away less heat, and the Earth will warm up. An increased covering by snow, or by clouds, means more heat reflected back to space, leading to a decrease in temperature.

Temperatures at the Earth's surface have not fluctuated wildly as a consequence of all these effects. At present the average temperature is 15°C (60°F); and during the 4000 million years of geological history, it has always stayed within 10 degrees of this figure. Climatologists are divided in their interpretation of this remarkable balancing act. One school sees it as a lucky coincidence. If the Sun were not warming up, we would now be frozen and life could no longer exist; if the Sun had its observed rate of heat increase, but Earth's greenhouse effect were not decreasing in step, then our temperature would now be approaching the boiling point of water—with equally disastrous consequences for life.

Other climatologists think our temperature maintains itself automatically 10 to 20 degrees above freezing point, like a thermostatically-controlled air conditioning system. Clouds are a good regulator, for example. If Earth warms up, more water evaporates, so the cloud cover increases and reflects more heat back to space. As a result, the Earth cools again.

But the fundamental balance is between apparently unrelated factors: the Sun's increasing heat and the destruction of carbon dioxide by life on earth. British scientist, James Lovelock, sees even this as a stable regulator. He regards our life-forms and atmosphere together as one complete living system, which he calls Gaia, after the Mycenaean earth goddess. Plants produce oxygen, which animals breathe in. Animals eat plants, and they combine this carbon matter with oxygen to make carbon dioxide, which is exhaled to be absorbed by plants again. It is a closed cycle. All scientists accept that this 'oxygen cycle' exists, but Lovelock goes further.

The balance of life forms, between plants and animals, is responsible for the carbon dioxide content of the atmosphere, and hence for the greenhouse effect. Lovelock suggests that live forms acting together, but completely unknowingly, always determine that the atmosphere's composition is just right to produce temperatures suitable for life. In other words, Gaia produces just the right conditions that she may survive.

ICE AGES

Lovelock's Gaia hypothesis is by no means widely accepted, but it does explain Earth's remarkably constant temperature. Even a fluctuation of a few degrees can produce fairly profound consequences, particularly for life, the most sensitive product of our planet. With a rise of a few degrees, the polar caps would melt and the rising sea level would drown millions of square kilometres of low-lying land; a cooling of a few degrees would result in the ice caps spreading down to now-temperate latitudes.

Geological records show that several such worldwide coolings have occurred, at intervals of a few hundred million years. Thick ice sheets have enveloped continents down to latitudes of 45° (about the present latitude of New York or New Zealand's South Island). There is no agreed explanation for these great Ice Ages. Part of the cause is undoubtedly continental drift—when there are land masses near both poles of the Earth, they accumulate snow and ice even during mild winters, when snow falling into the sea would melt. Hence large ice caps can build up, reflect more heat back into space and so chill the Earth and usher in an ice age.

But this may not be the only reason. Earth is not isolated in space, and over millions of years the influences coming to it from outside may change. The Sun's heat output is obviously a

We live in an Ice Age, during which the ice sheets have sometimes covered much of North America and Europe. Such worldwide coolings have occurred every few hundred million years (*see right*) and the cause is still in dispute.
The continent of Antarctica is buried under thick sheets of ice (*below*). Such continent-wide glaciations are typical of Ice Age periods, and outcrops of *tillite* on Table Mountain in Cape Town (*bottom right*) show that it suffered a similar fate to Antarctica some 300 million years ago—evidence for a previous great Ice Age. The recurrence of ice ages over such long periods may be due to influences from outside the Earth.

crucial factor. A few percent drop could bring on a major ice age on Earth. According to the standard theory of hydrogen fusion in stars like the Sun, such variations should not happen: the Sun is warming up gradually, but in a very even way.

There is, however, evidence that the standard theory is not entirely correct. The Sun should be producing huge quantities of fast, penetrating particles called *neutrinos*. From the Sun's light and heat output, we can

AREAS OF PRESENT AND FORMER GLACIATION IN THE NORTHERN HEMISPHERE
180°
90°
90°
0°
areas permanently covered by ice today
arctic circle
areas covered by ice in pleistocene period

calculate just how many neutrinos it should be producing—and experiments suggest that only one-third as many are being produced as predicted.

There are several explanations for the discrepancy (discussed in more detail in the next chapter). One is that the Sun's core can mix up its contents with the surrounding cooler gases every few hundred million years. The rate of nuclear reaction then declines temporarily and the outflow of neutrinos decreases abruptly, because they stream straight out from the core. On the other hand, the decrease in the Sun's light and heat will lag behind by a few million years, because this energy takes a long time to get from core to surface. The lag can explain the discrepancy between neutrino flux and the Sun's light and heat output at present. The proposed periodicity in the central mixing means that the Sun's heat output should vary periodically over hundreds of millions of years. If this is correct, it would explain the recurrence of the great Ice Ages on Earth.

Other astronomers argue that the Sun's actual output does not vary periodically, but that the amount reaching Earth can be affected by other factors originating outside the solar system. The Sun carries the solar system around our Galaxy of stars in an almost circular orbit, and every 200 million years the solar system passes through one of the *spiral arms* of the Galaxy. In these regions, stars are rather more closely bunched together, and between them are unusually dense clouds of gas and dust. If the solar system drifted into a very dense cloud, the dust could block off a substantial amount of the Sun's heat. Supporters of this theory point out that the period between major Ice Ages is quite similar to the interval between successive encounters with spiral arms.

Even within the arms, however, such very dense clouds are rare. It is unlikely that the solar system has ever encountered a cloud dense enough to block off an appreciable amount of the Sun's light and heat. On the other hand, each passage of a spiral arm is likely to bring an encounter with a relatively tenuous dust cloud, transparent to light and heat, but sufficiently dense to stop the solar wind from reaching Earth.

The solar wind is a stream of electrically-charged particles which continuously boil off the Sun's surface and spread out through the solar system. There is some indirect evidence that it affects weather on Earth, although meteorologists find it difficult to find a theoretical reason for this. One of its most dramatic manifestations may have been the 'Little Ice Age' of the seventeenth century, when average winter temperatures were a degree colder than they have been so far in the

twentieth century. Small though the difference sounds, it was sufficient to cause some of the harshest winters ever recorded in Europe, with rivers such as the Thames freezing solid for weeks at a time. Some of our greatest snowscape paintings date from this time, the forlorn conditions appealing to painters such as Pieter Bruegel the Elder and Jakob van Ruidael.

The solar connection is that the Little Ice Age coincided with a time when there were no sunspots on the Sun's usually mottled face. Sunspots normally come and go in an 11-year cycle, but during the late seventeenth century the Sun remained free of spots for over 50 years. Sunspots produce a strong solar wind, so many astronomers conclude that it was the unusually weak solar wind which caused the Little Ice Age.

The dust clouds that the Sun encounters while crossing a spiral arm would stop the solar wind near the Sun's surface, and prevent it from reaching Earth for a period of a few million years. Whatever meteorological process caused the Little Ice Age would then have operated with a vengeance, and the Earth would suffer a real ice age, with extensive glaciation.

Even this milder form of the dust cloud hypothesis has its critics, however. Effects of the solar wind on our weather are not absolutely proven. Many scientists have tried to show that weather patterns here follow the 11-year sunspot cycle, but none has been completely convincing. And even the famous case of the Little Ice Age has come under criticism recently. Chinese records indicate that oriental astronomers did observe sunspots during the seventeenth century, when Europeans recorded the Sun as spotless. If they are correct, then the solar wind cannot have been related to the Little Ice Age.

A passage through one of the Galaxy's spiral arms does, however, pose more direct dangers to Earth than the controversial effects of dust clouds. In the arms, many stars are born, and the heaviest quickly explode as *supernovae*. Each time the solar system traverses a spiral arm, there is likely to be a supernova explosion within 30 light years. The energetic gamma radiation from the supernova would hit Earth's atmosphere and weld nitrogen and oxygen atoms together as nitrous oxide, a highly reactive compound which would destroy the ozone layer. Since the ozone layer helps to keep the Earth warm (by absorbing solar ultraviolet radiation), its destruction would plunge the world into an ice age.

Although it is not possible to link each ancient Ice Age with a particular long-dead and unrecorded supernova, there is some intriguing indirect support for this idea. It comes from deep cores of ice dug from Antarctica. Deeper layers date back to successively earlier years, so scientists can study the snows of each winter for the past thousand years. The snowfalls of the years 1181, 1572 and 1604 all show a sudden increase in the amount of nitrates—compounds derived from nitrous oxide in the atmosphere. And, in each of these three years, astronomers observed supernovae in the skies. These exploding stars were all thousands of light years away so they could not have had any marked climatic impact on Earth. But the Antarctic cores are compelling evidence that their gamma radiation, arriving at Earth with their light, did produce some nitrous oxide in our atmosphere. Scale up the production to a supernova only a few dozen light years away, and the nitrous oxide would undoubtedly pose a grave threat to the ozone layer.

The Sun, and its family of planets, belong to a galaxy of a hundred billion stars which is spiral in shape, similar to the distant galaxy M101 (*above*). We pass through one or other spiral arm every 200 million years, an event which may produce ice ages on Earth.

The Milankovitch theory

Although we habitually speak of the great Ice Ages in the past tense, we are, in fact, in the middle of one now, and have been for 2,500,000 years. Man has been able to live in 'temperate latitudes' such as Europe and North America for the past 10,000 years only because the ice has temporarily retreated. But climatologists do not see this as the end of the Ice Age. The ice sheets have retreated at least seven times before during this particular Ice Age, and they will undoubtedly advance again; we live in the exceptionally pleasant climate of an interglacial period.

Unlike the still-raging disputes about the origin of the great Ice Ages themselves, the controversy over the ebb and flow of ice sheets within an ice age has been settled to most scientists' satisfaction. The now-accepted theory was first proposed by the Yugoslav scientist Milutin Milankovitch in the 1930s, but it is only recently that it has achieved respectability in scientific circles.

One reason for reluctance to accept the theory is that it does not invoke geological

The seasons occur because the Earth's axis points in the same direction during the course of a year, as Earth moves about the Sun (*top*). In June (*centre*), sunlight illuminates the northern hemisphere more directly, and it experiences summer; the weaker, oblique sunlight falling on the southern hemisphere leaves it cooler and hence in winter.

Over a long period of time, three factors can affect the contrast between summer and winter (*bottom*—all effects exaggerated). First the shape of Earth's orbit changes its ellipticity over a period of 100,000 years: a more oval orbit produces more extreme seasons. The tilt of the Earth's axis slowly alters, 'nodding' up and down over 40,000 years. There are extreme seasons when it has maximum tilt. Finally, the direction of tilt gradually swings round in a 26,000-year cycle. Seasons are most extreme when it points along the long axis of the Earth's orbit. The combination of these three factors causes slow but calculable changes in the seasons. According to Milutin Milankovitch, the seasonal changes in turn cause the succession of glaciations and interglacials during an ice age.

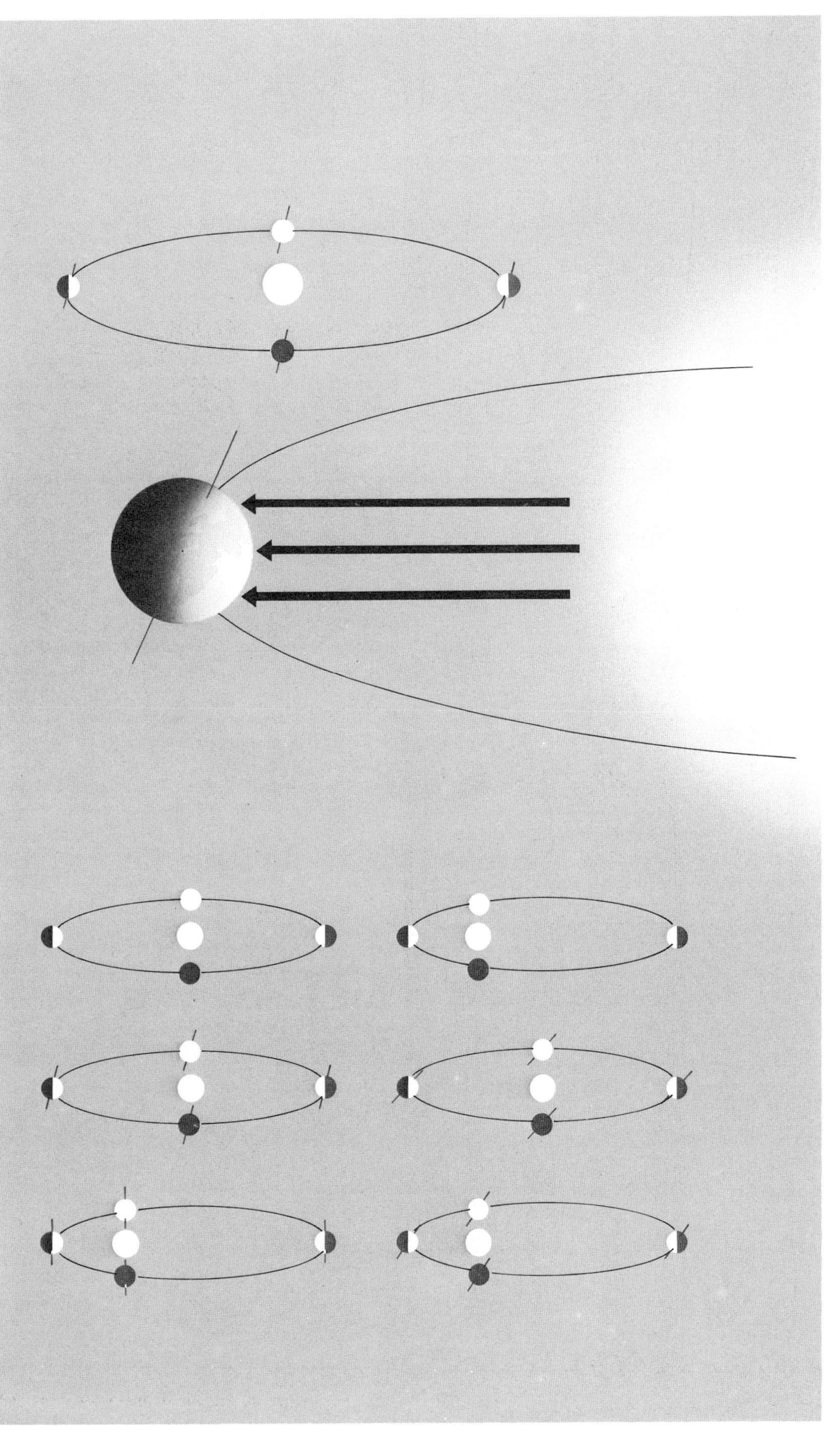

GEOLOGICAL TIME SCALE showing the Geologic Column with explanations of the names used

(dates indicate numbers of million years ago)

	ERA	PERIOD of time / SYSTEM of rocks	EPOCH of time / SERIES of rocks	Date at base
PHANEROZOIC obvious life	CAINOZOIC recent life	QUARTERNARY an addition to the old tripartite 18th Century classification	RECENT	
			PLEISTOCENE most recent	2
		TERTIARY third from the 18th Century classification	PLIOCENE very recent	7
			MIOCENE moderately recent	26
			OLIGOCENE slightly recent	38
			EOCENE dawn of recent	54
			PALAEOCENE early dawn of the recent	65
	MESOZOIC middle life	CRETACEOUS chalk		135
		JURASSIC Jura mountains in Europe		193
		TRIASSIC from the three-fold division of the period made at a locality in Germany		225
	PALAEOZOIC ancient life	PERMIAN from Permia province in Russia		280
		CARBONIFEROUS coal abundance	UPPER Pennsylvanian in USA	
			LOWER Mississippian in USA	345
		DEVONIAN Devonshire, England		395
		SILURIAN Silures, ancient British tribe		435
		ORDOVICIAN Ordovices, ancient British tribe		500
		CAMBRIAN Roman name for Wales		570
	PRE-CAMBRIAN	PROTEROZOIC		
		ARCHEAN		

Sudden widespread extinctions in the fossil record are used to define 'boundaries' in geological time, particularly those between the great 'eras of life' at 65 and 225 million years ago (*left*).

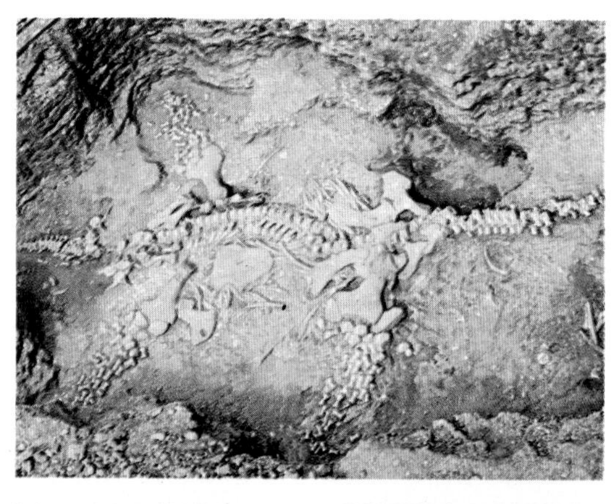

The 'Cretaceous catastrophe' of 65 million years ago is best known for its destruction of the dinosaurs—such as the *Plesiosaur* (*top*), the small carnivore *Coelophysis* (*above left*) and the reptile-bird *Archaeopteryx* (*opposite*). But most can be learned from the smaller sea creatures decimated in the catastrophe, such as the ammonites (*above right*).

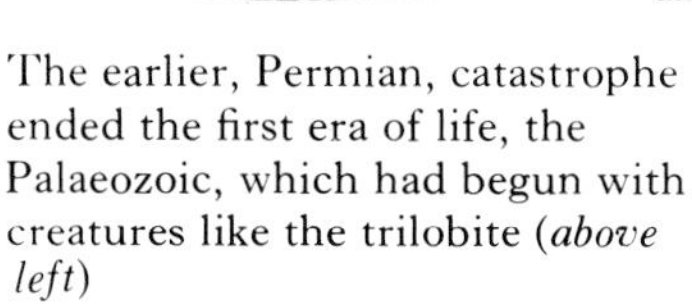

The earlier, Permian, catastrophe ended the first era of life, the Palaeozoic, which had begun with creatures like the trilobite (*above left*)

Geologists can estimate the scale of the catastrophe by its virtually total destruction of sea-creatures like the brachiopods (*above right*).

processes at all, but involves tiny changes in the Earth's orbit, and in our inclination in space. Not until recently have geologists accepted that forces outside Earth could have serious repercussions on terrestrial history.

Milankovitch explained the recurrence of glaciation and interglacials by the varying amounts of heat that Earth receives from the Sun at different times of the year and on different parts of its surface. The total amount of heat falling on Earth over a year is always the same, but both the shape of our orbit and the tilt of our axis of rotation affect the contrast between the seasons—and it is the contrast

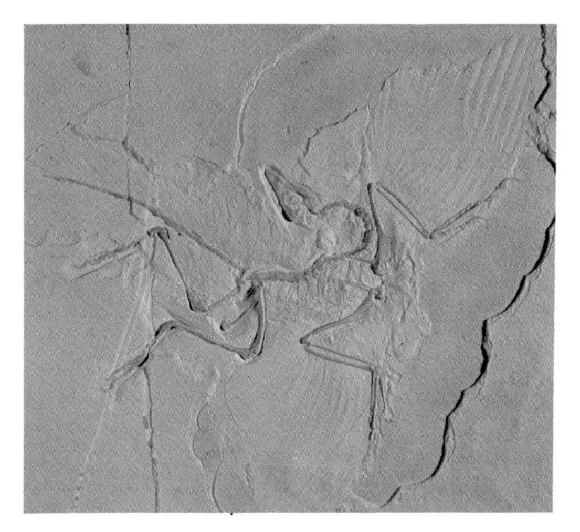

that causes the build up of ice sheets. In a period of cold winters and hot summers, for example, snow which falls at high latitudes melts again in the summer. But it is different in the opposite conditions of mild winters and cool summers. Although winter snowfall is less, the white snowy surface may reflect away most of the heat from the less intense summer Sun, and remain largely unmelted until the next winter. Ironically enough, a succession of years with little contrast between summer and winter will cause ice and snow to accumulate at the higher latitudes, and herald the outbreak of another period of glaciation.

The full theory is more complex than this outline, but with modern computers it is possible to calculate how our temperature *should* have changed. And geologists have perfected techniques for measuring not just past ice cover, but also the variation in temperature. The agreement between the temperature record preserved in ocean sediments and the theory is very close. The ebb and flow of the ice sheets does seem to be controlled by changes in the shape of the Earth's orbit over a 100,000 year period, and by alterations of its tilt, in both amount (40,000 year period) and direction (26,000 years). And so it is also possible to predict Earth's future climate. The forecast for the start of the next Ice Age puts it 10,000 years away.

WHAT KILLED THE DINOSAURS?

Of all the indicators on Earth's rocky record, the most sensitive are the fossils. Different types of living creatures and plants thrive in differing temperatures and locations on land and sea. Any alteration in local temperatures, in the distribution of continent and ocean, or in the amount of sunlight received, will produce significant changes in the fossil record. These changes will be quite sudden compared to the gradual changes that evolution brings about.

Geologists have located several such pronounced changes in the course of life on Earth, and have used some to draw boundaries between different eras of life, and between the subdivisions of geological periods. At such times, whole families of plants and animals die out almost at once, and their place is gradually filled by others during the next period. Some of these disappearances may have been caused simply by climatic change—perhaps the coming of an Ice Age with conditions too severe for some species to survive. Others were probably due to the effects of continental drift.

Some 225 million years ago, all the Earth's continents drifted together as one supercontinent, Pangaea. Land life from the various continents would have come into competition for the first time, and the strongest would have quickly destroyed the weakest—much as, in recent times, man's introduction of European animals has exterminated some of Australia's indigenous beasts. During a continental collision, the shallow seas bordering the continents are rucked up into mountain ranges, the water spilling off into the deep oceans. A single supercontinent has a much smaller area of shallow bordering seas than an equivalent area broken up into several continents. In the collision, many species of marine life, too, would have died.

At this time, the fossil record does indeed show a dramatic extinction of life of all kinds. Geologists have long used this event to distinguish the previous period, the Permian, from the succeeding, the Triassic. To take one example of the sudden extermination, the Permian had 125 separate genera of the shellfish, *brachiopod*: only two survived into the Triassic period. The change in fossils is so abrupt that the Permian 'catastrophe' is generally taken to mark the end of the first era of life on Earth, the Palaeozoic; and the gradual recovery during the Triassic as the beginning of the Mesozoic era.

This era ended as suddenly as its predecessor, at the close of the Cretaceous period some 65 million years ago. A planetwide catastrophe wiped out most of the living creatures on Earth. In the seas, the entire range of life was decimated, from microscopic plankton up to the large coiled-up ammonites. The latter had colonized the shallow seas for over 300 million years, apparently one of Nature's survivors, but they, too, were totally destroyed.

The most famous destruction in this carnage was that of the dinosaurs. These huge creatures had evolved from smaller reptiles some 135 million years earlier. Their massive bones are widely found in rocks from then on, and geologists have traced how one kind evolved into another to culminate in monsters such as the hundred-tonne Brontosaurus, a gentle vegetarian, and its predator, the fearsome carnivore, Tyrannosaurus.

For over a hundred years, scientists have puzzled over the sudden disappearance of the largest and most powerful land animals the world has ever seen. Early theories suggested that perhaps they could not cope with a cooling climate, or perhaps were struck down by a planet-wide epidemic. But neither theory could explain how the entire cross-section of life was ravaged simultaneously, from dinosaurs down to microscopic sea plankton.

Among the wilder hypotheses has always been the possibility of a nearby supernova explosion, or the impact of a huge meteorite on Earth. Most scientists have considered these ideas as little more than science fiction, but the past few years have seen evidence unearthed that suggests the dinosaurs were indeed wiped out by just such a cosmic cataclysm.

The evidence comes not from dinosaur fossils but from the more-easily interpreted sea-bed sediments of the time. At this point in the fossil record, sea life of course becomes very rare: below the dividing line in the rocks are large numbers of relatively large Cretaceous shellfish, above they are very few, and these Palaeocene shellfish are small and underdeveloped. The dividing line is unusually sharp and distinct, with new Palaeocene rock sitting on top of the older Cretaceous rock without any merging.

Dutch geologist Jan Smit has studied this boundary at outcrops in Spain, where the rate of sedimentation was so high that five to ten centimetres of new rock formed every thousand years. The boundary is so narrow that Smit concludes the extinction took place in just a few years—at most. The destruction must have been sudden, not just the result of slow climatic changes, or the gradual nuzzling together of continents.

In 1979 Luis Alvarez and his colleagues at the University of California at Berkeley analyzed the rock above and below this sharp boundary, taking their first samples not from Spain but from the Gubbio valley in Italy. To their surprise, they discovered that the first one centimetre thickness of the new Palaeocene rock was rich in the rare metal iridium. Subsequently, they have confirmed this high abundance of iridium in the first centimetre of Palaeocene rock everywhere in the world they have tested—including the United States, Denmark and the ocean beds.

These results show beyond dispute that the Cretaceous catastrophe coincided with a thin layer of iridium-rich sediment being deposited in oceans all over the world. There may be other explanations for this weird event, but the most reasonable so far put forward is a giant meteorite impact on Earth. Iridium is far more common in meteorites than in Earth rocks, and a powerful meteorite impact would readily spread dust particles evenly over the Earth. No other theory so far proposed can explain why the level of iridium instantly increases 25-fold across the Cretaceous-Palaeocene boundary.

Alvarez estimates that the meteorite must have been as large as a small asteroid (or the smaller of Mars's two moons), about 10 kilometres (6 miles) in diameter and weighing a million million tonnes. Its impact would have blasted out a 200 kilometre (120 mile) wide crater, and thrown up a hundred million million tonnes of fine dust. Mixed in with pulverized terrestrial rocks is the asteroid's material, which enriches the dust with

iridium. The dust pall must have hung in Earth's atmosphere for several years before settling out. It would have blocked out most of the Sun's light, turning the surface of the planet into a region of continuous night. And in this gloom, life would have all but died out.

All plants, from the simplest plankton in the sea upwards, rely on sunlight to live and grow. Without it, most plants would have died; and the animals that fed on them would have faced extinction, too. The dinosaurs were only the largest of the animals who starved to death as the plants wilted under the dust laden skies.

This theory explains all the facts known at present—including the peculiar thin layer of iridium-rich rock. It faces one embarrassment, though. We do not know of any crater in the world which is 200 kilometres (120 miles) in size—almost as large as Iceland—and only 65 million years old. Smit hypothesizes that it may lie hidden on the seabed.

Without this final piece of the jigsaw, many scientists will not be totally convinced that a wayward asteroid killed the dinosaurs. But from an astronomical point of view, it is quite likely. The number of old craters of various sizes and ages on Earth lets us calculate how frequently meteorites of a particular size have hit our planet. We can extend this law upwards in size to cover larger meteorites and asteroids, too. A 10-kilometre asteroid (the size that Alvarez invokes for the Cretaceous catastrophe) is likely to hit Earth every hundred million years on average. At least 40 should have smashed into our planet during its lifetime, five to six of these during the well-documented fossil record from the Carboniferous period onwards. From this viewpoint, we should *expect* to find several sudden changes in the fossil record as meteorites have smashed into our planet.

Geologists are now becoming more aware that Earth does not exist in isolation to space. Our world is subject to a wide variety of influences from outside. Some may cause gradual changes in climate, ushering the ice ages, for example, and controlling their ebb and flow; others may be sudden, and far more disastrous in their consequences. And the kind of cosmic catastrophes that have overtaken Earth in the past are events that will undoubtedly afflict our planet again in future.

A 12-metre (40-foot) crater blasted out by a relatively recent meteorite fall. It is one of the dozens of craters caused by fragments of the Sikhote–Alin meteorite which fell in eastern Siberia in 1947.

CHAPTER 6

BIOGRAPHY OF THE STARS

One November evening in 1572, Tycho Brahe took a walk before dinner. During his stroll he glanced up at the sky and was astonished to see a brilliant star, brighter than any other. Tycho was the greatest astronomer of his age, so he knew that there had never been a star in that position before, let alone one so bright. But he was so taken aback that he asked his servants, then passers-by, to confirm that a 'new star' was indeed blazing out amid the known stars.

At that time, men believed the Earth to be the centre of the Universe, and that while on Earth man was surrounded by change and corruption, up in the realm of the stars everything was perfect: no decay, no change could ever take place. Tycho's new star (*nova*, in Latin) shook men's belief in the old ideas, and prepared the way for the modern Universe where Earth is merely an insignificant planet orbiting the Sun, itself a very average star.

Astronomers now know that the stars do change. On the whole, the alterations are slow, spread over millions of years—outbursts like the one that surprised Tycho are rare. But if we watched over vast spans of time, the skies would be bustling with activity, as stars are born, experience the traumas of life, and eventually die.

Only very occasionally are we privileged to see such events in progress. Our view is a snapshot, encompassing generations of stars of all ages at once. And a confusing crowd it is: At one end of the scale, a few stars are almost a million times brighter than the Sun, incandescent blue-white furnaces five times hotter than the Sun's surface; while other luminous stars are far cooler, but are supersized—large enough to fill the orbit of Jupiter if they were placed in the solar system. At the other end of the scale are tiny red stars, miniature suns only one-ten thousandth as bright, and the *white dwarfs*, stars whose matter is incredibly tightly packed into a ball only one-hundredth the Sun's diameter.

Astronomers can, by modern techniques, actually see some details on the nearest giant stars, but even with the largest telescopes most are just points of light in the sky. A star's distance can, however, be judged by a variety of techniques, as can its true brightness. They only appear so much fainter than the Sun because their light spreads out over the huge reaches of space: even the nearest star is 250,000 times farther from the Sun than the Sun is from Earth—and that is about 150 million kilometres (93 million miles). Star distances are so great that they are not calculated in kilometres or miles at all, but by light years. Light travels at the speed of 300,000 kilometres (186,000 miles) per *second* so a distance of one light year is almost 10 million million kilometres (6 million million miles). On this scale the nearest star is four light years away.

The huge variety of stars baffled astronomers for many years but it is now accepted that the different 'types' are simply stars at different stages of life. The Sun, now in the prime of its life, will one day become a large giant star, and eventually a white dwarf. The run of temperature, brightness and mass ('weight') among stars tells us why they shine at all—and these properties agree well with calculations about what should be happening inside these huge gas spheres.

But they do still have their mysteries. Even the Sun is not completely understood, although it is a very ordinary star. Other stars behave much more strangely. Many of them, for example, vary in brightness. The cause of this changeability has been to some extent pinned down: some are pulsing in and out, while others are screened by moving clouds of soot. But others remain a mystery. We do now know that stars which appear suddenly as 'new' stars are, in fact, very old ones, temporarily becoming many times brighter. Ordinary novae have been revealed as hiccups of cannibalistic stars in a close partnership. Astronomers have also found super-explosions that mark the complete destruction of a star in a stupendous cataclysm. Such supernovae are imperfectly understood. Some are explosions of giant stars, but others defy explanation. Ironically enough, Tycho's star of 1572 was of this latter type: although it began the long process of enquiry into the nature of stars, it lies in the category of the still-unexplained.

THE SUN AS A STAR

Our Sun is a fairly average star. It is intermediate in mass, temperature and size. And it is the only star near enough to Earth to study in detail, so we know much more about it

Our Sun (marked by the cross) is an ordinary star, one of the 100,000 million that make up our Milky Way Galaxy. Although the stars look closely-spaced here, they are separated from one another by several light years—several tens of millions of millions of kilometres. The entire Galaxy is 100,000 light years across.

Properties and types of stars

Properties of stars		While using up hydrogen 'fuel' (as *main sequence* stars)	
Mass (compared to Sun)	Lifetime (millions of years)	Brightness (compared to Sun)	Surface temperature (°C and as colour)
25	3	80,000	40,000 blue–white
16	15	10,000	33,000 blue–white
6	100	600	17,000 white
3	500	60	9200 white
1.5	2000	6	6600 yellow–white
1	10,000	1	5500 yellow
0.8	20,000	0.4	4200 orange

than any other star. Yet our knowledge is far from complete.

No one disputes *why* the Sun shines. It is a huge ball of gases, some 1,400,000 kilometres (865,000 miles) across, made up mainly of hydrogen and helium, the commonest elements in the Universe. The Sun's surface is a searing 5500°C and within the core the temperature reaches 14,000,000°C, hot enough for the nuclei of hydrogen atoms to join together and make up the nuclei of the next lightest element, helium. As they do so, the reaction liberates huge amounts of nuclear energy, which makes its way up through the Sun's body, to emerge at the surface as sunlight and heat.

The Sun shines, therefore, because its centre is literally a hydrogen bomb—but, fortunately for us, a hydrogen bomb going off in slow motion. Although its core converts over 500 million tonnes of hydrogen to helium every *second*, it has such huge reserves of hydrogen 'fuel' in its massive bulk that it can keep going for some 10,000 million years. It is now about halfway through its life, having formed at much the same time as its family of planets, so its slowly-ticking core will keep us in sunshine for 5000 million years to come.

We cannot, of course, see the Sun's core directly, but hydrogen bomb reactions are the only source of energy that can possibly keep it shining for so long. And there is a whole 'family' of stars broadly similar to the Sun—astronomers call them *main sequence stars*. In this family, the heavier members are correspondingly hotter and brighter, the less weighty cooler and dimmer. All of them fit the theory that hydrogen is turning to helium in their central regions.

The Sun's powerhouse

A decade ago, physicist Raymond Davis of the Brookhaven National Laboratory decided to check on the Sun's nuclear furnace. The

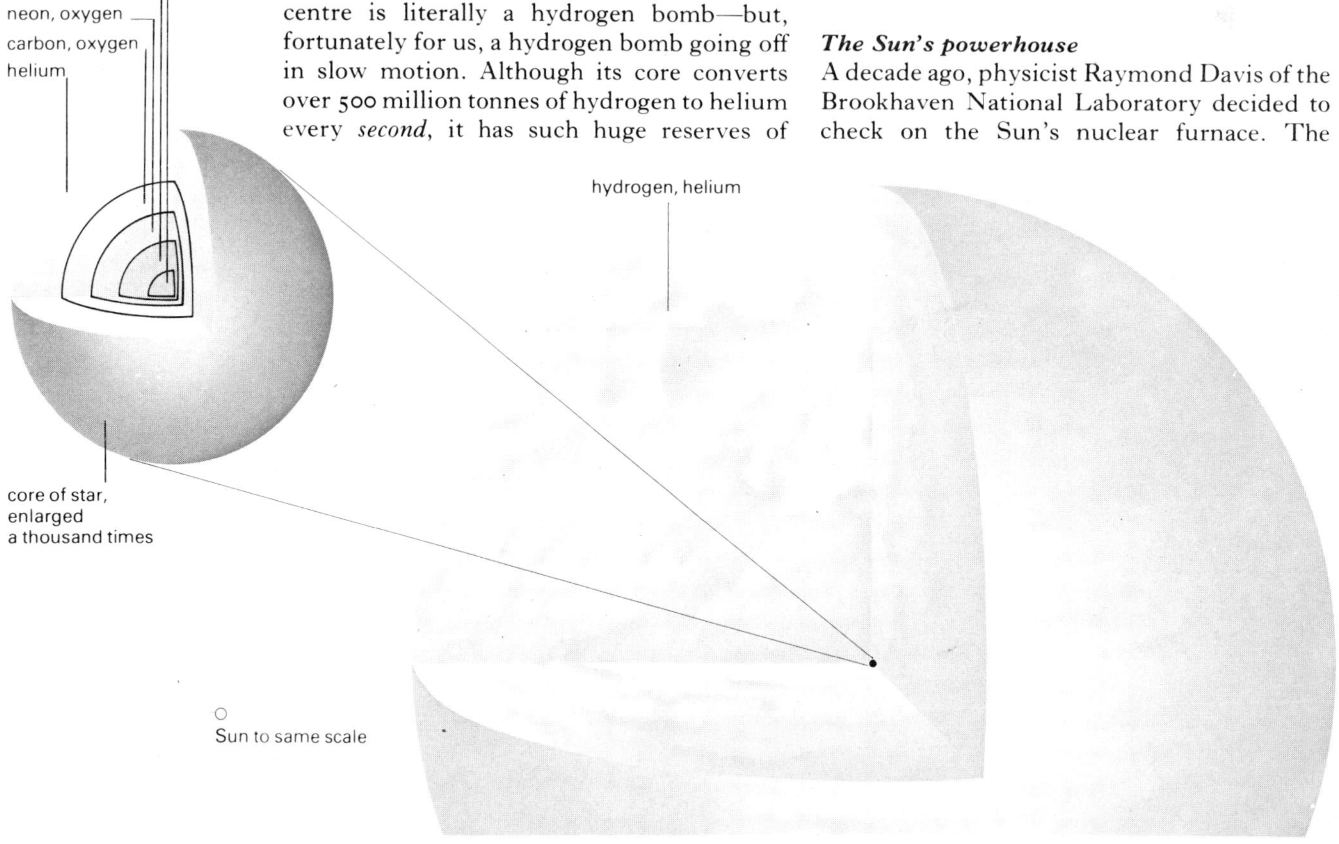

	Eventual Fate		
Example	Expands to become	Loses outer layers as	Core collapses to
zeta Ophiuchii	white supergiant	supernova	black hole
Spica	yellow supergiant	supernova	neutron star
Regulus	orange giant	supernova?	neutron star or white dwarf
Sirius	red giant	planetary nebula	white dwarf
Altair	red giant	planetary nebula	white dwarf
Sun	red giant	planetary nebula	white dwarf
epsilon Eridani	red giant	planetary nebula	white dwarf

As stars get older, they swell to become giants. The supergiant (*left*) represents the middle-age of a star 10 times heavier than the Sun. It has a complex structure. The outer layers have billowed out into a huge tenuous envelope, while its central core is tiny, but so compressed that it contains half the star's matter. Here successive nuclear reactions have created an 'onion-shell' arrangement of different elements. This star is ripe to explode as a supernova.

hydrogen in the core changes to helium in a number of steps, but the principle is simple. The nucleus of each hydrogen atom is a single positively-charged *proton*; the helium nucleus contains two protons and two electrically neutral *neutrons*. In the reaction, four protons add together successively, and two of them change to neutrons—which they do by shooting out lightweight positive particles called *positrons*, and (most important to Davis) very fast energetic particles called neutrinos. These ghost-like particles are difficult to stop—they will shoot through solid matter with the greatest of ease, and only rarely is one blocked on the way. So neutrinos are pouring out through the Sun's body from its centre: if our eyes were sensitive to them rather than to light, we would not see the Sun's surface at all, but instead the inferno at its core. But neutrinos pass right through us, at the rate of some 100 million million every second, and right through the Earth as well. A very small proportion are stopped by atoms in their path, however, and Davis planned to intercept some of them, and count them.

His neutrino 'telescope' is very odd: a huge metal tank filled with 400,000 litres (100,000 gallons) of dry-cleaning fluid, and situated 1.5 kilometres (1 mile) underground in a gold mine in the Black Hills of South Dakota. The fluid—perchloroethylene—contains a lot of chlorine atoms which are particularly good at 'catching' neutrinos. And the tank is underground to protect it from other, less penetrating, radiations from space that could confuse the results. When chlorine catches a neutrino, it changes to an atom of the gas argon. Davis periodically checks his huge tank for argon atoms, and so estimates the number of neutrinos he has caught. From laboratory experiments, he has calculated what proportion of the neutrinos passing through the tank are caught by the chlorine atoms, so he then works out the total number of neutrinos that have passed through his tank. From this he can calculate how many the Sun is producing.

But far from confirming conventional ideas on the Sun's central powerhouse, Davis has apparently contradicted them. According to his results, the Sun is producing only one-third as many neutrinos as had been confidently predicted. No one can doubt the scrupulousness of the experiment itself, bizarre though it seems at first sight: the Sun is simply not behaving as expected. And the 'solar neutrino problem' is now rated as one of the major mysteries in astronomy today.

One straightforward solution put forward is that the core is slightly cooler than expected. This could have happened if it contained fewer 'heavy' elements (elements heavier than helium) when it was born, and its surface composition has changed as it has been 'contaminated' when it has passed through gas and dust clouds in space.

Another theory has the Sun's centre periodically changing its temperature as it mixes in cooler gases from slightly further out in its body. In that case, the present central temperature is not high enough to make the Sun shine at its present brightness. This poses no real problem, however, for the neutrinos indicating a low central temperature shoot straight out and reveal the core's hot temperature *now*; but the energy making the Sun's surface shine has taken around 30 million years to percolate out from the centre. So the present brightness is actually telling us how active, and hot, the core was millions of years ago. On this theory, the low neutrino levels are warning us that the Sun is in the cooling part of a long-term cycle, a cycle of heating and cooling that may have produced the recurrence of ice age periods on the Earth. If so, then Davis's tank is providing a long-term weather forecast, of world-wide ice and snow millions of years ahead.

Other scientists think there is nothing wrong with our ideas about the Sun's interior; that it is the neutrinos themselves that are to blame. There are two or three kinds in nature, and some solar neutrinos may possibly change

The Sun's energy comes from a chain of nuclear reactions that convert hydrogen nuclei (protons) into helium nuclei.

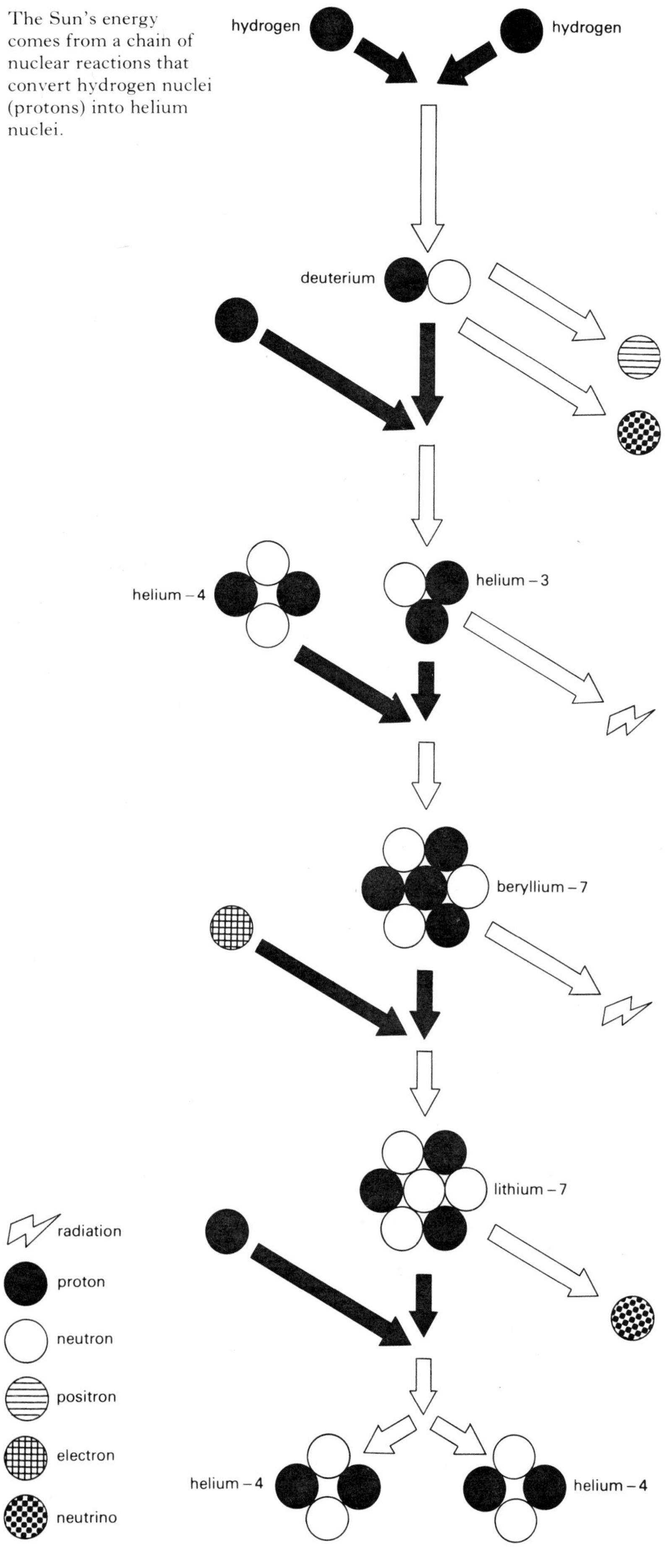

hydrogen
hydrogen
deuterium
helium – 4
helium – 3
beryllium – 7
lithium – 7
helium – 4
helium – 4

radiation
proton
neutron
positron
electron
neutrino

from one type to another during their flight from the Sun to the Earth; they could end up as objects that the 'neutrino telescope' cannot detect. According to physicists' present ideas of sub-atomic particles, such a change should not occur, but in 1980 Frederick Reines claimed that his laboratory experiments showed that neutrinos could, indeed, change type—and Reines's opinion is to be respected, for he first proved that the neutrino exists, some 30 years ago.

Meanwhile, astronomers wish to pin down exactly what is happening at the Sun's heart. The high-energy neutrinos that Davis detects come, in fact, from a side-reaction in the complex chain of hydrogen-bomb reactions, and there is a slight lingering suspicion that we may have got the details of these reactions wrong. We need to detect the lower-energy neutrinos from the fundamental first step, the joining of two hydrogen nuclei to make heavy hydrogen, and the best substance to catch them is the metal gallium—around 50 tonnes is required. Gallium is more difficult to obtain than cleaning fluid, and until recent years (when it has been used in the manufacture of electronic 'chips'), it was not available at all in commercial quantities. A gallium neutrino telescope is now under construction, next to Davis's tank, although at the present price of $500,000 (£250,000) per tonne, it will be a long time before all the gallium is installed! It is an important experiment, though, because it promises to solve once and for all the mystery of the Sun's missing neutrinos.

Blemishes on the Sun

The Sun's surface layers seem bland and uninteresting at first sight, a smooth ball of gas only occasionally marred by the black speck of a sunspot group. But special instruments show that it is a seething cauldron of activity, both above and below the surface.

Henry Hill at the University of Arizona has spent years measuring slight changes in the Sun's size and shape, and he has found that the Sun wobbles like a jelly. The up-and-down movements are minute in size—about 100 metres, on a globe over a million kilometres across—and wobble with several periods at once, from once every five minutes to as slow as once an hour. Other astronomers have confirmed the wobbles, by studying the surface gas near the centre of the Sun's disc, and measuring its speed towards us or away from us as it moves up and down. They have also proposed that the Sun may oscillate at the much slower rate of once every $2\frac{2}{3}$ hours as well, a still controversial suggestion.

Wobbles in the Sun are not just for fun. The periods of vibration of the different oscillations depend directly on how the density and composition (relative amounts of hydrogen,

The Sun's surface is not a bland featureless disc when viewed with special techniques. Observations in light of just one wavelength show the rippling gas clouds making the 'surface', and spikes of hot gas standing above the Sun's apparent edge (*right*). X-rays from the Sun show up only the 'active' regions in the solar atmosphere (*below*): the Sun's disc (outlined) does not produce any X-rays.

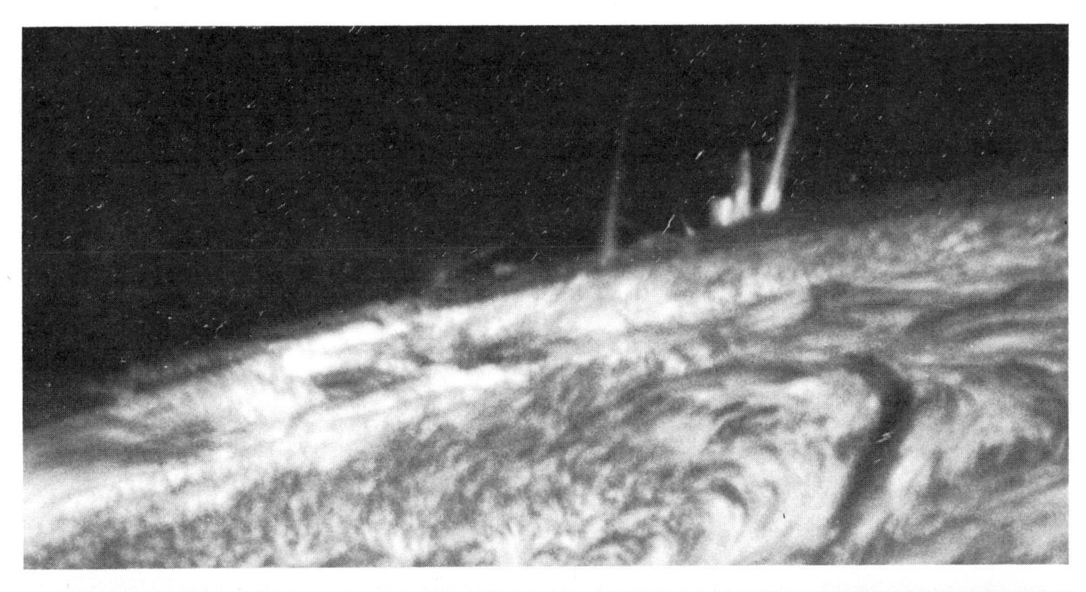

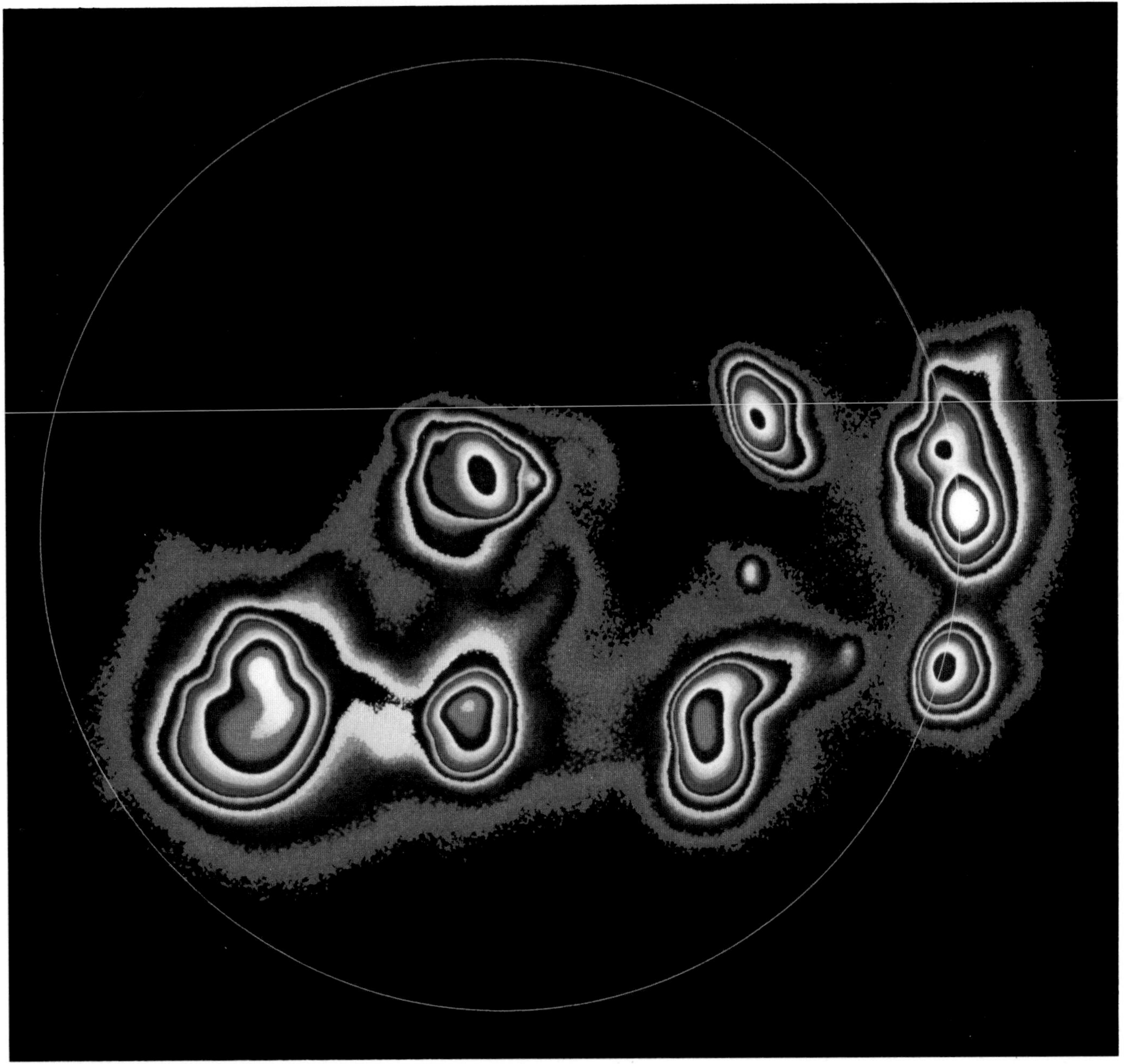

helium and heavy elements) change from the surface down to the centre. Apart from the embarrassing neutrino results, solar wobbles are the only way astronomers can find out about the Sun's interior. Once the current controversy over which vibrations are real has been settled, we will be able to use the results to interpret the Sun's structure, rather as earthquakes and moonquakes tell us of the structure of these planets.

The obvious blemishes on the Sun are *sunspots,* small dark regions on the shining disc. They are 2000°C cooler than the rest of the surface, and are dark only by contrast: an individual spot is shining as brightly as the Moon appears in our skies. Astronomers agree that sunspots are caused by a strong 'bunch' of magnetic field poking up through the Sun's surface; and that a sunspot is accompanied by peculiarities of 'weather' in the Sun's atmosphere, provoked by this magnetic field. The reasons for the magnetic activity are not known.

Sunspots live for only a few weeks, but the number of spots comes and goes over a period of 11 years. This is an average, in fact, for to confuse the issue, this 'cycle' can vary in length from 7 to 17 years. A maximum number in the current cycle was reached in 1980, and numbers will now decrease until the mid-1980s, then increase to another maximum around 1991. The most popular theory to explain the existence of these cycles is that the Sun's weak pole-to-pole magnetic field (no stronger than Earth's) becomes wound up around its equator by the fast rotation of the gases there. After 11 years or so, the magnetic field is so strong that it becomes unstable, and bursts up through the surface in the isolated patches we see as sunspots. When this happens, at sunspot maximum, the field can dissipate and die down. The surface activity then declines until eventually the winding-up becomes serious again. Even so, it is difficult to explain why the magnetic field is so very strong—up to 10,000 times the Earth's field—or even why it cools down the Sun's gases where it breaks through the surface.

Above the sunspot, loops of magnetic field hold up filaments of glowing gases in the atmosphere, filaments which appear at the Sun's edge as beautiful arched prominences.

A magnetic region's *tour de force* is a solar flare. When two opposed magnetic loops come together within a complex sunspot region, they annihilate each other in an explosion that shoots gases out into space at enormous speeds. These gas particles race across interplanetary space and, when they hit the Earth, they disrupt the upper atmosphere. Tremendous displays of Northern and Southern Lights appear above the north and south poles, and long-distance radio communication is wrecked. Space-faring nations are wary of sending astronauts above the shelter of the Earth's atmosphere when large solar flares are likely. But, despite their practical importance to us on Earth, it is not yet known why opposed magnetic fields should nullify so dramatically, instead of draining each other's strength gradually. And there is no sure way to predict the appearance of flares, except the knowledge that they are most likely every 11 years, at times of sunspot maximum.

The Sun probably has not always had a regular cycle. There were very few spots recorded for 50 years in the late seventeenth century (the so-called Maunder minimum) when it seems that the Sun 'stuck' at minimum activity. American solar expert Jack Eddy has related such historically-recorded 'spotless' periods to variations in the amount of radioactive carbon (carbon-14) in the Earth's atmosphere; and by testing the carbon preserved in even older wood, he has

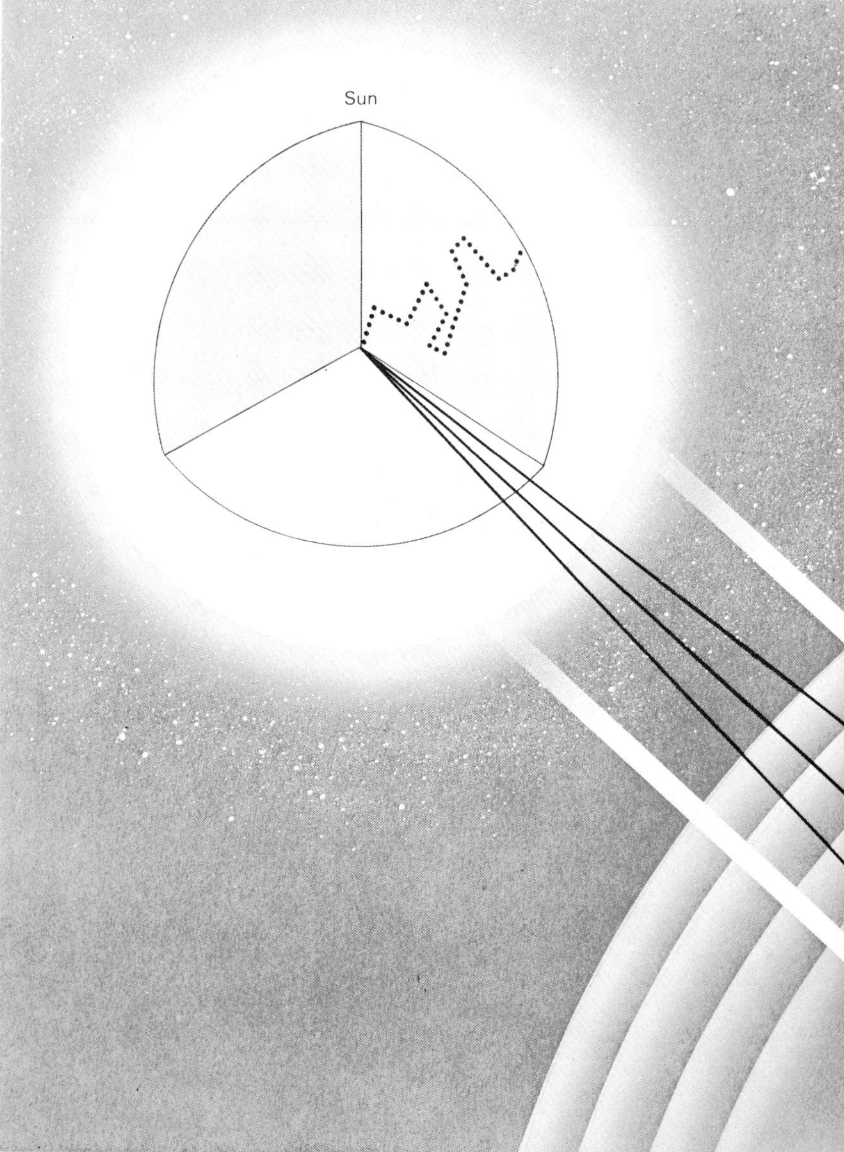

The neutrino 'telescope' (*left*) is a huge tank filled with cleaning fluid, situated down a gold mine where it is protected from less-penetrating radiations from space.

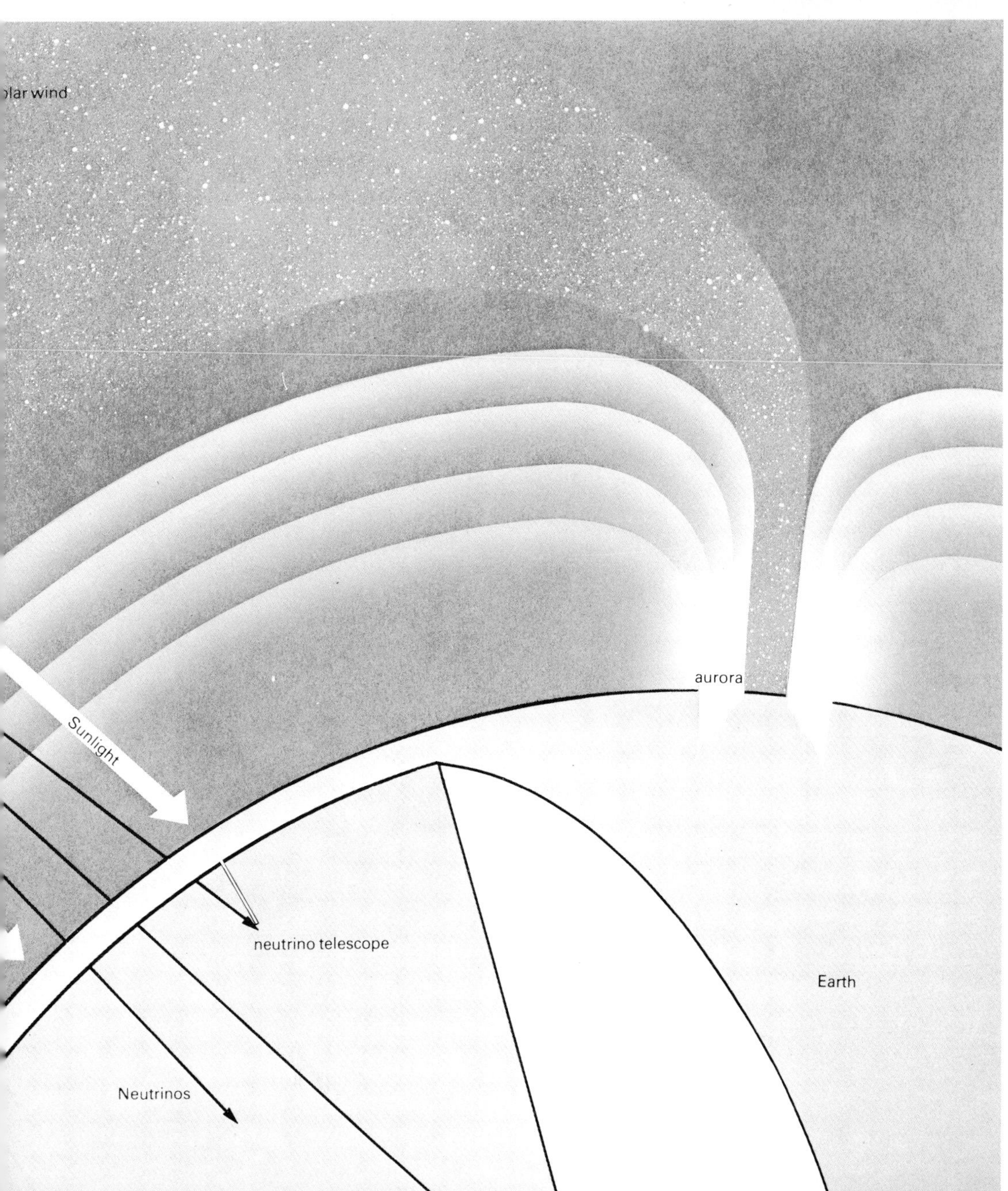

The Sun's huge energy output reaches us in many forms. The most important by far is radiation, which travels outwards from the solar core, taking millions of years to reach the surface because it is impeded by the Sun's dense gases (dotted line). From the surface, it shines out as light and heat. Neutrinos from the central nuclear reactions stream out completely unimpeded, and pass through the Earth, too. A few may be captured by the neutrino 'telescope', located in a mine in South Dakota. Active magnetic regions on the solar surface produce the solar wind of protons and electrons, which are deflected by Earth's magnetic field to the polar regions where they produce aurorae (Northern and Southern Lights).

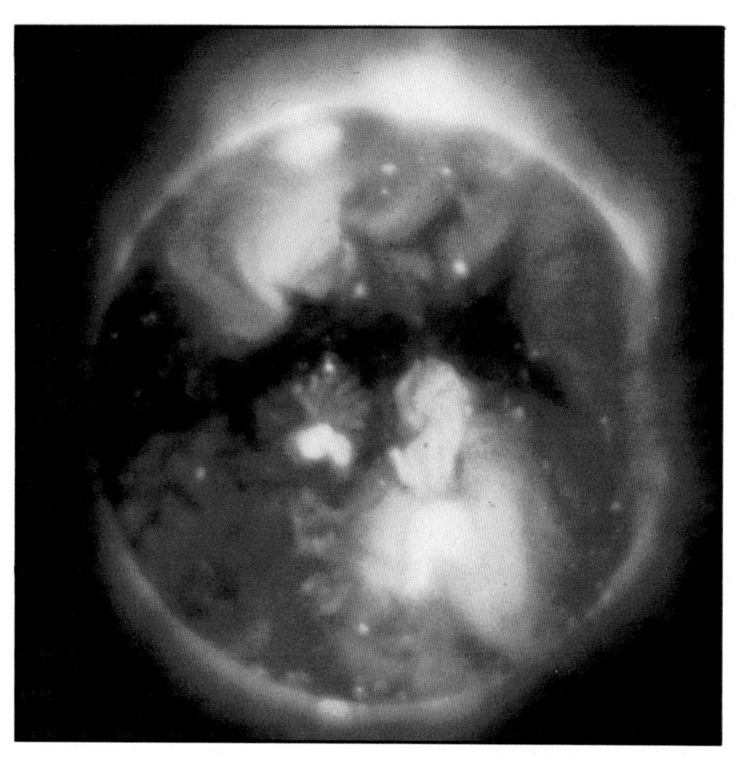

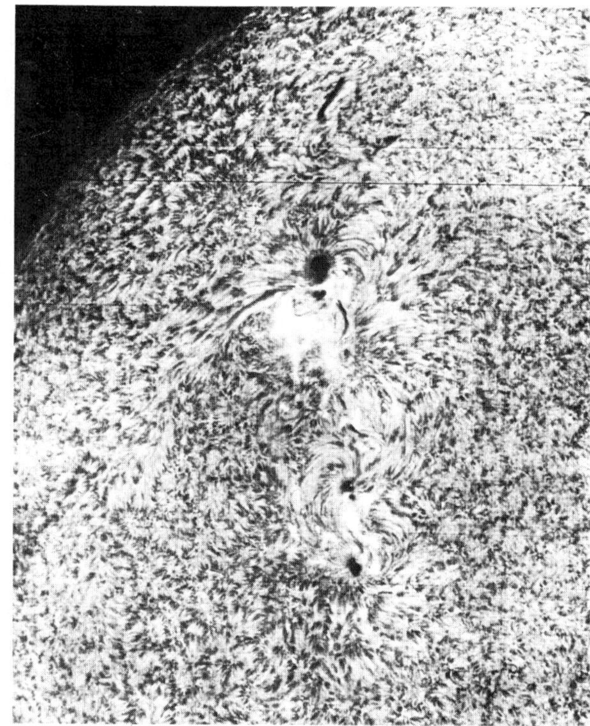

concluded that the Sun has, by and large, been unspotted for the past 3000 years. Only for odd periods has it had magnetic activity on its surface. We happen to be living in one of these unusual times, and are privileged to see sunspots, flares and all the related paraphernalia of the active magnetic regions. This 'evidence' is not beyond dispute, however: Chinese records of sunspots seen during the Maunder minimum have recently been discovered, which may undermine his conclusions.

Above the Sun's visible 'surface', the photosphere, is its more tenuous 'atmosphere'. The outer atmosphere, the '*corona*' is a faintly glowing region only visible to the eye when the brilliant photosphere is hidden during an eclipse. Oddly enough, it is far hotter than the Sun's surface. Our star's gases are around 14,000,000°C at the centre, dropping to 'only' 5500°C at the surface, and then becoming hotter again above the photosphere. The corona has a temperature of over a million degrees, although the gas is so tenuous that it does not hold much heat-energy. This high temperature cannot be due to heat coming directly from the surface because heat cannot flow spontaneously from a cooler to a hotter region. Astronomers believe the energy comes, not from the heat content of the photosphere, but from gas motions within and below it. Seen in detail, the photosphere has hot streams of gas rising, and cooler gas descending, a pattern of convection which carries energy outwards in the Sun's outer layers.

There is more than enough energy in the rough and tumble of the convection layers to heat the corona—but how does the energy get up there? It could be by means of magnetism again, or by sound waves. The convecting gases must make an incredible din in the Sun's lower atmosphere and as these reach the thinner, outer gases of the corona, the sound energy can perhaps change to heat.

ACTIVITY ON OTHER STARS

If astronomers have problems in solving the secrets of the Sun's magnetism and its hot corona, these pale into insignificance beside the difficulties of explaining the even more intense activity on the surface and in the atmosphere of other stars. The Sun is mild-mannered indeed when compared to other members of its immediate family.

Some of the heavier main sequence stars have far stronger magnetic fields than the Sun. In fact, their light varies slightly, because they are so disfigured by 'starspots' that their output drops noticeably as their rotation turns their spottier side alternately towards us and away from us.

But it is the lowest mass stars which are the magnetic exhibitionists. Lightweight stars are progressively more common in space, and those of around one-tenth of the Sun's mass are two-a-penny. They are also very dim, shining with only one-ten thousandth of its light and so, despite their number, we can see only those nearest to us. (The closest star to the Sun, Proxima Centauri, is just such a dim main sequence star.) In spite of their paltry light output, some of them, the *flare stars,* can stage several flares a day, each far more brilliant than

Activity in the Sun's surface and atmosphere is controlled by magnetic fields. Sunspots (*above*) mark the regions of most intense magnetism, and they twist the surrounding hydrogen clouds into elongated shapes like the pattern of iron filings around a bar magnet. The Sun's hot outer atmosphere, the corona, is spectacular when photographed in the X-ray it produces (*top left*). The corona is probably heated by magnetic fields extending up from the surface. Gas in the corona can cool and condense, flowing back to the surface down the magnetic field lines, as seen in the ultraviolet photograph (*right*). Such detailed solar studies can provide clues to the much more dramatic activity which other stars experience.

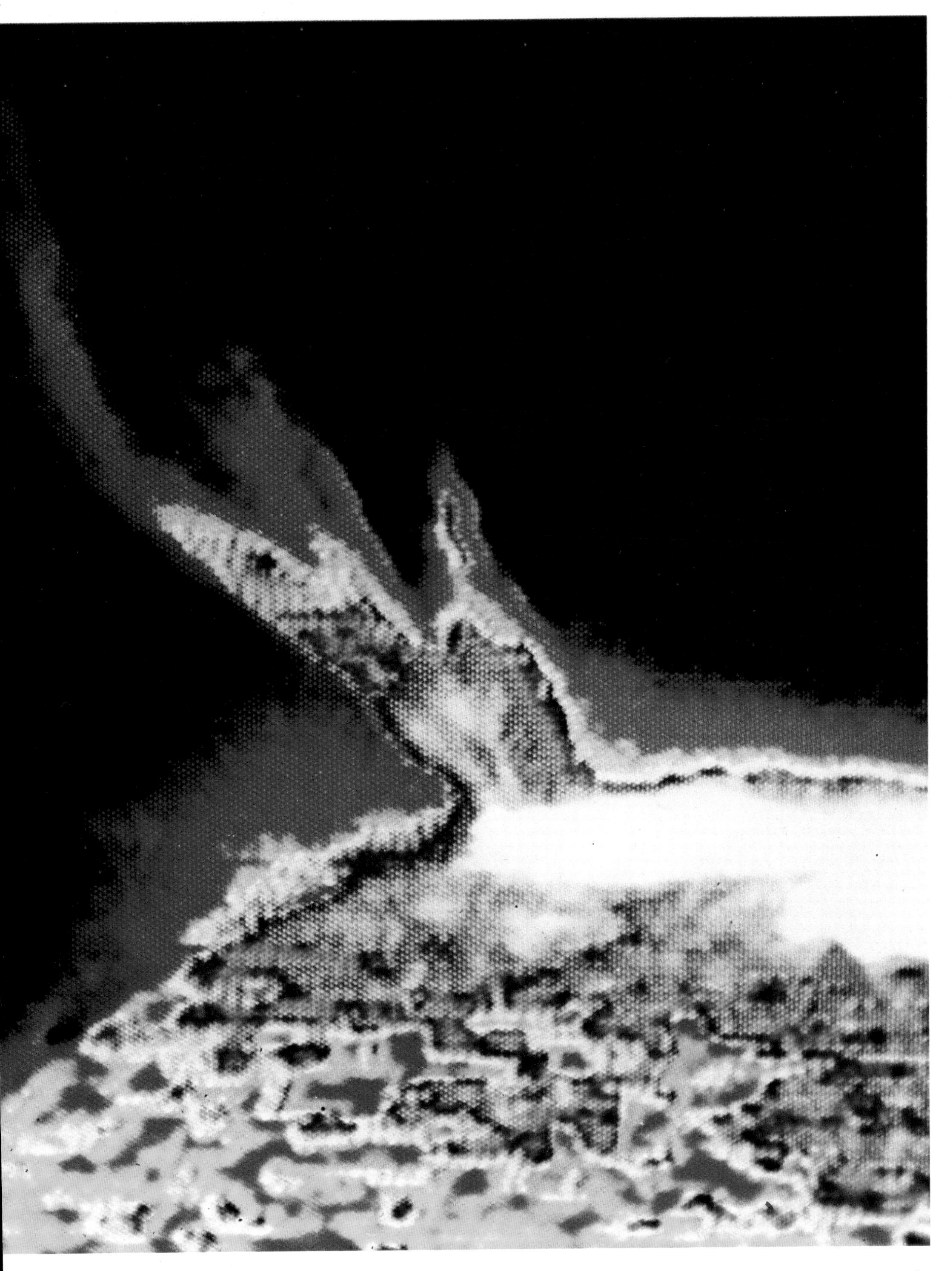

even a large flare on the Sun. And in contrast to the Sun, where the light flash from a flare is hardly noticeable, a flare star's outburst can be a hundred times brighter than the star itself.

These low-mass stars also had a surprise in store for astronomers using the orbiting X-ray telescope on the Einstein satellite which began observing in 1978. It was expected that stars like the Sun would have coronas hot enough to emit X-rays, and if their X-ray output was, like the Sun's, about a millionth of the light output from the surface, Einstein should have detected them. Indeed it did. But to everyone's astonishment, it also showed that the lightweight main sequence stars have coronas, too—and far more powerful ones than the Sun. In these stars, fully one-tenth of the total energy shines out as X-rays from the corona.

Like the tremendous flares on these apparently insignificant stars, their X-ray output indicates that an incredible amount of energy is channelled into their atmospheres. If there are problems explaining the Sun's flares, its magnetic activity as a whole, and how its corona is heated, they are dwarfed by the problems of ascertaining what is going on in the dim small stars. To put it into perspective, if the Sun emitted one-tenth of its energy from its corona, the radiation would destroy life on Earth; while scaling up the flares in proportion, a single flare a hundred times the Sun's brightness would in a flash wipe life off the Earth's surface. The answers to many of the Sun's riddles may well come from studying its faintest, but outrageously active, relatives.

Star-birth

The diversity of main sequence 'Sun-like' stars results from their different masses. Any star with the same weight as the Sun will shine with the same brightness, glow with the same surface temperature, and form into a sphere with the same diameter. Lightweight stars are cooler, dimmer and smaller in proportion, while heavy ones are brilliant superhot beacons in space. And the heavier a star, the more profligate it is with its hydrogen 'fuel': a star 10 times the Sun's weight will rip through its central hydrogen store in only 10 million years.

But we can see such stars in the sky, stars which must have been born within the past few million years—in the very recent past, astronomically speaking. Even our Sun is not that elderly, being born when the Universe was well over half its present age. So star-birth must be going on all around us, if we had eyes to see it. While we know roughly how stars are born, however, star-birth is a private matter, young stars only appearing when they are almost fully-formed.

The original material of stars is the very tenuous gas that fills space, mainly hydrogen and helium. Although it is rarefied, there is enough to make one new star for every 10 existing ones. Left to its own devices, interstellar gas is content to remain as it is. It needs to be given a push before it becomes interested in making stars: it must be rounded up into much denser 'clouds' of gas.

This first, pre-natal, stage, can result from any of several influences, and which are most important is still a matter for dispute. In a *spiral galaxy* (such as the Milky Way), interstellar gas is wheeling around the galaxy's centre; and when it passes through one of the spiral arm patterns the extra gravity of the stars there pulls on the gas and bunches it together. There is no doubt that far more stars are born in the arms than anywhere else in a spiral galaxy. But the push from the gases thrown out

Names of stars and other celestial objects

The brightest stars in the sky have individual names, usually bestowed by Arabic astronomers a thousand years ago. Thus the red star in Orion is Betelgeuse ('the armpit of the holy one'), and the end star in Leo (the lion) is Denebola ('tail'). In 1603, Johann Bayer systematically labelled the twenty-four brightest stars in each constellation with Greek letters: Betelgeuse is alpha Orionis, and Denebola is beta Leonis (note that the constellation names change to the genitive form—*of* Orion, *of* the lion).

With the invention of the telescope, astronomers could see far more stars, and the constellation patterns became less meaningful. Stars became labelled by numbers in specific catalogues, each resulting from a different systematic survey of the sky. The most generally useful is the Henry Draper catalogue, containing all stars down to a limit ten times fainter than the unaided eye can see. It lists not just their brightness and position, but also the type of spectrum their light has. The first list of 225,300 stars was published in 1918–24, and gives stars 'HD' numbers; an extension to include 47,000 fainter stars came a decade later, and stars in this have 'HDE' numbers. Thus the star identified with the X-ray source Cygnus X-1 (chapter 9) had already been catalogued as HDE 226 868.

There are many more specialized catalogues. For example, C. B. Stephenson and N. Sanduleak searched the sky for stars emitting strong spectral lines. Their 'SS' catalogue was known only to specialists in their branch of astronomy for 12 years, until one of their catalogued objects—SS 433—turned out to be a startingly bizarre kind of star pair (chapter 9).

Variable stars have a nomenclature of their own, going by constellation. The first variable known in each constellation is called 'R'—for example R Coronae Borealis, a giant star with a sooty atmosphere. The sequence runs S . . . Z, then RR . . . RZ. For later stars, the system becomes tortuous: SS . . . SZ, TT . . . TZ, . . ., WZ, YY, YZ, ZZ, AA . . . AZ, BB . . . BZ, . . . QZ. Hence the designation FU Orionis for a star which flared up in 1937. This system covers 334 variables in each constellation. The next discoveries are simply labelled V335, V336 . . . with the constellation name. So the bright nova which appeared in Cygnus in 1975 is V1500 Cygni, and a possible black hole companion in Scorpius is V861 Scorpii.

Nebulae and galaxies share a common system of names. Although they are very different kinds of object, they both appeared 'fuzzy' to early astronomers. The most prominent have their own names: the Orion Nebula, the Crab Nebula, the Andromeda Galaxy. Over a hundred were catalogued by the eighteenth century French astronomer Charles Messier, and therefore bear 'M' numbers. M87 in Virgo is a giant galaxy which modern research indicates harbours a massive black hole. Many more are included in the New General Catalogue of 1888, the source of 'NGC' numbers. The Orion Nebula, for instance, is also M42 and NGC 1976, while the Andromeda Galaxy is M31 and NGC 224.

Radio and X-ray astronomers have catalogued their sources too. In the early days of each subject, when positions were known only approximately, such sources were labelled by constellation. The strongest radio source in the constellation Taurus is Taurus A—it turned out to be the Crab Nebula, already well known to optical astronomers. The strongest X-ray source in Cygnus is Cygnus X-1. Systematic catalogues then followed: the third Cambridge catalogue of radio sources gave many '3C' designations, such as 3C 273, the nearest bright quasar. The most distant known quasar, OQ 172, was picked up at the Ohio State University in a series of surveys designated OA, OB, . . .

Many recent catalogues, at all wavelengths, simply list objects in terms of their position in the sky. This is a very convenient system when one is studying the same object at different wavelengths, when it would otherwise be known by different names to X-ray, optical and radio astronomers. The binary pulsar (chapter 8) is PSR 1913+16—'PSR' meaning pulsar, the first number the sky equivalent of longitude (right ascension) and the second number the equivalent of latitude (declination—the '+' indicating it is north of the celestial equator). Its position, in fact, puts it in the constellation Sagitta. The 'double quasar' (chapter 10) is known as 0957+561 and lies in Ursa Major (the great bear). Despite their utility, the new names certainly lack the romance of the old!

in a star explosion—a supernova—can compress gas very effectively, too; and supernovae occur more frequently in the spiral arms. Another influence is the powerful radiation from intensely hot, heavyweight stars, which can push away nearby gas and ram it into farther-off gas to build up a denser cloud. So gas in space leads an eventful life; it is constantly disturbed from its natural peaceful state, and much of it is forced into denser clouds floating amid thinner surrounding gas.

Once it is in a cloud, the gravity of every gas atom pulls more strongly on every other atom, and the cloud begins to shrink. At its centre, the gas compresses itself so much that it is no longer tenuous at all, but reaches the density of gas in a star—and, during this compression, the central gas heats up so much that nuclear reactions can begin. A star is born.

But although that account is true in principle, it is a tremendous over-simplification. For a start, stars are not born singly: a star-brood is a clutch of hundreds born at the same time from the same cloud of gas. Although gravity shrinks the cloud down towards its own centre, it does not make just one huge central star, but splits up into a multitude of fragments, each of which then becomes a star in its own right. The fragmentation presumably begins because the cloud is full of motion, streams and eddies of gas intermingling. Their motions prevent these whirls from sinking to the centre, and they form separate fragments which condense on their own account into stars.

But we are still struggling to understand how a gas cloud breaks up like this. The swirls must happen more or less at random, like wisps of vapour in the Earth's clouds, but here the chance occurrence of a particular whirl can make the difference between a star forming—with perhaps planets and living beings on them—or the gas just dispersing back into space. The way the fragments break up from one another must control how many stars of different masses are born. The fact that the stars around us today are predominantly small, with the Sun above average in weight, must be a result of these obscure and little-understood processes.

Most of our problems stem from the 'mistiness' of interstellar gas. Mixed in with it are tiny dust grains, each about the size of the particles in cigarette smoke and, like cigarette smoke, they dim our view of things beyond. When a gas cloud forms in space, the dust is concentrated too, and the dust grains block off so much light that there is no way we can see what is happening at the centre where the stars are forming.

Some astronomers have turned to calculation, and have worked out in detail how a star can form from a collapsing fragment of gas cloud—but only if it is not rotating, and has no magnetic field. But there are certainly magnetic fields in space, and the real gas fragments are undoubtedly rotating, so the calculations are far too difficult to solve completely, even with the assistance of a powerful computer.

Observations can help, however. Although we cannot see stars during their birth, they do emit other radiations, and infrared and radio

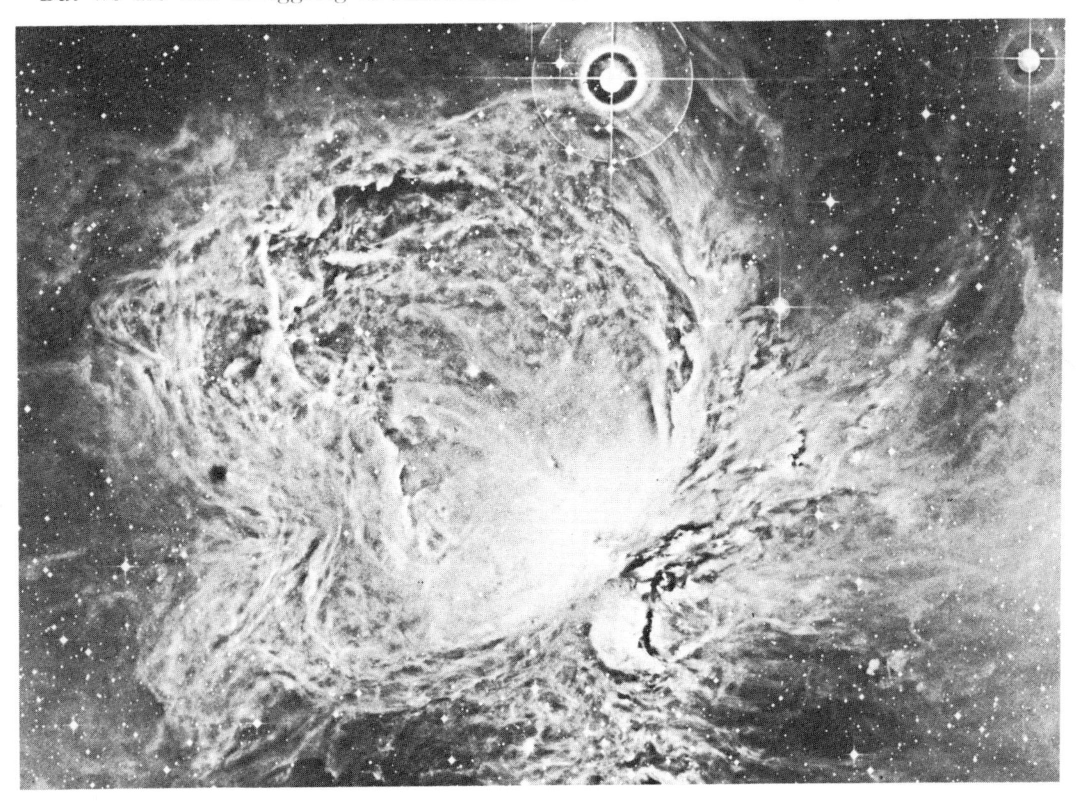

The Orion Nebula is a huge gas cloud lit up by the radiation from newly-born stars within it. The contrast-enhanced photograph emphasizes the twisted strands of gas and dark dust lanes in the outer parts of the nebula.

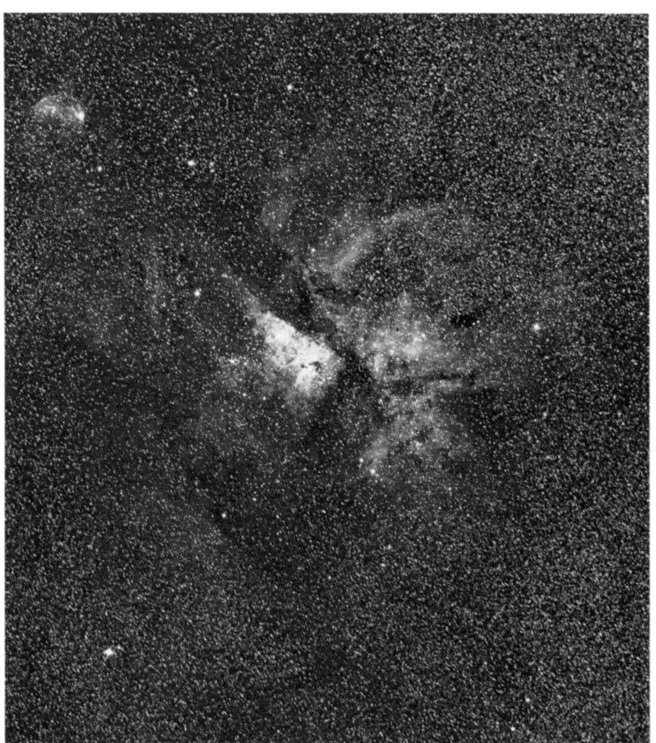

Optical telescopes show the Carina Nebula (*top right*) as a glowing gas cloud lit by young stars within. Many of the young stars are, however, hidden by thick dust clouds in the nebula. X-rays from these stars can penetrate the dust, and the young stars show up prominently on a photograph of the central regions of the nebula, taken in X-rays by the orbiting Einstein Observatory (*left*).

waves can penetrate the dust clouds which envelop the stars. Astronomers tuned into these radiations have picked up signals from the hatchlings in the centres of the dark gas and dust clouds, pinpointing where stars are forming. Infrared is 'heat' radiation, and it comes most strongly from the concentrated dust around a star that is shining for the first time and heating up its neighbourhood. Radio astronomers have two strings to their bow. As a young star continues to heat up its surroundings, the gas atoms around it are split up (each atom losing its electrons), and as such hot gas emits radio waves, the slightly more developed hatchlings can be 'seen' within the cloud.

They can also follow the first stages of starbirth, the clumping-together of the gas fragments before they become stars at all. The unexpected discovery of radio emission characteristics of *molecules* proved to be the key. Only 20 years ago, it was thought that such groups could not exist in space, for the ultraviolet light from stars would split them up into individual atoms. But in the dense clouds, shaded from ultraviolet by the fog of dust grains, there are indeed molecules. Some are just two atoms joined together, like carbon monoxide; but others are fairly complicated, including the nine-atom molecule of ethyl alcohol. Astronomers have now identified the characteristic radio wavelengths of over 50 types of molecule in space.

Although molecules are evidently protected from disruption in the dust clouds, no one is sure how they came to form out of individual atoms in the first place. Astronomers have had to turn for help to chemists, whose speciality is the reactions of atoms and molecules—although out in the cold and near-vacuum of space 'chemistry' must be very different from the test-tube and bunsen burner of the laboratory! Some atoms may have teamed together as gas atoms floating freely in the cloud; others must have come together on the surfaces of the tiny dust grains. The grains may, in fact, be coated with even more complex molecules. British astrophysicist Fred Hoyle suggests that the chemicals of life formed on dust in interstellar clouds, not on the early Earth. His view is not generally accepted by astronomers and biologists, but even more unorthodox is the theory put forward by his colleague Chandra Wickramasinghe. He has proposed that actual living cells have formed in space, and that the dust grains—far from being inanimate chips of rock, metal and carbon 'soot'—are actually 'freeze-dried' bacteria and viruses.

Returning from speculation, radio astronomers can 'read' the molecules' broadcasts to work out the gas density, speed and temperature within a dark cloud about to fragment into stars, a vital insight into the moment of star-conception. There are also intensely strong molecule broadcasts from near the youngest hatchlings (those revealed by the infrared astronomers), radio signals so strong, in fact, that they must be coming from natural masers. (A maser is the radio equivalent of a laser: a blindingly-intense source of radio waves rather than light.) Here we have laser-power radio waves coming from a region the size of the solar system. Astronomers believe that the maser is powered by the strong infrared radiation around in these parts of the cloud, but something of a mystery hangs over the way nature converts this radiation into the strange balance of wavelengths that is necessary to keep the maser shining.

The energetic young hatchlings eventually disperse the dense cloud around them, and at last reveal themselves to astronomers studying ordinary light. The heaviest of these young stars are so hot that their ultraviolet radiation lights up the residual gas around them, to create the beautiful glowing *nebulae*, some of the most beautiful sights in the sky.

The young stars are not yet stable, though. Their light flickers over a period of days or weeks, while streams of gas flow off their surfaces into space. Some astronomers think it was during this unsettled period—the T Tauri phase—in the Sun's early life that gas streams carried off its angular momentum, and left it a slowly-turning star. Occasionally a T Tauri star brightens up, in an unexpected show of youthful enthusiasm. In 1937, the star FU Orionis suddenly became a hundred times brighter. At the time it was believed it had suddenly 'switched on' its central nuclear reactions; but FU Orionis is now gradually fading again, and other stars have been seen to perform a similar feat, too many for this to be a once-in-a-lifetime event. Although no one really knows why this should happen, it seems likely that such enthusiastic bursts represent necessary adjustments of the young star's outer layers, during the traumatic first stages of growing up.

Stars in middle age

The Sun has long since passed the traumas of birth, and the readjustments of adolescence, as have most of the stars in the sky. Some have now moved on to middle age and senility, and by studying them we have a time machine to show what the Sun's fate will be. Once a star like the Sun has converted its central hydrogen to helium, something has got to change.

In fact, the central core—now just the helium 'ashes' of its extinct reactor—shrinks in size and, to compensate, the outer layers inflate grossly to something like a hundred times the star's previous size. When this eventually happens to our Sun, it will swallow up Mercury, Venus and perhaps Earth; even if our planet survives, the vast luminous Sun filling half our skies will boil off the oceans and scorch to cinders any living thing.

Many of the brightest stars in the sky are just such red giant stars, destroyers of any planets they might once have had. Their distended outer layers have grown so much that their gases are extremely tenuous. The Sun's gases now are packed to a density about one-third that of water, while a red giant's outer gases are no denser than the air we breathe. These loose layers of gas can pulse in and out, not as minute changes in size like the Sun's pulsations, but colossal heartbeats in which its size alters by half over a year or so. As it pulses, its light output changes in step. From the Earth, we see many of these giants as *variable stars*, detecting their light changes much more easily than the underlying pulsing. The star Mira is, at its brightest, the most prominent star in the constellation Cetus, but it regularly fades to a thousandth of this brightness and is then invisible without a telescope.

The most regular pulsators are bright, heavy giants called Cepheid variables. They are particularly useful to astronomers because their period of pulsation reveals their true brightness, and their distances can then be calculated from their apparent brightness. These variable giants are the basis of our knowledge of the distances to other galaxies. By comparing the average brightness of Cepheid variables in the Andromeda Galaxy with Cepheids in the Milky Way, we can calculate that the Andromeda Galaxy is two million light years away. But among giant stars, variability is the rule rather than the exception. Some can even change in brightness without pulsing in and out: carbon stars, for instance, whose atmospheres are thick with carbon gas, can suddenly fade a thousand

A table of some of the more important and unusual molecules discovered in space. They radiate at radio wavelengths, unless otherwise indicated.

Interstellar molecules

Year	Name	Formula	Wavelength
1937	methylidyne	CH	430 nanometres (visible)
1940	cyanogen	CN	387 nanometres (visible)
1941	methylidyne ion	CH^+	375–423 nanometres (visible)
1963	hydroxyl*	OH	2.2, 5.0, 6.3, 18 cm
1968	ammonia	NH_3	1.20–1.26 cm
1968	water*	H_2O	1.35 cm
1969	formaldehyde	H_2CO	2.0, 4.0 mm 1, 2.2, 6.2 cm
1970	carbon monoxide	CO	2.60 mm
1970	hydrogen	H_2	110 nanometres (ultraviolet)
1970	hydrogen cyanide	HCN	3.38 mm
1970	formyl ion	HCO^+	3.36 mm
1970	methanol*	CH_3OH	0.3, 1.2, 36 cm
1970	formic acid	HCO_2H	18.3 cm
1971	carbon monosulphide	CS	2.04 mm
1971	silicon monoxide*	SiO	2.30 mm
1971	acetaldehyde	CH_3CHO	28.1 cm
1972	hydrogen sulphide	H_2S	1.78 mm
1972	methyleneimine	CH_2NH	5.67 cm
1973	sulphur monoxide	SO	3.49 mm
1974	ethynyl	C_2H	3.43 mm
1974	dimethyl ether	CH_3OCH_3	3.47, 9.6 mm
1974	hydrodinitrogenyl ion	N_2H^+	3.22 mm
1975	cyanamide	NH_2CN	2.98, 3.73 mm
1975	silicon sulphide	SiS	2.75, 3.30 mm
1975	ethanol	CH_3CH_2OH	2.8, 3.3, 3.5 mm
1975	sulphur nitride	SN	2.60 mm
1975	sulphur dioxide	SO_2	3.46, 3.58 mm
1976	formyl	HCO	3.46 mm
1976	cyanodiacetylene	HC_4CN	2.80, 11.28 cm
1977	ketene	CH_2CO	2.94, 3.00, 3.67 mm
1977	cyanohexatri-yne	$HC_2C_2C_2CN$	2.95 cm
1978	nitric oxide	NO	1.99 mm
1978	methane	CH_4	3.91 mm, 1.4 cm

* These molecules can radiate very intensely as natural masers (the radio equivalent of lasers)

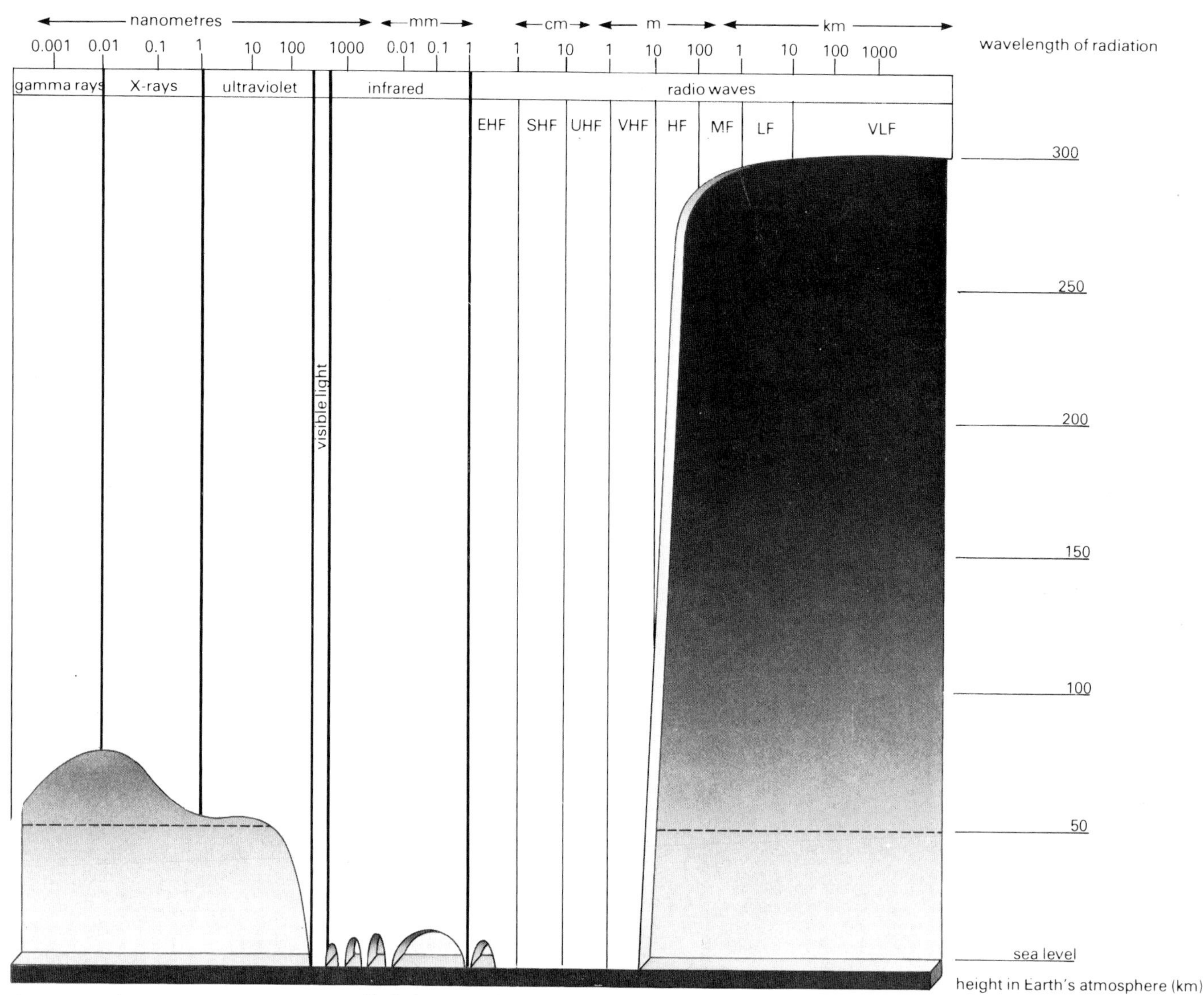

times, as the transparent gas expelled from the star suddenly condenses all around it as a thick shell of soot, which hides it from our sight.

Eventually, a giant star will shed its bloated outer gases altogether. Without the fuss and fury of an explosion, the gas is pushed gently out into space in all directions as a hollow shell surrounding the star. After a succession of these episodes, the shed gas forms a neat shell around the now-bare core of the star, and the intense radiation from the still incredibly hot core lights up the shell to make it appear as a softly-glowing iridescent ring. Some of the most beautiful 'nebulae' in the sky are this type (miscalled 'planetary nebulae' because the ring looks rather like a planet's disc when seen in a small telescope), and if we could wait some 5500 million years we would see the dying Sun swathe itself in just such a beautiful funeral shroud.

White dwarfs

All through the red giant stage, the star's central core has been very small and dense. In fact, its shrinking act heats up the centre so much that the helium there starts to react with itself to build up heavier elements. Just as the simplest element, hydrogen, had earlier combined to make helium, so at these higher temperatures helium can combine to make carbon and oxygen by further nuclear reactions. And while this is going on right at the centre of the core, there is a region immediately around the core where hydrogen from further out is still turning to helium. These reactions both produce energy to keep the star shining—and indeed produce it even more rapidly than before, so the star also becomes brighter as it expands to become a giant.

But these reactions must ultimately stop. The core is bared as the outer layers depart into space as a planetary nebula, and it is left as a closely packed ball of 'ashes'. It is searingly hot, from the heat of its earlier nuclear reactions, but with no further source of energy it must now start cooling. From white-hot at first, it will fade and cool through yellow-, orange- and red-hot, until at last it stops

Astronomers now use the entire range of radiations coming from space to interpret the Universe. Most of this wide spectrum is absorbed in Earth's atmosphere, and astronomers must use satellites to carry their 'observatories' above the atmosphere. The atmosphere's opacity has several causes. Gamma rays and X-rays are stopped by collisions with atoms high in the atmosphere, while ultraviolet is absorbed by the ozone layer. Water vapour and carbon dioxide in the lower atmosphere absorb most infrared wavelengths, but some infrared observations can be made from high dry mountain peaks. Long wavelength radio waves are reflected back to space by the ionosphere. Only visible light and short radio waves find the atmosphere to be 'transparent', and penetrate down to the ground.

shining altogether as a dark relic of a former star.

Since the most luminous of these hot ash-balls are the white-hot ones, these stars—no matter what their actual colour—are called *white dwarfs*. And dwarfs they certainly are. They consist of about two-thirds as much matter as the Sun, but compressed into a globe no larger than the Earth. They are so compacted that a matchboxful of matter from a white dwarf would weigh several tonnes.

STAR PARTNERSHIPS

Astronomers have thus charted the course of a single star's life, from the cradle to old age. But our Sun is an oddity in that it is a single star. Most stars form partnerships with another star—over three-quarters of all stars occur in such doubles. We do not see many of these partnerships when we look at the sky simply because most double stars are very close together, and appear single without a telescope. Some double stars pair up with other stars too, to make multiple star systems—the bright star Castor is actually six stars in close proximity, three pairs in mutual association.

In most star partnerships, each star lives its life as if it were a single entity. The heavier partner is more profligate, ripping through its hydrogen faster and dying first; the lighter carries on for longer. As a result, it is quite possible to find stars of different types together, for example a white dwarf and an ordinary main sequence star orbiting each other, even though both partners must have been born at the same time.

Astronomers have a special interest in double stars, for they constitute a natural weighing machine. There is no direct way to measure the weight of an isolated star, which is unfortunate because it is mass that determines all other behaviour. The stars in a partnership, however, hold on to each other by their gravity, and since gravity is a property of mass ('weight'), the masses can be found by working backwards from the size of the orbit and the time taken for the stars to revolve once around each other.

The heaviest known stars turn out to be about 50 times heavier than the Sun. This is probably a natural weight limit, because a more massive star would be so bright that its intense radiation would push off its own outer layers at birth. A few astronomers do dispute that there is such a limit, though. Perhaps we do not see any heavier stars in doubles simply because they are extremely rare. There is, indeed, some indirect evidence that a few nebulae have just hatched real heavyweight stars in their centres: the central object in the Tarantula Nebula, for example, may be a star at least 200 times the Sun's weight, and shining over a million times brighter than our star.

At the other end of the scale, there are stars only one-twentieth the Sun's weight, and dim to match. Undoubtedly, smaller fragments must occur in the gas clouds of the stars' maternity wards, but in these the central temperature never becomes high enough for the nuclear reactions to commence. 'Stars' lighter than this limit will simply become heavy planets, like Jupiter on a larger scale, but possibly pursuing their own paths without being held captive by another star.

Contact binary stars

When two stars lie close enough together, they do influence each other's lives. As one expands to become a giant, for example, it may shed gas onto the other and so alter its life-cycle. But even before this stage of development, they can influence each other. In a contact *binary system* (sometimes called W Ursae Majoris stars after their proto-type in the constellation of the Great Bear), two main sequence stars have been born so close together that they are literally touching as they spin around each other and sharing their outer layers of gas like a pair of cosmic Siamese twins. A typical contact binary has one star as heavy as the Sun, the other half this mass. If they were normal stars, the heavier should be 20 times brighter than the lightweight star, but in a contact binary it is only twice as bright. The central nuclear reactions cannot be affected by the stars' lying in a common outer layer of gas: the extra heat from the heavier star must spread through the outer layers of gas so that this star radiates less than normal while the other radiates more. The outer gas layers should be moving up and down in convection patterns like the Sun's topmost layers below the photosphere, and this is an efficient way to transfer the heat; but even so, it is difficult to explain the exact temperatures and luminosity. The Einstein Observatory has added a further surprise: in their X-ray output, contact binaries shine a million times brighter than would either star alone; a large fraction of the star's energy output must go into heating their surrounding corona, rather than appearing as light and heat.

The later life of these stars may be even odder. As each swells up in middle age, the double star system may be swallowed up entirely within the outer layer of the expanding partner. The double may become effectively a single giant star with two centres, revolving around each other deep within it, and hidden from us by the huge distended outer layers of gas. We do not know if such coalesced giant stars do exist, although the peculiar pulsing stars called dwarf Cepheids may be this type of two-cored giant. When double stars are farther apart, their middle age traumas can give rise to the spectacular explosions called novae.

Dwarf novae

We have known of novae since medieval times, but it is only within the last 20 years that we have discovered that the stars which explode as novae are never single: they are always close doubles, with one star a white dwarf. American astronomer Robert Kraft's study of the novae's near relatives, *dwarf novae*, laid the foundations for understanding these star explosions.

Dwarf novae are very faint stars which suffer fairly regular, sudden brightening. The star S S Cygni, for example, increases in brightness 40 times in just a few hours, and then fades to its original state over the next two weeks. These outbursts are repeated roughly every seven weeks. But dwarf novae do not actually throw out gases in an explosion. They consist of a small red star (low-weight main sequence star) and a white dwarf, orbiting around each other.

The red star is beginning to expand to become a giant; as its outer gases move away from its centre, they come closer to the white dwarf, and under the influence of its gravity. Caught in the gravitational tussle between the two stars, streams of gas are pulled down towards the white dwarf, but they cannot fall straight onto its surface. All the time, the two stars are whirling about each other, taking only a few hours to complete an orbit and, as a result, the gas streams have a sideways motion too. They head for the white dwarf at a tangent, and get spun out into a disc of gas circling the white dwarf star.

Where the infalling streams of gas freshly ripped from the red star hit the existing disc of gases, the collision heats the gas to incandescence in a glowing 'hot spot'. This bright, flickering spot is so intense that the 'star' which is seen between the outbursts of a dwarf nova is not either the red star or the white dwarf but the hot spot shining more brilliantly than either.

In a dwarf nova outburst, the whole of the disc around the white dwarf brightens up, without ejecting any gas into space. The outburst is, in fact, caused by gas moving *inwards*, spiralling in from the outer part of the disc down towards the dwarf's surface. There is still some dispute as to whether the repeating performances occur because the red star is unstable, and periodically dumps larger amounts of gas into its companion's disc, or whether the gas is continually building up in the outer part of the disc, until there is so much that the rotating disc becomes unstable, and the gases fall inwards, brightening up as they do so.

Nova eruptions

So much for the dwarf novae, a long-running enigma now near its solution. But these are faint objects visible only with a telescope. Much more impressive are the true novae, stars which explode with such violence that they fling out clouds of gas at tremendous speeds, gas to a total of 30 Earth-masses. Every decade or so there is a nova bright enough to appear to the unaided eye, sometimes a starburst that rivals the brightest stars in the sky.

Astronomers patiently studying the stars left after a nova explosion have concluded that these, too, consist of a small red star and a white dwarf in close orbit. But here the white dwarf is stripping gas off its partner much more quickly and the disc of gas around the white dwarf is permanently lit up and glowing brightly. Most of the light from an old nova (and presumably from a nova before its outburst) comes from this disc: and as we might expect, it is all the time about the same luminosity as a dwarf nova at maximum brightness.

Gas in the disc is, however, always spiralling gradually inwards, its orbital energy lost by friction with other gas atoms. So the white dwarf slowly accumulates a layer of this gas on its surface—like a cannibal feeding on its partner's substance. By some kind of cosmic retribution, such cannibalism gives the white dwarf severe hiccups. The gas accumulated on its surface is very hot, and it is highly

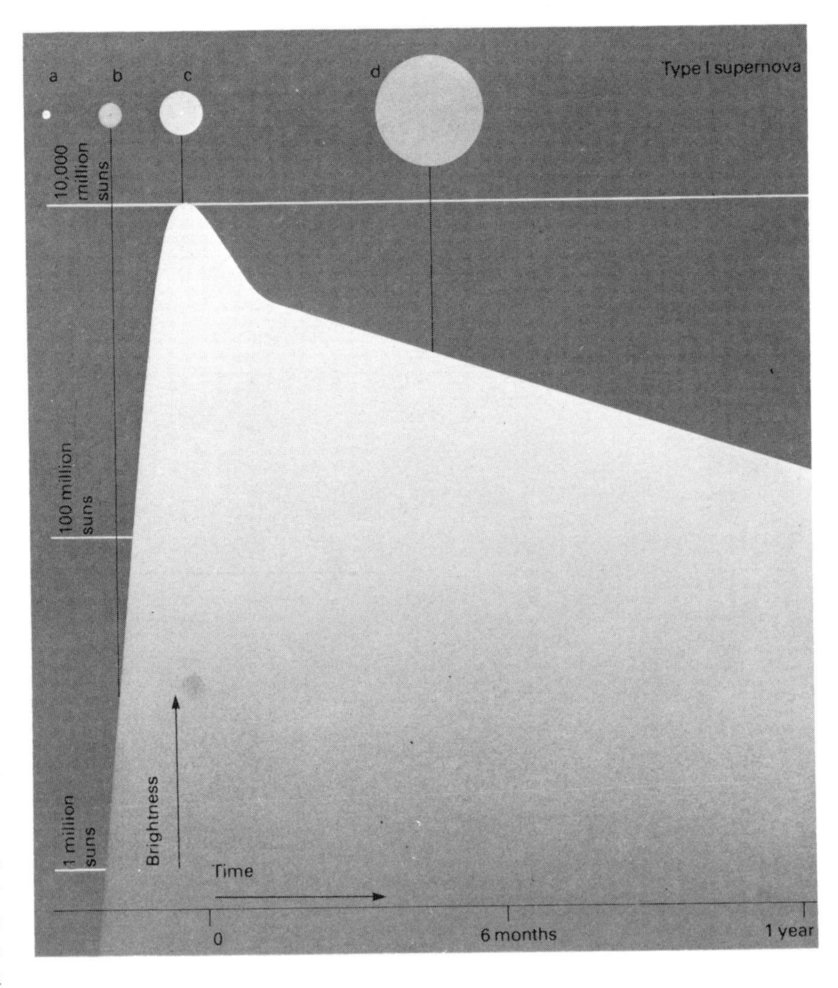

Light curves of the two types of supernova (*above*). Type I have an extremely uniform fading after the initial peak. Light from Type II drops more quickly; the residual brightness probably comes from gases left near the centre.

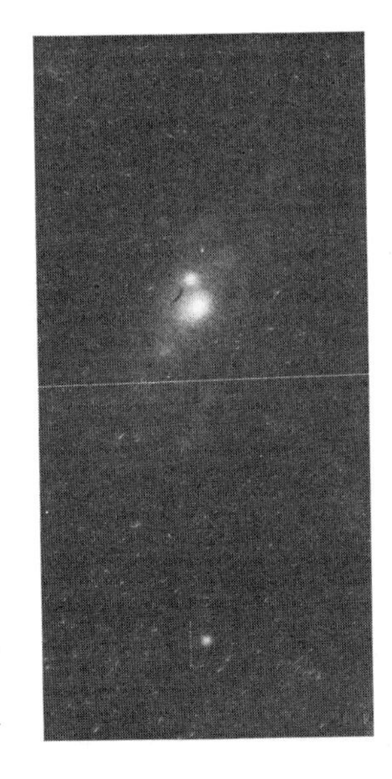

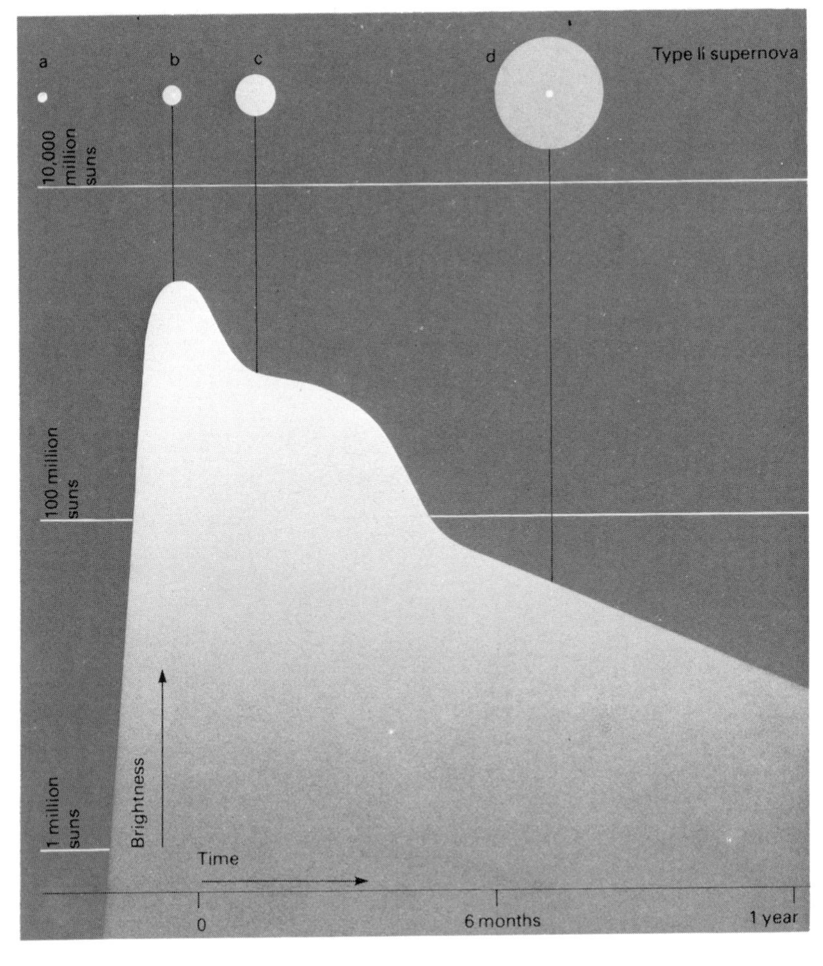

compressed by the dwarf's gravitational pull, which is mainly hydrogen. Eventually there is enough to trigger an explosion, literally a cosmic-scale hydrogen bomb.

The eruption rips off the gas on the white dwarf, and that in the disc, and ejects them into space at 5,000,000 kilometres per hour (3,000,000 miles per hour). The 'flash' of light from this enormous bomb takes months, sometimes years, to fade and, at its brightest, a nova shines with the light of 100,000 Suns.

But all this tumult does not affect the two stars much. As the gases eventually dissipate into space, the tug-of-war begins again, and the white dwarf starts accumulating hydrogen once more. After some 10,000 years, the layer of accumulated hydrogen on the white dwarf is thick enough to react, and it explodes again—although astronomers have not yet been around long enough to see a nova explode a second time.

Although some details remain to be confirmed, novae are no longer the major puzzle they once were. After a long search, and a painstaking reconstruction of evidence with the help of the world's largest telescopes, forensic astronomers can rest assured that both novae and dwarf novae are eruptions in an unusually close and cannibalistic double star system.

The Type I supernova which appeared in the galaxy NGC 4414 in 1974 (*bottom right*) almost outshone all the stars of the galaxy (*normal appearance bottom left*).

SUPER-EXPLOSIONS

The same cannot be said of the most violent and spectacular of all star explosions, the supernova. In a supernova outburst, a star literally blows itself apart. In the process, it flings out more matter than the Sun contains, at the incredible speed of 15,000 kilometres per second (10,000 miles per second), and at its brightest outshines a thousand million Suns.

Such extravagant displays are not common. Tycho's star of 1572 was a supernova which erupted several thousand light years away, and even so it was so brilliant that it outshone everything in the sky (apart from the Sun and Moon). For a short time it could even be seen in the bright blue daytime sky. Tycho's pupil, Johannes Kepler, recorded another only 32 years later, but since then no supernova has blazed out to shame the other stars.

We know of earlier supernovae, though, because the Chinese have kept a faithful watch on the sky since 221 BC. They were not astronomers so much as astrologers, searching for omens that would portend calamity for the Emperor. Unlike western astrologers, the Chinese were unconcerned with the gradual and predictable motion of the planets through the signs of the zodiac; it was objects which appeared *unexpectedly* that were, to them, signs for good or evil. They have left records of spots on the Sun, Northern Lights, comets, novae and supernovae (they called both 'guest stars') and even loud noises in the night. With careful detective work and a knowledge of the Chinese constellation patterns (very different from ours), it is possible to identify several other supernovae of the past two thousand years. The most famous appeared in the constellation Taurus in AD 1054 and was another supernova so bright that it was visible during daylight for three weeks.

Such historical accounts do not tell us *why* stars should blow themselves apart in such a spectacular way, although they are useful in guiding modern astronomers looking for the aftermath of supernovae explosions. Supernovae are so bright that a telescope can easily show one millions of light years away, in a distant galaxy way beyond our own. Indeed, at its brightest, a supernova can outshine the entire galaxy of stars in which it was formerly just one ordinary insignificant member. Nearly 500 distant supernovae have been discovered over the past 50 years, and by investigating them with modern instruments, astronomers have found they fall into two categories. Of these, those called Type I remain an enigma; Type II are partly understood.

Type II supernovae are seen only in the spiral arms of the spiral-shaped galaxies, and they eject several times as much matter as the Sun contains. This gives immediate clues to

Stars in their millions, of all sizes, ages and temperatures appear in this view towards the centre of the Milky Way Galaxy. The bright and dark gas clouds mark the birthplaces of stars.

the type of star it was before its suicide. To throw out so much matter, it must have been several times heavier than the Sun and, as such stars are short-lived, they bustle through their lives and die without moving far from the spiral arm where they were born.

A star 10 times heavier than the Sun, for example, rips through its central hydrogen and expands to become a red giant in only 10 million years—a heartbeat on the cosmic timescale. It has a small compact core of helium, and a huge flaccid body of tenuous gas. The centre of the core now begins to turn helium to carbon, just as a Sun-mass star does when it becomes a red giant. But a heavy-weight star does not stop at carbon once its helium core has an inner core of carbon, the weight of overlying layers crushes the centre more, until it is hot enough for carbon to turn to silicon. Later, the silicon at the very centre will turn to iron.

The star is now like an onion in structure. Its very centre is iron, surrounded by a shell of silicon, a shell of carbon, a shell of helium; and outside this small dense core of heavy elements, is an enormous layer of original gas, mainly hydrogen. The outer hydrogen layers have now swollen up so much the star is a *supergiant*, big enough to fill the orbit of Jupiter. And this superheavy, superlarge giant star now explodes. It rips itself apart as a supernova, of Type II.

The reason is basically that the central iron core is unstable. It is squeezed and heated more and more by the overlying layers, but unlike the preceding cores of helium, carbon and silicon, iron cannot make up heavier elements *and* give out nuclear energy at the same time. So it gets hotter and more compressed, until it is at a temperature of 1000 million degrees. Now conditions are so extreme that the iron breaks down into the simpler nuclei of helium. But this is exactly the reverse of all the other reactions that have gone on, and the situation is so unstable that the core is crushed in a matter of seconds to a tiny remnant, made of an entirely different kind of matter—the neutron stars and black holes of the chapters which follow.

Why should the outer layers of the star blow outwards as the centre collapses? There is still disagreement on the cause. The layers of the star nearest the core must certainly begin to fall in when the core collapses on itself. Perhaps some of the infalling layers bounce back when they hit the surface of the newly-formed neutron star at the centre and rip off the outer layers; or the elements in these inner layers may suddenly erupt in new nuclear reactions as they are so rapidly compressed. But the most popular theory for the past few years has been that an instantaneous flood of neutrinos pours outwards from the core, lifting off the star's outer layers and blowing them out, like a high-pressure air line attached to a balloon.

Certainly the final catastrophic breakdown of iron to helium when the star concedes defeat to its own relentless internal pressures, will produce a vast number of neutrinos. But we do not yet know enough about these elusive particles to say if they will indeed blow the star apart. Early calculations, in fact, seemed to show that they would simply fly through the star's outer layers without impeding their infall, let alone lifting them outwards again. Some later theories, on the other hand, indicate that the neutrinos would actually get trapped right in the centre, in the neutron star core, and would never get out into the rest of the star.

Type II supernovae are stars that astronomers do understand—heavy supergiants—exploding for reasons that are half-understood; Type I supernovae are a total enigma. They are actually more colossal explosions still, rivalling at their brightest 10,000 million Suns, and throwing out the Sun's mass of gas at an even higher speed. But this gas contains virtually no hydrogen. And all Type I supernovae follow almost the same pattern of brightening and fading away at the same rate—and in a very regular way, the brightness dropping by half for every two months elapsed.

These could be the explosion of star-cores, like Type II supernovae with their outer hydrogen layers stripped off—perhaps by a companion's gravity. But Type I supernovae are found even among the oldest stars in a galaxy, and it looks as though they must be old, lightweight stars that have been sitting around for a long time before getting around to exploding. One possibility is that they are white dwarf stars which have a natural weight limit, about one and a half times the Sun's mass. Although a white dwarf is normally stable, if it greedily pulled off enough gas from a companion star to take it above this limit, it would start to collapse, and then all hell would break loose. Weight for weight, the material of a white dwarf is the most explosive substance known in the Universe. If a white dwarf started to collapse, the carbon and oxygen in it would suddenly turn to iron, in an uncontrollable nuclear runaway reaction. The collapse would turn into the most powerful kind of star explosion that the mind can envisage.

Some astronomers think that the uncannily steady decline of brightness after a Type I explosion may be caused by radioactive atoms in the gases left over. Radioactivity always decays in a steady way, and so the fading can be explained if the energy of radioactivity can be turned into light—like a fluorescent watch dial. Radioactive cobalt produced in the explosion could be the culprit, decaying to iron in the months after the explosion.

But it is not universally agreed that a Type I supernova is an exploding white dwarf: the true answer may still await discovery. Although the traumas of an average single star's life have been sorted out, there are still mysteries left. Many discoveries of recent years have turned out to be due to the unexpected antics of close star partnerships—novae are just one example—and there is still a great deal more to be understood about how such close double stars grow up together, the games they play, and how they affect each other's later life and death.

When Tycho Brahe saw the supernova of 1572 flare up and fade away a couple of years later, and returned his attention to the ordinary stars in the sky, he could not have realized that later generations of astronomers would be able to write the biography of the stars. Modern astronomy has pieced together the broad outline from their birth, through vigorous life, to death—quiet for lightweight stars, dramatic for others—but still the stars hold secrets to tease future generations.

The Helix Nebula is a planetary nebula, the gently expanding outer layers of a red giant star. This beautiful nebula has a macabre relevance for us, because our Sun will eventually die in a very similar way—without the excitement of a supernova explosion.

CHAPTER 7

LIFE IN THE UNIVERSE

Radio telescopes all over the world are tuned in to the natural broadcasters in space: clouds of gas, ticking pulsars, distant quasars. Now and then, one is pointed at a nearby star. Stars do not emit radio waves. The quarry in this case is something else: a planet circling that star, not emitting radio waves naturally but as an artificial beacon to alert astronomers elsewhere in the Universe to the presence of intelligent life.

Picking up such a signal is a dream shared by many radio astronomers all over Earth. In 20 years, however, although literally hundreds of stars have been investigated, in no case has a message been picked up. Despite this, many astronomers are confident that our planet cannot be the only one in the Universe to harbour life. Their reasons are sometimes vague—It happened here, so why not elsewhere?—or sometimes precise: a numerical estimate of how many civilizations like ours there *should* be in our Galaxy of 100,000 million stars. On the other hand, there is no evidence of alien life in our Galaxy, and the belief in the possibility has ebbed and flowed in the past 30 years, as various arguments have come into vogue, then fallen into disrepute.

LIFE ON EARTH

But before investigating the possibility of life elsewhere in the cosmos, let us look at life on Earth. Our planet is the only place where we can study living things, where we can hope to understand how life evolved in the first place from the non-living matter of the Universe.

Life as we know it is a complex series of chemical reactions, occurring in small, jelly-like bags called cells. The simplest living things, plant and animal, are individual cells, while a more complicated organism, like the human body, is a collection of huge numbers of cells working together. The chemicals involved in the construction of cells, and in their internal reactions, are made principally of carbon atoms. The other vital ingredient is water, which fills out the cells and allows the dissolved carbon compounds to move around and react. Every living cell is three-quarters water: so our 'solid' bodies are, in fact, mainly composed of liquid.

In recent years biologists have come a long way in unravelling many of the mysteries of the living cell, probing the secrets of life itself. But the greatest mystery is still unsolved. How did life arise on Earth? The first living things were single freely-floating cells, like present day algae. A single cell is an incredibly complex system of interlinked chemical reactions that allow it to move, digest food, grow and reproduce. Somehow the simple, inanimate materials on the early Earth must have come together to make up the complexity of the first living cells—a feat as apparently miraculous as taking a random selection of electronic components, wiring them together at random and ending up with a computer.

The first steps in the climb from inanimate matter to living cells are rather less mysterious than the later stages. The processes occurring within a living cell depend on 'organic molecules', long strings of carbon atoms to which are attached other atoms such as hydrogen and nitrogen, and scientists are beginning to understand how small strings of atoms, at least, could have come about.

The Earth's primeval atmosphere must have been composed of noxious gases, either exhaled from volcanoes or brought in from space by comets hitting Earth: carbon dioxide, water vapour, sulphurous gases and cyanides. The water condensed into huge clouds enveloping the Earth, which poured down rain in torrential storms, drenching the surface for hundreds of millions of years and filling the hollows between the shifting continental blocks with oceans. Lightning rent the early clouds. They were lit on top by raw ultraviolet light from the fierce young Sun, and the impact of infalling planetesimals smashed sonic booms through the atmosphere. In these harsh conditions, life's fragile carbon strings began to assemble. During the past 30 years, scientists have subjected gas mixtures resembling the Earth's early 'air' to electric flashes, radiations and shocks in laboratory flasks. And their apparatus has gradually accumulated a tarry deposit of organic molecules, many of them the basic 'building blocks' needed to form the complex molecules of life.

Within a few hundred million years, Earth's early seas must have dissolved huge amounts of such chemicals—to achieve the consistency of a foul-smelling thin soup. And from ancient fossils, geologists know that living cells appeared within the next thousand million years. The real mystery is how these simple organic molecules built up into the enormous and precisely-tailored molecules needed for life's processes, and imprisoned themselves within the shelter of enclosed cells. It is not yet certain whether this was an inevitable progression, built into the way that organic molecules can react with one another, or whether there was a series of lucky chances that culminated in a living cell. On the latter view, all life on Earth must have descended from just one cell that happened to form; and the chances are slim that that lucky combination would have occurred on any other planet in our Galaxy.

Not all scientists are this pessimistic. Some argue that life started extremely quickly on Earth, when all the necessary reactions are considered, and this indicates that living cells are an inevitable consequence of the chemistry of carbon compounds. Recent discoveries have shown that simple carbon compounds are widespread in the Universe, even in the hostile environment of space, and this greatly increases the chances of their reacting to make life's chemicals. Some meteorites falling to Earth are rich in carbon compounds. These carbonaceous chondrites appear to be the oldest unaltered matter in the solar system, and they contain some 'organic' molecules which must have formed in the early days of the solar system. These are as complex as the chemicals produced in the experiments which mimic the Earth's early atmosphere. So it is possible that the raw materials of life were not synthesized in the chemical industry of the Earth's turbulent and noxious atmosphere, but instead were made during the formation of the planets and landed on our world in the form of meteorites.

Some of these carbon-rich meteorites have traces of tiny enclosed sacs, which look very

'Origin of life' experiments

In 1953, American chemist Stanley Miller simulated the effect of lightning on the gases of the Earth's early atmosphere. His laboratory experiment (see diagram) involved passing a high voltage discharge through a mixture of hydrogen, methane, ammonia and water vapour.

After a week, the apparatus contained a tarry mixture of chemical compounds, some of which were molecules essential to the structure of living cells, various types of amino acids which make up proteins. Miller had shown that 'organic' compounds could be produced from simple inorganic gases that served Earth as its first atmosphere.

Other experiments have used different gas mixtures, energized by different sources of energy. In every case, organic molecules are produced. Some are amino acids, others the constituents of the DNA and RNA molecules that pass on the hereditary knowledge from cell to cell.

There is no doubt that Earth's early oceans were rich with the simpler molecules needed for life to arise. But no one has yet demonstrated in the laboratory how these simple ingredients came to work as a living cell. The process probably took hundreds of millions of years on the early Earth, and there is no way to speed it up in the laboratory.

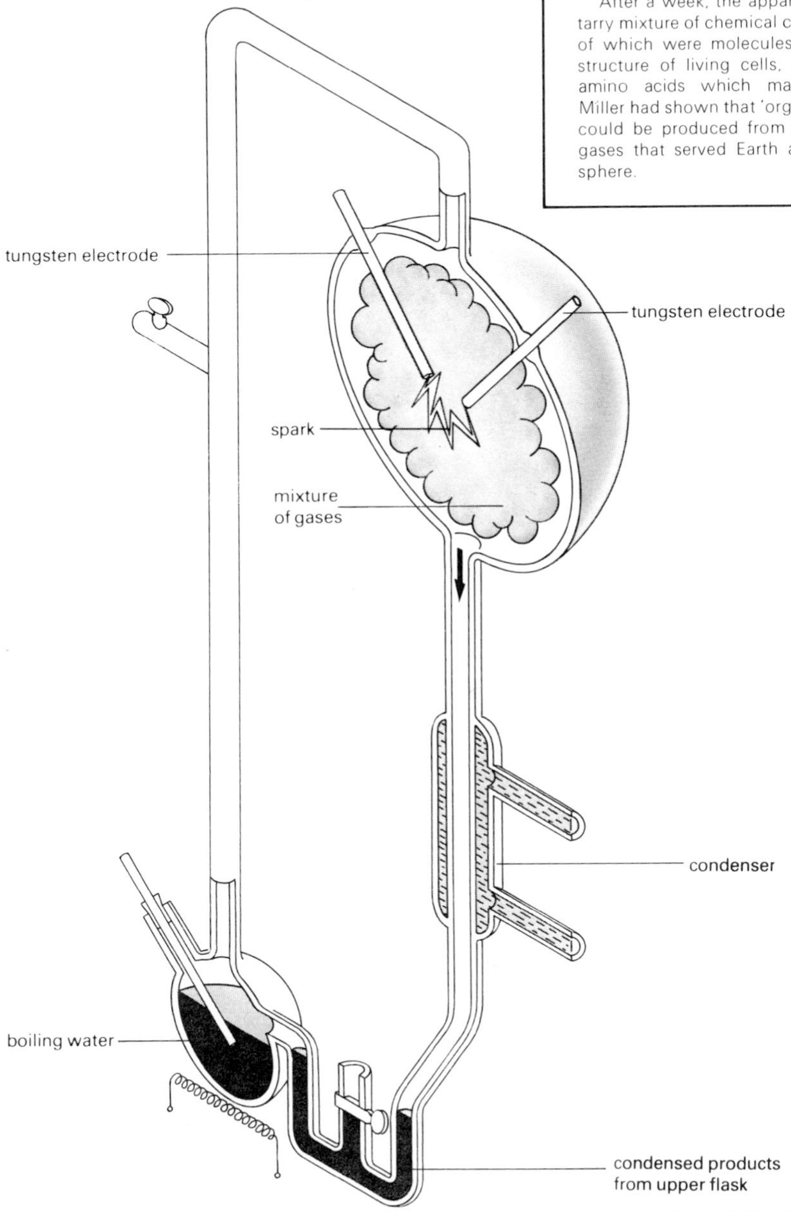

Starting gases	Sources of energy – simulates	Important products		
Water vapour H_2O Carbon dioxide CO_2 Carbon monoxide CO	electric spark—lightning ultraviolet—irradiation from Sun shock waves—meteor shocks	glycine $C_2NH_5O_2$ alanine $C_3NH_7O_2$ aspartic acid $C_4NH_7O_4$	amino acids: constituents of proteins	
Ammonia NH_3 Methane CH_4	fast protons—cosmic rays heat—contact with lava			
Hydrogen H_2 Hydrogen cyanide HCN Formaldehyde H_2CO	freezing—winter conditions	adenine $C_5N_5H_5$ guanine $C_5N_5H_5O$ cytosyne $C_4N_3H_5O$	bases	constituents of genetic molecules DNA & RNA
Acetylene C_2H_2		ribose $C_5H_{10}O_5$ deoxyribose $C_5H_{10}O_4$ glucose $C_6H_{12}O_6$	sugars	

1 carbon dioxide, water and ammonia in atmosphere

2 glycine and adenine dissolved in pools

3 DNA molecule

4 si
or

5 oxygen and nitrogen in atmosphere

Life has evolved from the inorganic matter of the early Earth, and has in turn altered the planet and its atmosphere. The diagram shows the likeliest route from non-life to life, although many details are still uncertain and controversial. Volcanoes gave the Earth its first atmosphere of carbon dioxide, water vapour and ammonia (1). Lightning and other violent activity broke up these molecules, and they reformed into simple organic molecules such as glycine and adenine (2). Washed into pools, these combined into the large molecules which are essential in living cells, including the molecule of heredity, DNA (3). Within a billion years, the first single-celled creatures and plants had formed (4). Plants photo-synthesized the carbon dioxide atmosphere to oxygen (5). High in the atmosphere, oxygen formed the ozone layer, absorbing damaging ultraviolet radiation from the Sun and allowing both plants and animals, now large and multi-celled, to invade the land.

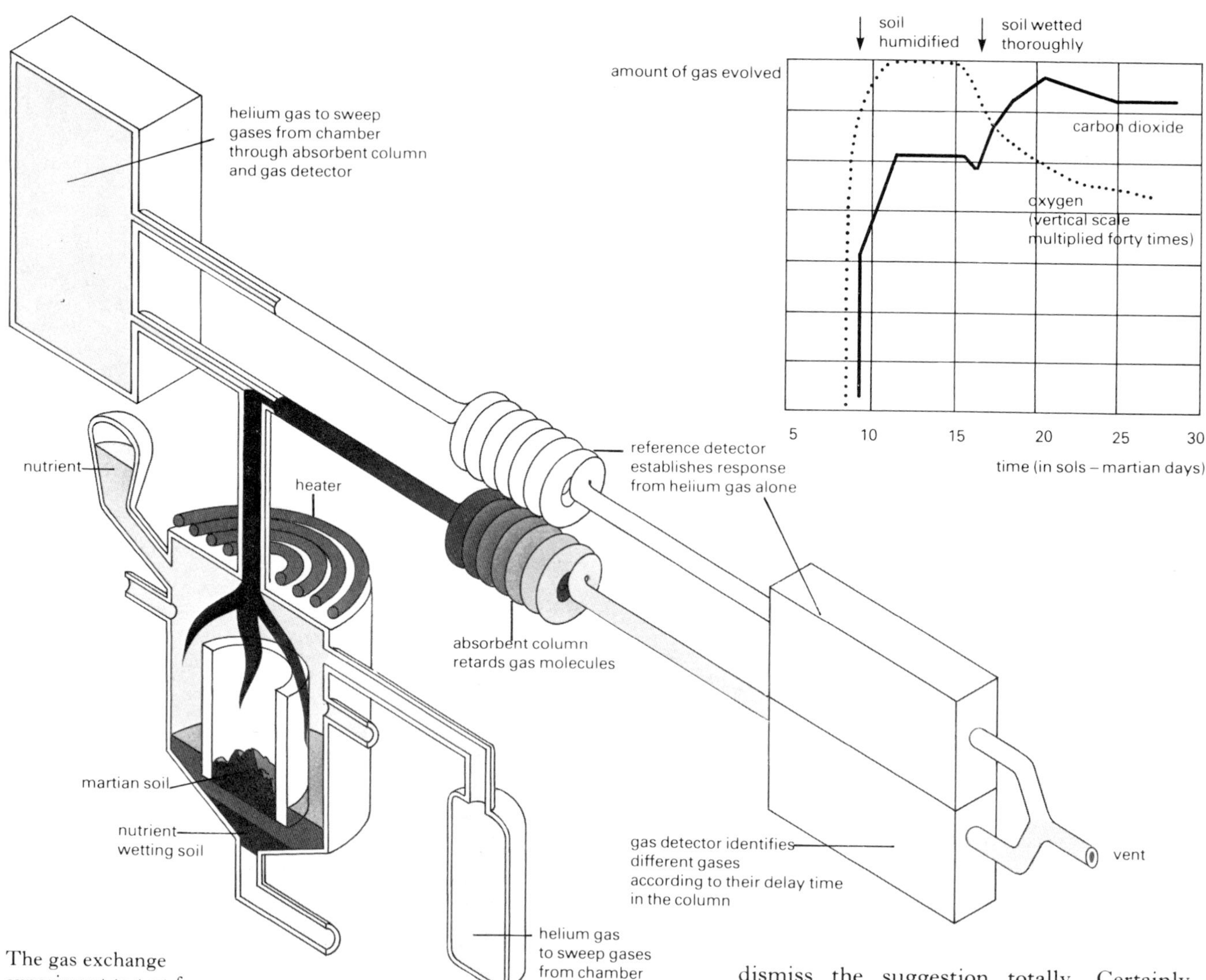

The gas exchange experiment to test for Martian life proceeded in two stages. Nine Martian days after the landing (on Sol 9 in Viking jargon), a small amount of nutrient was added to the test chamber to humidify it without wetting the soil. Carbon dioxide and oxygen immediately appeared (see graph). On Sol 17, more nutrient was added to wet the soil thoroughly, and the oxygen then began to decrease, as it was absorbed by one of the nutrient compounds. The carbon dioxide production increased gradually, then also started to decline. This was a tell-tale sign of chemical reactions between nutrient and soil.

like living cells. They are almost certainly seeds which have entered the meteorites as they have lain on Earth's surface after their fall, but these 'organized elements' did at one time provoke speculation that living cells themselves came from beyond the Earth. This idea has recently been revived in a different form by astronomers Sir Fred Hoyle and Chandra Wickramasinghe. They believe that comets are the ideal site for life's first stages. They argue that icy comets must be thick with carbon compounds, and if some of the ice deep within them melted, it would make a small enclosed 'test-tube' where dissolved carbon compounds could build up into a living cell. The heat needed to melt the ice could come from the chemical reactions themselves: life would have warmed its own incubator. During Earth's early history, they continue, such a life-bearing comet hit our planet, and life left its interplanetary icy craft to colonize a whole world.

Few astronomers—or biologists—would go along with this view, but since we know so little about life's origins, it would be unwise to dismiss the suggestion totally. Certainly, many scientists far more conservative than Hoyle and Wickramasinghe are convinced that life will be found wherever there is a suitable environment. In that case, there may be other homes for life even within our solar system.

Down below Jupiter's frozen cloud layers, heat coming up from that giant planet's core must produce a layer where water is liquid. American astronomer Carl Sagan has envisaged a race of huge gas-bag 'fish' swimming about in the enormous ocean. Marine life might—just possibly—be disporting itself in seas below the frozen ice-crust of Jupiter's moon, Europa. Some scientists had regarded Saturn's giant moon Titan as a likely site for life, if its atmosphere produced a greenhouse effect to warm its surface from the −180°C (−290°F) temperature of this part of the solar system to one where water is liquid. But Voyager 1 dashed these hopes in 1980, when its instruments found that the surface temperature is indeed −180°C (−290°F) and that its atmosphere is mainly nitrogen, a gas that does not produce a greenhouse effect. Any 'oceans' on Titan must therefore be composed of liquid nitrogen.

LIFE ON MARS

All astronomers would agree that the most likely place for Earth-like life in the solar system is the planet Mars. With its volcanoes, canyons and deserts, Mars looks like Earth. And it has a lot of water. Although most of this is frozen into the soil at present, the Mariner 9 and Viking orbiter probes discovered old, dried-up river beds. In the past, therefore, Mars must have had liquid water flowing over its surface. And, in that case, life could have evolved there, life which has now gone into hibernation, or is deep-frozen, during the current Ice Age which grips the planet.

The two Viking lander probes which soft-landed on Mars in the summer of 1976 were designed to find out. They could not return samples to Earth for a rigorous biological analysis, so instead they were equipped with simple laboratories to test Martian soil for life in three supposedly unambiguous ways. The laboratories were designed to 'wake up' any life forms dormant in the soil. An extendable arm scooped up samples of soil and deposited them in the tiny test-chambers, each scarcely larger than a pen-cap. The dry desert dust was warmed and moistened to revive any cells that might be present.

In the first test, the water added to the soil had a mixture of organic and inorganic compounds dissolved in it. The experimenters reasoned that if life of any kind was present it would 'eat' some of the compounds, and produce gas as a result. Sure enough, gases did come from the soil. But the instant surge of oxygen died away within two hours; carbon dioxide emitted more gradually also faded over 11 days. If life was there, it should have started off processing only a small amount of nutrient and, as the cells grew and divided, more and more gas should have been produced—not less and less.

The scientists handling this 'gas exchange' experiment by remote control from Earth reluctantly concluded that the added water and nutrients were reacting *chemically* with the soil of Mars. Although Mars's red rocks look very much like Earth's deserts, they evidently contain compounds that are highly reactive, substances that fizz if liquid water is added. Chemists think that these are super-oxides, produced when the Sun's ultraviolet light strikes the exposed surface of the planet.

This unexpectedly reactive soil caused headaches when it came to interpreting the results of the other biology tests, since the experimenters had to assess whether the results could simply be chemical reactions.

The second experiment looked for photosynthesis, the signature of plant-type life. Plant cells on Earth use the energy of sunlight to absorb carbon dioxide gas, and incorporate carbon atoms from this gas into their structures. In the Viking experiments, the Martian soil samples were incubated for five days under an artificial light to mimic the Sun, and with an artificial carbon dioxide atmosphere in the chamber. In this gas, the carbon atoms were radioactive. If Martian cells incorporated the carbon atoms from the gas by photosynthesis, the cells themselves would end up slightly radioactive. The experimenters tested for this by removing the original atmosphere in the chamber, then cooking the soil at 640°C (1180°F). This heat would have broken up any cells present into a tasty aroma of gases, which would have carried the tell-tale radioactive carbon atoms into detectors designed to look for them.

Sure enough, another positive result. Carbon dioxide *was* being incorporated into the soil, and driven out again when it was cooked. But was this a sign of life, or just another odd chemical reaction? The ex-

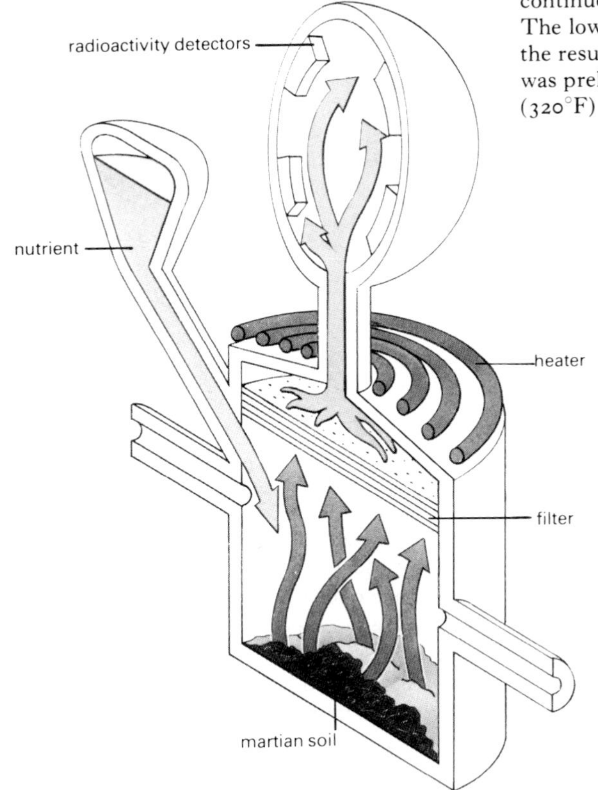

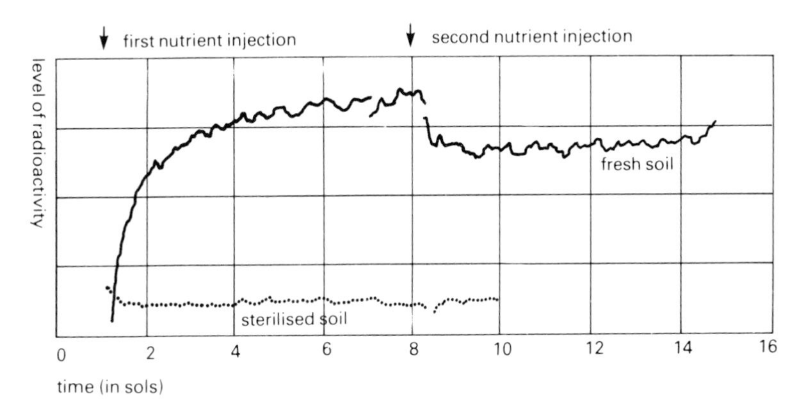

The Viking 'labelled release' experiment tested for animal-like life that consumes nutrient and exhales carbon dioxide. The nutrient contained radioactive carbon atoms, so living cells would release radioactive carbon dioxide gas, to be detected in the upper chamber. The first stage (during Sol 1) was to humidify the chamber and wet part of the soil: the soil immediately produced carbon dioxide (see graph). At Sol 8, a larger nutrient addition wetted the soil thoroughly, and gas continued to appear. The lower curve shows the result when the soil was preheated to 160°C (320°F) to sterilize it.

perimenters attempted to sterilize the soil by heating it to 175°C (350°F) for several hours before adding water and nutrient. But even after this treatment, the soil could still assimilate carbon dioxide from the atmosphere: the sterilization had not killed off whatever was active in the soil. The Viking scientists concluded that chemical reactions were once again responsible.

It was the third experiment that still causes dispute. This was almost the opposite of the carbon assimilation experiment which tested for plant-like life. Here the soil was fed with nutrients that contained radioactive carbon atoms: any *animal*-like cells would ingest these and breathe out carbon dioxide gas, in this case radioactive. Radioactive gases were indeed emitted, from the moment that the nutrients were added to the soil. In this experiment, moreover, the gas output did not quickly reach a peak and die away, as would have happened if a chemical reaction were occurring. The soil produced more and more carbon dioxide over the entire length of the experiment, which lasted almost seven weeks.

Biologists were not entirely convinced that this was the work of living cells. They had predicted that activity would very quickly get out of hand, as the cells grew and divided into more and more cells, each turning nutrient into carbon dioxide. But this doubt stemmed from theoretical expectations of how cells should behave, not on experiments. Viking experimenters Gilbert Levin and Patricia Straat responded by actually testing soil from the Californian desert under simulated Martian conditions. The cells in these soils produced a *gradually* increasing amount of gas, just like the Martian samples.

What was most intriguing, however, was that heat treatment *did* sterilize the Martian soil in this experiment. Whatever was turning nutrient to carbon dioxide was 'killed' by a temperature of 160°C (320°F) for three hours; and partially incapacitated by heating to 50°C (120°F). This is much what would happen to Earth cells, and it is difficult to imagine a chemical compound whose potency would be cut by these quite mild temperatures.

If the Viking landers had carried out only these three tests, scientists would probably now be saying: 'There may well be life on Mars, but we cannot be certain until we understand the chemistry better.' But today's attitude is different. The official line is that Mars is a dead world.

The experiment which has been most decisive in reaching this conclusion was not designed as a biology experiment. The gas chromatograph mass spectrometer (GCMS) sampled Martian soil for different types of chemical compound. And much to everyone's surprise, it found absolutely no trace of any carbon-containing compounds. Since life is based on carbon the GCMS experiment seemed to rule out life on Mars entirely.

Most scientists take this view. But those who consider the question still open point out that the 'successful' biology experiment would require only one million bacteria per gramme of soil, and that this is far below the level that the GCMS can detect. But they have to concede that it is extremely odd that there are not a whole lot of carbon compounds around, either as the remains of old cells, or as the 'food' for the cells that they claim exist.

Weight of opinion therefore now says that Mars has no life—but a fascinating soil chemistry. Before the issue is finally settled, though, we need more than a few ambiguous results from simple (though highly ingenious) experiments on an automated probe some 100 million kilometres away. The mystery of Martian life will only be completely resolved when scientists can test the soil of the planet in

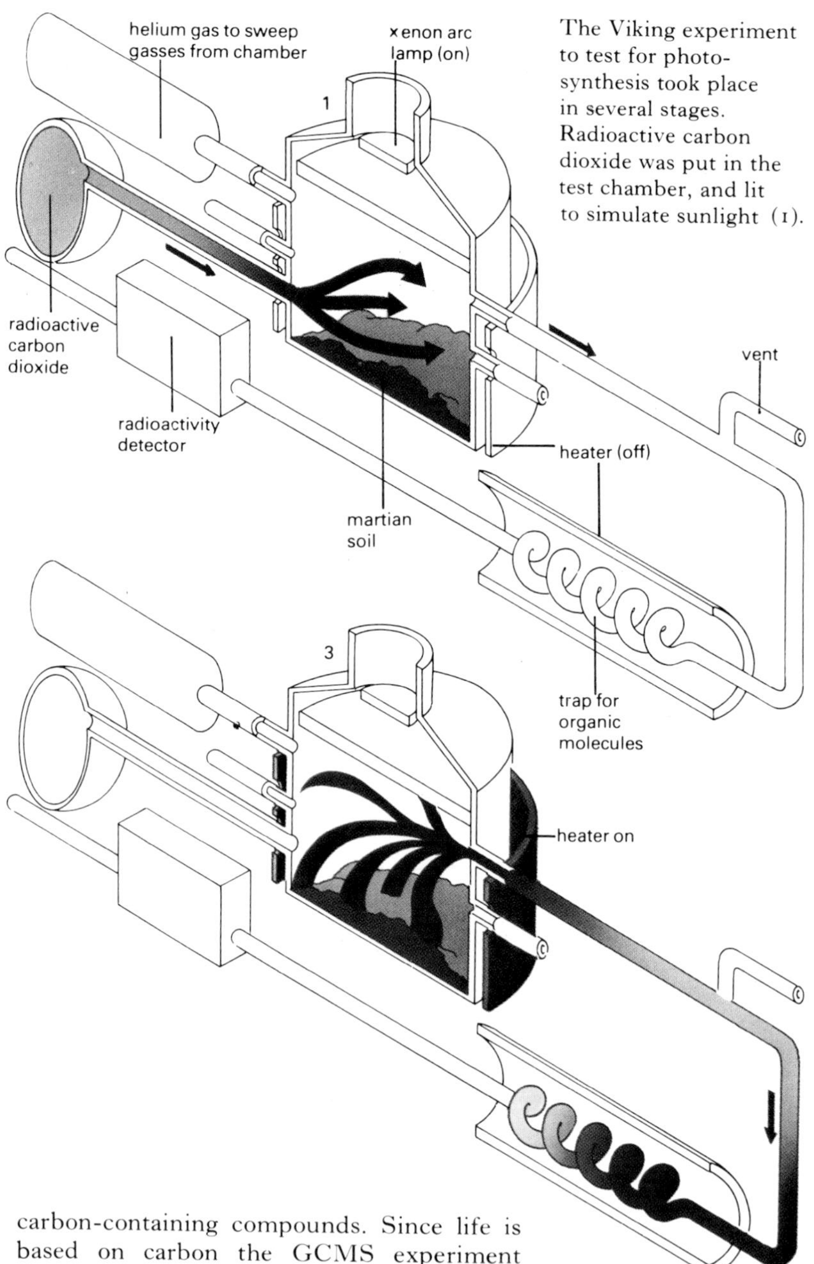

The Viking experiment to test for photosynthesis took place in several stages. Radioactive carbon dioxide was put in the test chamber, and lit to simulate sunlight (1).

Heaters around the chamber decomposed any organic molecules or living cells, and any resulting fragments were swept into a column which trapped them (3), while any of the original carbon dioxide was swept through. On heating the trapping column (4) the organic fragments (if any) would be oxidized and swept into the radioactivity detector, where their radioactive carbon atoms would give them away.

fully equipped laboratories, soil either brought back to Earth, or tested on Mars itself by a manned expedition.

PLANETS OF OTHER STARS

The quest for life elsewhere is now reaching beyond the solar system. Certainly, if there is *intelligent* life elsewhere in the Universe, it is not on one of the Sun's planets. But do other stars have planetary systems like the Sun's? If not, there is nowhere else suitable for life, and we may as well resign ourselves to interstellar solitude.

Unfortunately, the formation of planets is still disputed territory. Most astronomers do agree, though, that the planets condensed out of a rotating disc of gas and dust surrounding the early Sun; and that most stars are born with such a whirling halo. Astronomers studying the infrared (heat) radiation from newly-born stars are actively looking for signs of such pre-planetary discs. There was a false alarm in 1977. Infrared waves from the vicinity of the star MWC 349 in the constellation Cygnus were interpreted as coming from such a disc, but later observations revealed that the radiation actually came from a companion star so large and cool that its energy shines out as heat rather than as light. It will probably be only a matter of a few years, however, before the first nascent planetary system is discovered.

From these theoretical ideas, we should expect at least one star in ten to have planets. Astronomers have, indeed, been trying to detect planetary companions to nearby stars, but it is an incredibly difficult task because the planets are small and dim, shining only by the reflected glory of their star. Compare Jupiter's brightness with the Sun's; then imagine that the Sun was so far off that it appeared as faint as a typical star in the sky. Jupiter would then be so dim that even the largest telescopes could not pick it up, especially in the dazzle of the star itself.

When telescopes are lofted above Earth's atmosphere, it is just possible that they may be able to spot other stars' planets directly. In the meantime, roundabout means must be used to infer that a star has a planetary family.

The key factor is that a planet does not, strictly speaking, orbit a star: both planet and star orbit around their mutual centre of gravity—the balance point. Because the star is by far the heavier body, the balance point lies very close to it. The star consequently swings around in a very small orbit as it keeps on the opposite side of the balance point from the planet. Astronomers measuring star positions accurately are familiar with stars which seem to circle perpetually for no apparent reason. Such a star 'wobbles' about its average

If plant-like-cells were present in the soil, they would have incorporated the radioactive carbon into organic molecules within the cell. After five days, the radioactive atmosphere was removed (2).

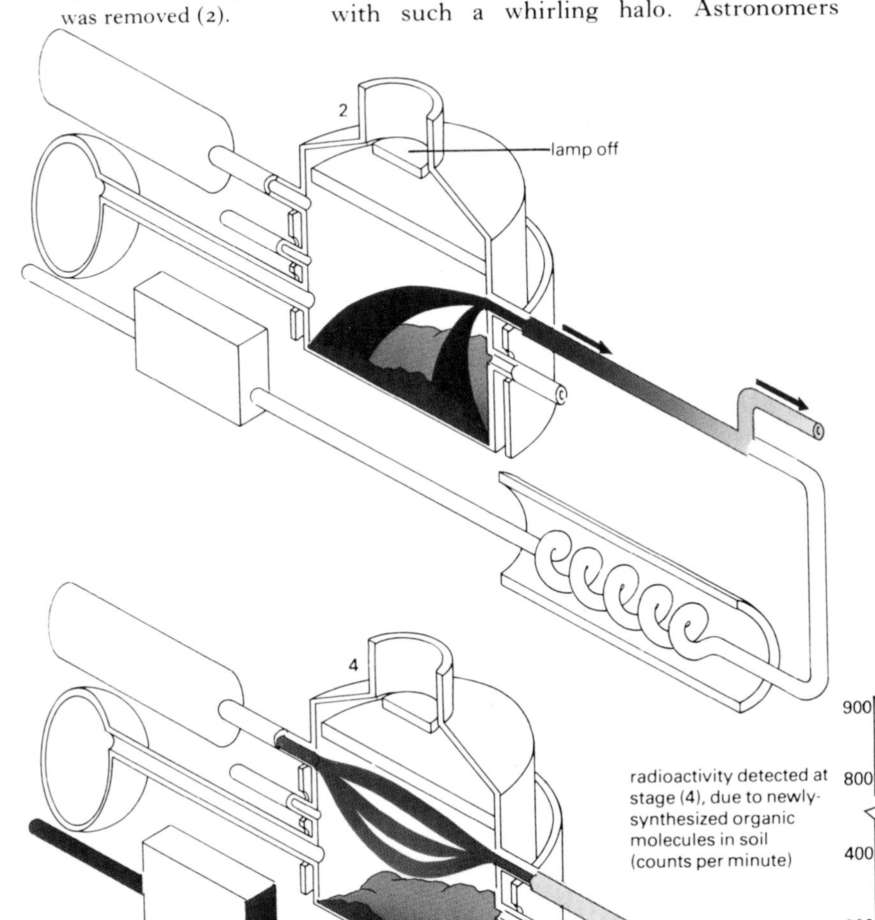

radioactivity detected at stage (4), due to newly-synthesized organic molecules in soil (counts per minute)

900
800
400
300
200
100
0

0 100 200 300 400 500 600 700

radioactivity detected at stage (3), due to artificial atmosphere (thousands of counts per minute)

Results are shown as a graph of radioactivity detected at stage (3) against stage (4). The small dots are tests on sterilized samples on Earth; almost all the Martian results (large dots) lie above the dashed line, which means the samples incorporated carbon dioxide from the artificial atmosphere.

position because it is not solitary, but has a fainter companion star which cannot be seen directly from Earth.

One of the pioneers of this kind of measurement has claimed that some of the 'unseen companions' he has unearthed are far too light in weight to be stars. Peter van de Kamp has produced evidence for companions only the weight of Jupiter. Any astronomical body lighter than 50 Jupiters (one-twentieth the Sun's weight) cannot shine in its own right as a star, and is a planet, dependent on its star for light and heat. In van de Kamp's reported planetary systems, it is safe to assume that if there is a Jupiter-like planet circling a star, there is a whole system of lighter-weight planets, too.

Barnard's Star

Van de Kamp and his colleagues at the Sproul Observatory, Pennsylvania, think they have found planetary systems accompanying several nearby stars. In 1975, they claimed to have evidence that the star BD 68° 946 has a planet 26 times heavier than Jupiter; BD 43° 4305 one between 9 and 23 Jupiters; epsilon Eridani a planet only 6 times as heavy as Jupiter. But the Sproul astronomers' strongest case was Barnard's Star. According to van de Kamp this star—at only 5.9 light years distance, the second-closest to the Sun—has two planets: one is similar to Jupiter in mass, the other about half as massive.

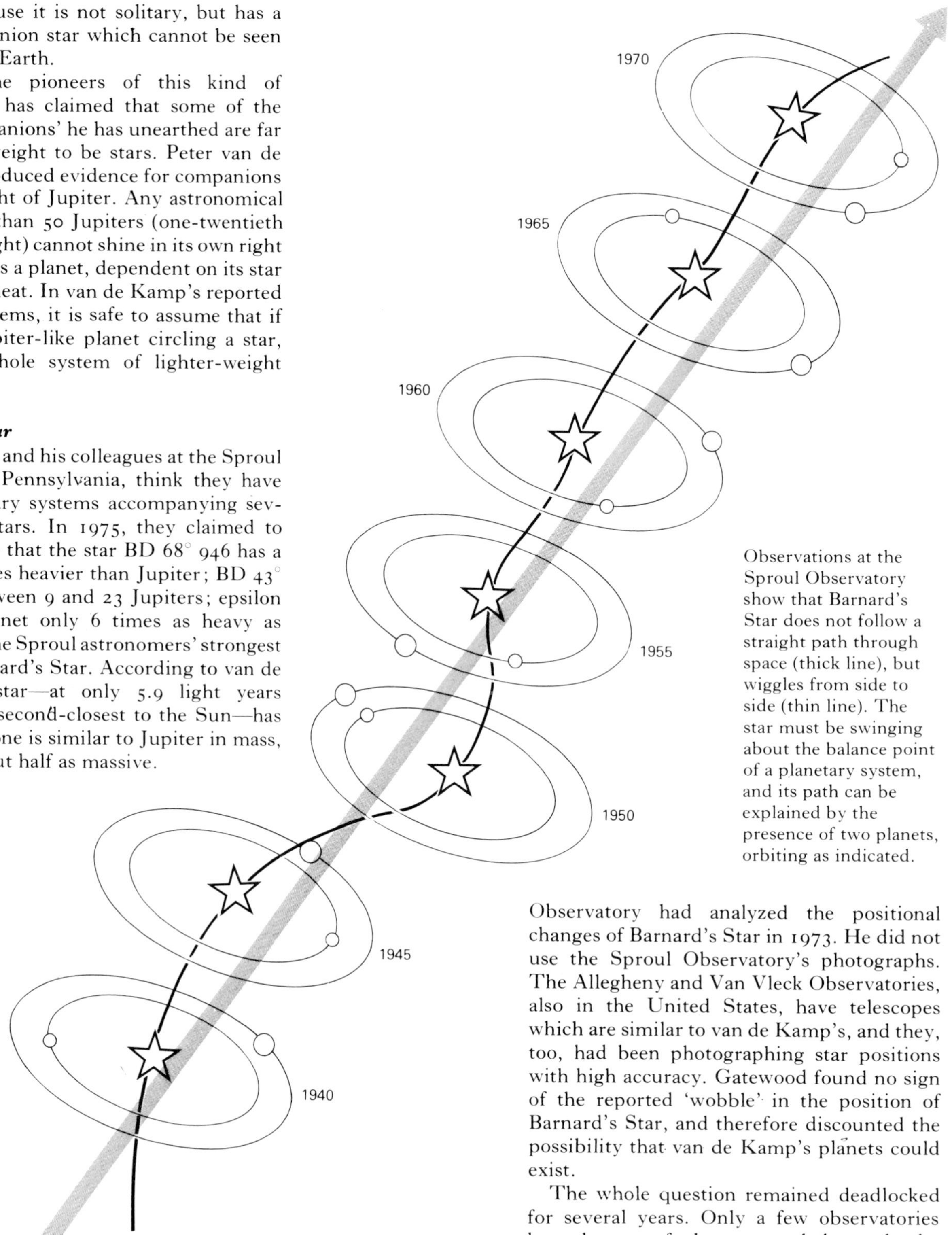

1970
1965
1960
1955
1950
1945
1940

Observations at the Sproul Observatory show that Barnard's Star does not follow a straight path through space (thick line), but wiggles from side to side (thin line). The star must be swinging about the balance point of a planetary system, and its path can be explained by the presence of two planets, orbiting as indicated.

This, indeed, seems a second solar system. Our own planetary system has one Jupiter, with the second weightiest planet, Saturn, about one-third its mass. Astronomers generally took van de Kamp's results as confirmation that planetary systems are common throughout the Galaxy.

But even at the time, some had their doubts. George Gatewood of the Allegheny Observatory had analyzed the positional changes of Barnard's Star in 1973. He did not use the Sproul Observatory's photographs. The Allegheny and Van Vleck Observatories, also in the United States, have telescopes which are similar to van de Kamp's, and they, too, had been photographing star positions with high accuracy. Gatewood found no sign of the reported 'wobble' in the position of Barnard's Star, and therefore discounted the possibility that van de Kamp's planets could exist.

The whole question remained deadlocked for several years. Only a few observatories have the type of telescope needed to make the measurements—ironically enough, the old fashioned, long refracting telescope with a lens at the top end, rather than the wide-mirrored reflecting telescope now in favour for its ability to see very faint objects. And the measurements are tricky. They involve determining positions on a photographic plate with an accuracy better than a thousandth of a millimetre. The motions that van de Kamp and Gatewood were seeking were right at the limit of measurement; on these microscopic differences rested the existence of whole planetary systems.

There are subtle statistical reasons for believing that van de Kamp's measurements were the less reliable. Wulff Heintz, an astronomer working in Cambridge, England, demonstrated this resoundingly in 1976. He took the Sproul Observatory's published data on 11 low-mass unseen companions (planets or lightweight stars), and showed that in all 11 cases we apparently see the orbit practically edge-on; and in 9 of the 11 the orbit is orientated in the same direction in space. It is virtually impossible that this could really be so. Heintz concluded that very small measuring errors were affecting all the star positions and so the apparent motions indicating the presence of planets were not real. He backed up his analysis by showing that all these stars, lying at different distances in various directions, are apparently swinging backwards and forwards in step with one another as the years go by, if the Sproul results are taken at face value.

On the observational side, Gatewood spent three years studying Barnard's Star with special care at Allegheny. By 1980, he had shown that Barnard's Star does *not* swing as van de Kamp had claimed, and other astronomers have been convinced by Gatewood's use of modern, sophisticated techniques, far more reliable than the traditional methods still in use at Sproul. The Jupiter-like planets of Barnard's Star seem to have vanished into the same limbo as the canals of Mars.

In astronomy, it never does to be too dogmatic. Although present opinion is firmly against van de Kamp's results, only more accurate measurements at a variety of observatories will finally settle the issue. And Gatewood's measurements do not mean that other stars are necessarily barren of planets. Barnard's Star itself may indeed have planets quite a bit lighter than Jupiter; present techniques would not reveal a planet as light as Earth, for example. The search is now on at Allegheny and other observatories: the search first for the best, most modern way to measure star positions to the last decimal place of accuracy, then the search for planets to other stars.

Life on other planets

For the moment it must be back to theory. Many astronomers have played the numbers game of extraterrestrial life. Answer in turn a chain of questions, and we can calculate not just how common life is likely to be, but whether we stand any chance of conversing with intelligent aliens: How common are

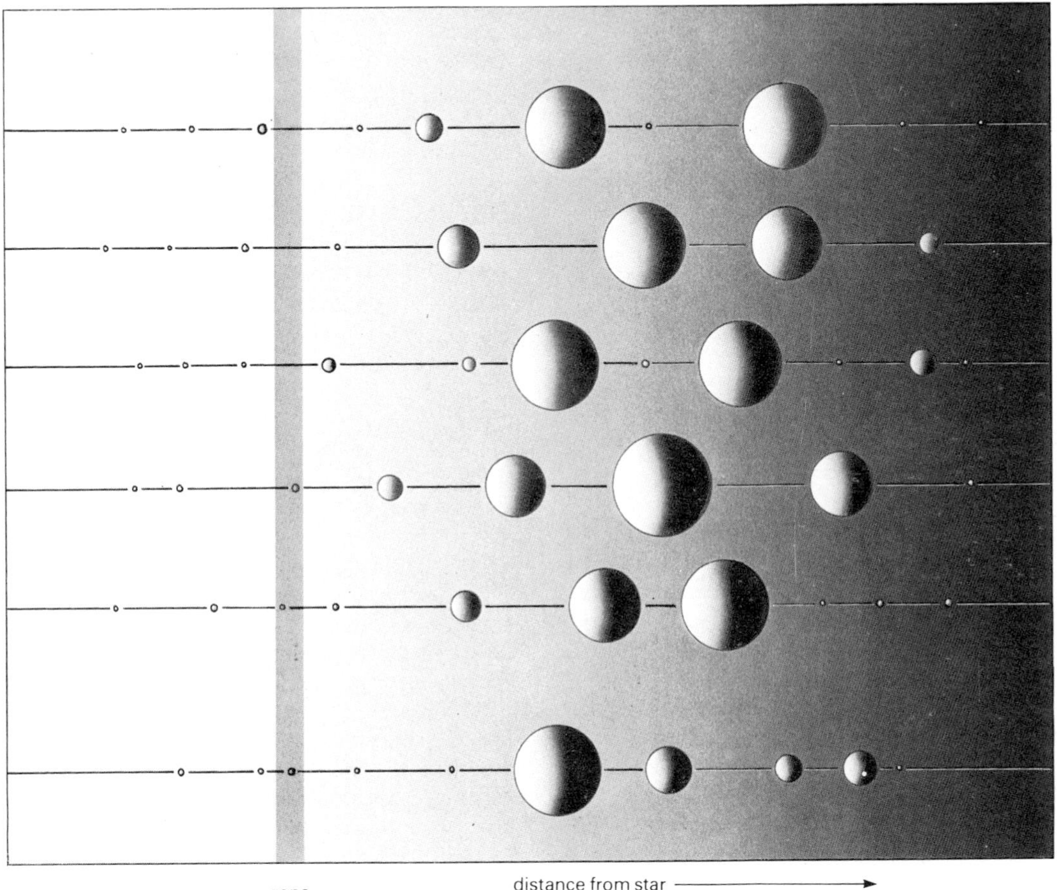

Computer simulations of planetary systems, based on widely-held theories of planet formation, produce results which are remarkably similar to the solar system (which is included as the lowest of the systems shown here). The planets' sizes are drawn to scale; their distances from the star have been compressed in an ever-decreasing (logarithmic) way towards the right. The zone where the temperature would be suitable for life is shaded. By investigating the proportion of simulated systems which have an Earth-sized planet lying in this region, astronomers can estimate what proportion of stars could have a planet where life may have evolved.

planetary systems? How likely is there to be an Earthlike planet in such a system? What are the chances that life would evolve there? Would it become civilized, survive nuclear war, and be interested in making interstellar contact?

If we could give definite answers to all of these questions, we would have a very good idea of how many alien cultures are trying to get in touch with us, and even the likely distance of the nearest. But at each stage down the chain, the numbers become more tentative, less estimates than 'guesstimates' and, in the final instance, nothing more than pure guesswork.

It is possible to make a stab at the purely astronomical questions: if there is a planetary system, the planets nearest to the star will probably be Earth-like, those farther out more Jupiter-like—gaseous rather than solid worlds. For life to form and evolve, a planet very like Earth must lie at the right distance from the star. A planet which is too heavy will retain too much atmosphere; a light planet not enough. The distance from the star must be such that the planet's temperature lies between 0° and 100°C, in other words, the right degree of warmth for water to be liquid. This zone around a star is called the *ecosphere,* region of life. It is surprisingly narrow. If Earth were only a few percent closer to the Sun, it would have suffered the kind of runaway greenhouse

Man's first intentional message to the stars was transmitted in November 1974, from the huge radio telescope at Arecibo, Puerto Rico (*above*). The message was a string of 'on-off' pulses, 1679 in all, in the sequence shown (*left*, where '1' represents 'on' and '0' 'off'). The key to decoding the Arecibo message lies in recognizing 1679 as the product of 23 and 73, so the pulses can be arranged in a two-dimensional picture. (*right*). Here the 'off' pulses are drawn black, and the 'on' pulses bright. The intended interpretation is shown on the far right. The top row, reading right to left, represents the numbers 1 to 10, in the binary code used by

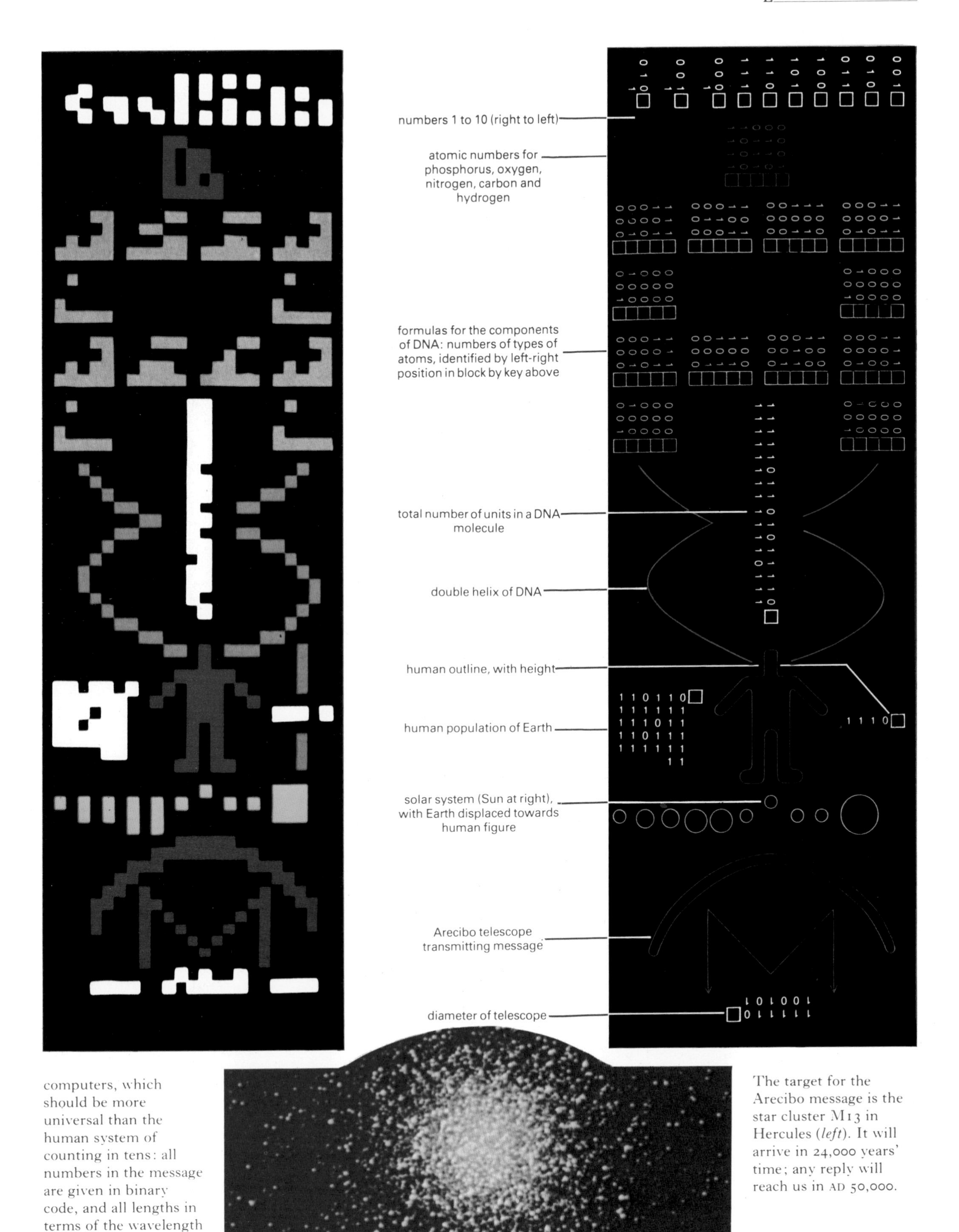

computers, which should be more universal than the human system of counting in tens: all numbers in the message are given in binary code, and all lengths in terms of the wavelength of the radio transmission (12.6 cm).

The target for the Arecibo message is the star cluster M13 in Hercules (*left*). It will arrive in 24,000 years' time; any reply will reach us in AD 50,000.

effect that has made Venus the hottest and most hostile planet in the solar system. Only one or two percent farther away, and our periodic Ice Ages would have gained an irreversible grip, and left us a totally frozen world like Mars.

Although astronomers still argue about the details of planetary formation, some have produced computer simulations that 'generate' planetary systems surprisingly similar to the solar system. If these are any guide, the chance of an Earth-like planet lying in the ecosphere of a sunlike star is between 1 in 10 and 1 in 100. Not all stars are like the Sun, of course; the more massive ones are hotter and brighter, and they have wider ecospheres. But these stars get through their lives quickly. A star twice the Sun's weight lasts only one-tenth as long. There would not be time for life to form on any planet around it, certainly not enough time for intelligent life to evolve.

Lightweight stars have narrower ecospheres. And when we take into account all the factors that can cause a runaway greenhouse effect or a permanent ice age, it turns out that a star less massive than 85 percent the Sun's mass has no ecosphere at all. All the planets of a very lightweight star are either parched or frozen.

So the stars which might have life-bearing planets are only those very like the Sun. These make up about 1 in 100 of the stars in our Galaxy. Two-thirds of these are partners in double-star systems, and here the gravitational tug of the second star would disturb a planet's orbit, making it very unlikely that it could stay permanently in the ecosphere region. Only single Sunlike stars are good candidates.

Slotting these particular figures together, the numbers game comes up with an answer of about a million planets in our Galaxy where life could exist. That is only a rough estimate, however, and probably on the pessimistic side. Even with these purely astronomical calculations, there are so many uncertainties that different astronomers have come up with remarkably different figures. Sir Fred Hoyle works out that there are 2000 million 'habitable' planets in our Galaxy; the astrophysicist Shiv Kumar reckons there must be less than a million, and possibly only one—the Earth. Most astronomers carry around in their heads an answer somewhere between 1 million and 1000 million.

As we have seen earlier, scientists disagree very much more over the probability that chemicals on such a planet would get together to form living cells. So estimates of the number of planets in our Galaxy which have living things growing, creeping or flying on them are even wilder. And by the time we ask if intelligent life would evolve we really are into the realms of guesswork. Although many books and magazine articles discuss in apparently scientific terms the arguments for and against various types of civilizations, their ability to survive nuclear war, and their inclination to get in touch with other civilizations, such discussions are little more than stabs in the dark. To get an idea of what intelligent aliens would look like, an evening at the local science fiction cinema is as good a guide as a scientific seminar, and a lot more enjoyable!

Interstellar communication

Although our Earth *may* be the only populated planet in the Galaxy, there is a good chance that it isn't. And, taking Earth as a guide, there may well be intelligent being up there looking into their night sky, and—like us—wondering if they are alone in space. There may indeed be civilizations spread throughout the Galaxy, attempting to contact one another by radio waves beamed across the void.

With this thought in mind, radio astronomer Frank Drake pointed the big 26 metre (85 foot) diameter radio telescope at Greenbank, West Virginia, in the direction of two nearby Sun-like stars, back in 1960. He called his search for artificial radio signals from beyond the Earth, Project Ozma, after the princess of the mythical land of Oz in L. Frank Baum's children's story. Drake picked up no signals; nor have other radio astronomers who have 'listened in' to hundreds of other stars in the past two decades.

But interstellar signalling is a tricky business. Whoever may be out there will be sending his or her message in a tight beam of radio waves, like a searchlight beam, to conserve its energy over the light years of intervening space. So he must be pointing it at likely stars in turn, not spending long on signalling to any particular star. And radio astronomers on Earth are also 'listening in' to other stars in turn. There is only a remote chance, therefore, that our radio telescopes would be pointing at a particular star at the same time as an alien broadcaster is beaming a message towards the Sun.

Another problem is tuning-in. Radio waves come in all wavelengths, from a few millimetres up to several kilometres. Broadcasting stations on Earth transmit at particular wavelengths: to pick up our favourite station, we must tune in exactly on the receiver dial. In the case of interstellar broadcasts we do not know the wavelength we must tune into. Frank Drake assumed that any alien civilization would choose a wavelength of 21 centimetres; the wavelength that hydrogen gas produces naturally. Hydrogen is both the simplest kind of atom and the commonest element in space, and because it has a truly universal significance, 21 centimetres does seem the

Nuclear-powered starship Daedalus passes Neptune and its large moon Triton, *en route* for Barnard's Star. The British Interplanetary Society believes that this unmanned spacecraft could be built with today's technology. Its propulsion system consists of billions of small hydrogen/helium pellets, stored in the large spheres, and exploded in turn within the large bowl behind. The succession of small nuclear explosions kicks Daedalus forward at an ever-increasing speed.

obvious wavelength for civilizations to choose when they have no other guide. This wavelength does, however, have one drawback. All the hydrogen gas in space is emitting it, too, making it the 'noisiest' wavelength in the spectrum. Other radio astronomers have suggested looking for artificial radio waves at the natural wavelength emitted by water vapour, 3.5 millimetres, since water is such a vital ingredient of living things. Another possibility is the range of wavelengths between that of hydrogen (21 centimetres) and the unstable molecule hydroxyl which broadcasts at 18 centimetres. This wavelength region is known as the 'water hole', because hydrogen and hydroxyl together make up a water molecule—and where better in the radio spectrum for water-based life to meet?

Scientists on Earth have thought of several 'obvious' wavelengths in only 20 years, but it is more than likely that an alien civilization millions of years more advanced than ours has come up with a totally different 'obvious' wavelength for communication. It may not even be a radio wavelength. Intense laser beams could carry messages in the form of light or infrared pulses; interstellar signals could even be winging their way across space as X-rays. From their basis of advanced knowledge, an alien civilization may use a form of communication we have not yet thought of.

Sticking to radiation of kinds that we do know, it is no use if everyone is just 'listening'. Someone must broadcast a message. Man has already sent his first communication: in November 1974 the world's largest radio telescope, the Arecibo radio telescope in Puerto Rico, sent a message from Earth to the stars. The transmitter was programmed to beam its powerful radiation towards a distant

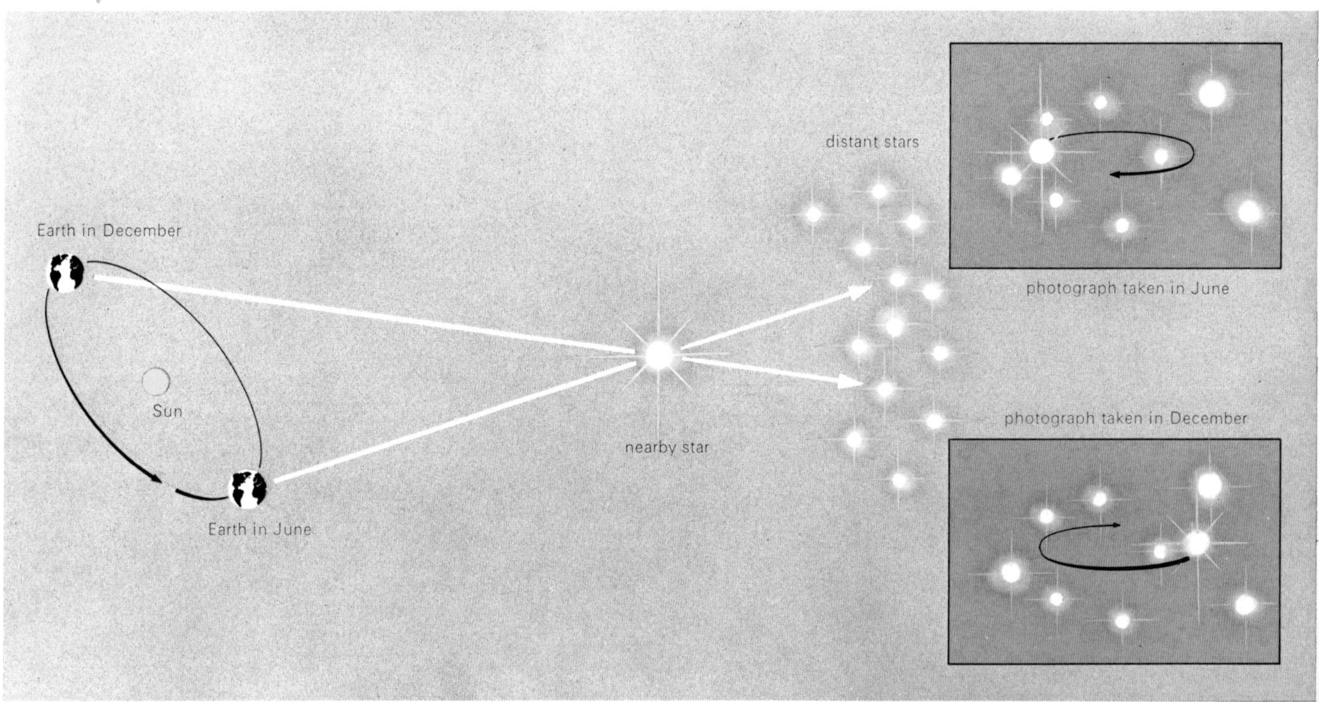

Measuring star distances

Astronomers measure the distances to nearby stars by the parallax method (see diagram). As the Earth moves around its orbit, the nearest stars appear to swing back and forth against the background of those situated farther away. The star's distance from Earth can be calculated by measuring the small angle of swing, because the distance of the Earth from the Sun is accurately known (p 41).

Even the nearest star, Proxima Centauri, is 40 million million kilometres (25 million million miles) away, and a larger unit than the kilometre must be used to express these huge distances concisely. Many professional astronomers use the *parsec*, a unit related to the mathematics of the parallax method, but more familiar is the *light year*. One light year is the distance a light beam travels in a year, and equals 9.5 million million kilometres (6 million million miles). Proxima Centauri is therefore 4.2 light years away; the bright star Rigel in Orion lies 800 light years from us; and our entire Galaxy of stars is 100,000 light years from edge to edge.

The parallax method becomes less reliable for stars more than 300 light years away since the small, apparent swing then becomes too tiny to measure accurately. A variety of methods is used to calculate distances of the more distant stars, like Rigel. One simple approach is to compare their properties with nearby stars whose distance is known. If a distant star seems identical with a nearby star in all its properties (temperature, types of spectral lines, and so on), it is presumed to be just as bright intrinsically and the apparent difference in brightness then shows how far off the more distant star lies.

Other methods rely on studies of star motions. Detailed analysis of the movements of individual stars in a cluster can reveal the cluster's distance. An investigation of similar types of star spread over the sky can again reveal the intrinsic luminosity of that star type, and so the distances to individual stars. The spectral changes in stars which pulsate regularly in size tells of their intrinsic brightnesses, and so leads to their distances.

Astronomers always check the results of one method against another, for example comparing the distance to a pulsating star as derived from its spectrum with the distance to the cluster of stars it lies in, determined from the motion of stars in the cluster. The agreement is always to within 10 percent or so, showing that star distances right out to the edge of our Galaxy are fairly reliable.

An alien starship leaves its home planet and heads for our Galaxy, in this artist's impression (*right*). The location of the Sun is shown by the green square, and an expanded view of the location of our nearest-neighbour stars is given in the green box. (Barnard's Star is unlabelled, the nearer of the two stars at the 11 o'clock position from the Sun.) Manned space travel may never become reality, because of the immensity of interstellar distances. It takes light several years to reach the nearest stars, and even the fastest present-day spaceprobes would take 100,000 years to cover such a distance. Interstellar exploration may instead rely on automatic spacecraft.

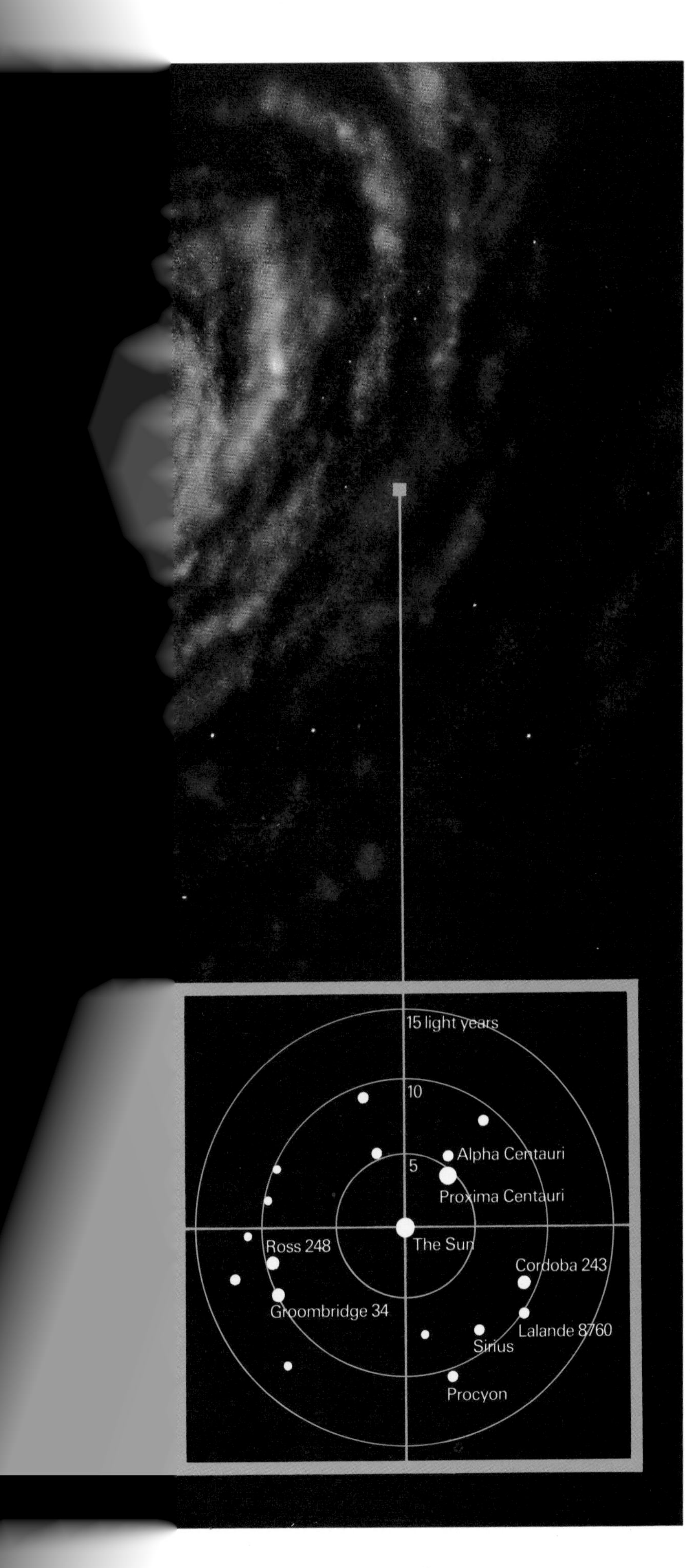

cluster of stars. Encoded in the waves was a message, telling of the solar system, of Man, and of the vital chemicals of life. The Arecibo communication will travel through space for 24,000 years before reaching its destination, a cluster of a million stars called M13, lying in the constellation Hercules.

But even if radio astronomers are scanning the skies for a planet of a star in M13, there is only the remotest chance that their telescopes will be pointed in Earth's direction during the brief three minutes of the message. The broadcast is also at a wavelength of no significance: 12.6 centimetres was simply a convenient wavelength for the radar research being carried out at Arecibo at the time. After M13, Man's first utterance to the cosmos will carry on out into the endless depths of space beyond our Galaxy.

ROBOT MESSENGERS

Galactic conversation must be conducted in monologues. Even the nearest habitable planet is likely to be so far off that the gap between question and answer would be centuries or even millennia. Interstellar messages are not likely to be 'please reply if there's anyone there,' but instead a coded recital of *Encylopedia Aliena*. Because of this, radio astronomer Ron Bracewell thinks it is far too wasteful to keep beaming powerful radio beacons at other stars continuously for centuries or more. Instead, he feels it would be far more economical to record the *Encyclopedia,* enclose the recordings in small space probes, and despatch one to each nearby star which has planets. The probe would stay dormant until it picks up artificial radio signals from one of the planets, then it would make its presence known, playing back all its recorded information.

Man has already sent out his interstellar visiting cards, although they are by no means as sophisticated as Bracewell's probes. The Pioneer 10 and 11 spacecraft investigated Jupiter and Saturn in the 1970s, and are now heading out of the solar system. Each carries a small plaque showing the figures of a man and a woman, and a chart showing the Sun's location in the Galaxy, referred to nearby *pulsars.* The more sophisticated Voyager 1 and 2 craft carry disc recordings—like ordinary long-playing records: their musical selection includes tribal songs, Beethoven and Chuck Berry! There is also a 'sound essay' depicting the evolution of life, from volcanoes and ocean sounds (to indicate life's beginnings) to rocket launches, and a selection of pictures encoded into the record grooves. Spoken messages include greetings in sixty languages, and a message from President Carter referring to 'our wish to become a member of the galactic community.'

These probes are not aimed at any particular star: even if they were, they would take 80,000 years to reach the nearest one. And it is very unlikely that even a super-civilization would notice them in the immensity of space. Their significance is that Man's emissaries have left not just the Earth, but our solar system, too. When Carl Sagan designed the Pioneer plaques, it was not only a message to anyone 'out there,' but also to Earth, to remind Man that he is now a creature of the cosmos.

Optimistic scientists see Man himself venturing out into interstellar space within the next couple of centuries. Unfortunately, however, interstellar travel is not as easy as science fiction suggests. The yawning gulfs that separate the stars are immense beyond human comprehension. It would strain present technology to make a man-carrying spaceship that could travel as fast as the Voyagers; yet even this craft would take a thousand human lifetimes to reach the nearest stars.

One answer is a 'space ark'. Like Noah's original, it would be a self-contained colony afloat in a hostile environment. The couples on board would all be human; living, bringing up children, and dying, as their ark ponderously floated on its slow journey to the stars. Eventually the descendants of those who left planet Earth would step out onto a new home, colonizers of what would really be a new world for Man.

In practice, it would probably not work out like that. The interplay of plant and animal life on Earth which keeps our planet's surface fit for both is not entirely understood, and some scientists feel that in creating a simplified Earth-environment some vital but obscure ingredients may be left out, and the ark would then gradually become poisoned, perhaps by a slow build up of noxious gases. And a hundred thousand years is anyway a very long time for the humans concerned: the evolution of Man from the Stone Age to today took only one-tenth of that time. So the descendants of the space pioneers would undoubtedly think very differently from their Earth-born ancestors, even if they survived the inevitable disputes and conflicts within the frail skin of the ark. The psychological pressures of living in a small, completely-enclosed 'world' cannot be estimated in advance. But it is plausible that, at journey's end, the supposed colonists would peer in terror at the wide-open spaces of a planetary surface, so frighteningly open to the heavens above—and return to the snug comfort of their familiar enclosed home.

Cutting the travel time by high-speed rocket flight may just be possible, but we do not yet know precisely how it could be done. Scientists and engineers have drawn up plans for fast spaceships, but all have their drawbacks. If the ship carries its own fuel, it needs a vast amount to speed up its payload to a high enough speed. The British Interplanetary Society has proposed a spaceship powered by hydrogen bombs. Called Daedalus, it could reach Barnard's Star in 49 years. To do this, it needs 50,000 tonnes of fuel—and even so it could not carry a payload big enough to contain the life-support system that astronauts would need. Daedalus would be a purely unmanned expedition. And note that even Daedalus, which will probably strain the technology of the year 2000, takes most of a lifetime to reach the second-closest star.

Much higher speeds are possible in principle, despite the engineering and practical difficulties that put them beyond our present capacity. But Nature does impose an ultimate speed-limit to even the most futuristic trans-galaxy craft: the speed of light. So the journey to Barnard's Star, 5.9 light years away, can never be cut to less than 5.9 years of travel time. (Although one of the odd predictions of relativity is that time seems to pass more slowly to the traveller on board the spaceship than to those on Earth; so the spaceship pilot will be less than 5.9 years older when he arrives.) There can never be any hope of zapping at will around our 100,000 light-year Galaxy.

Science fiction spaceships get around the cosmic speed limit by the clever knack of dipping in and out of hyperspace, which gets them across light years of space in a matter of seconds. Present-day scientists are rather less convinced about hyperspace than science fiction writers. Since Einstein formulated his theory of relativity at the beginning of this century, it has been known that space and time are inextricably linked together: cutting corners in space by hyper-drives and the like would also mean we could cut corners in time. In other words, if hyper-drives could exist, they could be used for time travel, too.

And time travel is a logical impossibility. If an 'effect' can happen *before* the 'event' that caused it, then the Universe would be total chaos. Stable atoms, planets, stars would not exist—nor could life. To take a simple example: Suppose you travelled back in time, and while doing so you shot your parents when they were children. Since they could never have grown up, you could not have been born, and therefore you could not have travelled back in time and shot them. But if they were *not* shot, then they *did* grow up and you *were* born; and in that case you do exist to travel back in time and shoot your parents as children.... It is not just a trick argument: time travel poses real logical impossibilities.

Even so, a civilization a million years more advanced than ours may use means we cannot envisage for interstellar exploration. It may,

Carved stones from Palenque, Mexico, display highly-stylized symbols. With sufficient imagination, it would be possible to make out 'spacemen'—or anything else one liked!

perhaps, have constructed unmanned interstellar probes with computers so advanced that the probes themselves are intelligent beings. If these probes were able to repair themselves they could last practically for ever, and make trips across the Galaxy even at the low speed of present-day probes without getting impatient.

Computer pioneer John von Neumann has conceived the idea of machines which can reproduce themselves from naturally occurring materials. A probe based on this principle could colonize the entire Galaxy.

A hundred years from now, say, we send out a single von Neumann probe. After a 100,000-year journey, it gently cruises into the vicinity of Barnard's Star. Even if this star does not have planets, there are bound to be rocky fragments such as asteroids. The probe dissembles itself into a factory in space, smelting asteroids for their metals, extracting their carbon compounds for plastics and their silicon to make new electronic 'chips', as well as hydrogen for fuel. From these materials, it constructs several copies of itself. One stays in the Barnard Star system, sending its tales of exploration back to Earth by radio. The others head off to more distant stars. Here they, too, construct new probes; and so the exploration continues. Avenues to farther and farther stars spread out at an ever-increasing rate as the number of probes multiplies. Within 300 million years, von Neumann probes could be circling every star in our Galaxy. And while this may seem a long time to mere humans, it is only a moment in the history of the Universe.

The thought is staggering, but quite believable. The rockets and computers needed for von Neumann probes are already on the drawing board; they are likely to be reality by the year 2000. The cost would only be the expense of building the first probe: American mathematician Frank Tipler estimates it would be only one-tenth the cost of the Apollo Moon programme—and for that price we could explore the entire Galaxy.

But the concept of von Neumann probes only adds to the mystery of intelligent life elsewhere in the Universe. If other intelligent, technological civilizations exist, why haven't their probes visited us?

VISITORS TO EARTH?

We certainly do not have any evidence that aliens are here, or have ever visited our planet. Erich von Däniken and others may have claimed to produce evidence for visits by 'spacemen' within the past few thousand years, but when these claims have been investigated, they have always failed to stand up to scrutiny.

Von Däniken, for instance, insists that the Egyptians could not have built the Great Pyramid without extraterrestrial help. Yet tomb paintings exist showing exactly how it was done. And why the need to invoke alien technology at all for the buildings of our ancestors, when we know that the greatest artificial structure in the world—the Great Wall of China—was built by Man with only primitive means?

Another piece of 'evidence' produced by von Däniken consists of statues and pictures which look like spacemen in spacesuits or rockets. These fare no better when investigated. All cultures have stylized representations of human beings or gods; search hard enough and there are bound to be a few with details of clothing that look like a spacesuit, or (in some central American carvings) sit in a vaguely 'spaceship-shaped' building. Von Däniken's strongest contender for representations of aliens are clay figures from north Japan. Both he and Russian science fiction writer Aleksander Kazantsev have studied these Dogu statuettes, and they point out 'obvious' details of spacesuits: a slit-goggled helmet, overlong sleeves ending in mechanical manipulators, and details resembling rivets, buckles and small inspection hatches. But according to Ron Bracewell, a leading American investigator of the possibility of extraterrestrial life, these claims have been examined by a professional Soviet ethnologist, S. Arutyunov, and dismissed.

If the statuettes had been realistic portraits of arriving astronauts, the space-suited figures should appear without any relation to previous sculpture. Arutyunov has examined a whole series of Dogu statuettes from the earliest primitive clay figures 11,000 years old to the highly ornate 'spacesuit' figures which date back only 3000 years. There is an obvious gradual transformation from the crude representations of head and limb, through realistic portrayal, to the latest over-stylized forms. There is no sudden appearance of 'spacesuit' features: eyes *evolve* towards a stylized representation like goggles with slits; heads to dome-shapes; and hands to mechanical-looking stubs. And Arutyunov points out they could not strictly speaking be space*men*: all the Dogu statuettes are female, with breasts protruding through a gap in the 'spacesuit'. Is it likely that visiting astronauts should all be female—or that, having donned elaborate pressurized spacesuits complete with complex helmets, they should then expose part of their anatomy?

The UFO mystery

In no sense are the von Däniken-type claims a mystery: they are misinterpretations, distortions, and in some cases sheer fabrications. But many people believe that UFOs (unidentified flying objects) are spacecraft from another world; and the evidence for UFOs is not so easy to dismiss.

There is a lot of confusion as to what a UFO is. Most of us have seen objects in the sky that we cannot identify—bright lights at night, or distant 'hovering' objects by day—but that does not necessarily make them UFOs. The majority of UFO reports can, in fact, be explained by natural phenomena: weather balloons, bright meteors (shooting stars) and the bright planet Venus are commonly reported as 'UFOs' for instance. The person who sees them reports his sighting in good faith. After an experienced investigator has interviewed him, visited the site, and questioned other people around at the time, the object can usually be identified. If the investigator cannot find any natural cause for it, the object is a genuine UFO. Those seen at night are usually erratically-moving lights in the sky, while daytime sightings are often metallic-looking discs, hence flying saucer.

Networks of lines cross the Nazca desert in Peru, tracing out the outlines of spirals, zigzags and even birds and animals. Since the figures can only be made out from the air, some writers have speculated that they were constructed as signals for UFOs; archaeologists, however, point out parallels with other huge religious constructions, such as the 'serpent mounds' found in the United States.

Many scientists would say there is no such thing as a genuine UFO. They maintain that even in cases where it is impossible to work out what caused the sighting, there is a natural explanation which would be obvious if we had more information. This was the conclusion reached by Edward Condon in 1969. The distinguished physicist had been called in by the US Air Force to study the UFO reports it had received over the previous 20 years.

The Condon report was a milestone in UFO investigation, hailed at the time as proving scientifically that they did not exist, and that all sightings could be explained as misinterpretations of known types of natural objects. But Allen Hynek, long-standing scientific adviser to the USAF on UFOs, points out that this is not so. Condon began the report with his pessimistic conclusions. But reading through the rest of it, you find that he and his team of experts were actually unable to explain over one-quarter of the cases they reviewed—despite intensive investigations. The bulk of the report apparently contradicts Condon's conclusions.

Hynek is a professional astronomer, and was sceptical of UFOs when he first started to advise on sightings. It was the weight of inexplicable sightings that changed his mind. Many of the strangest reports came from respectable and experienced people, including policemen and air traffic control operators, whose judgement we trust implicitly in everyday life. Scientific researchers, including professional astronomers, have also sent in UFO reports.

All of these people had nothing to gain by filing such a report; in fact, since the whole business has now become so associated with 'little green men' and other crackpot ideas, responsible people have everything to lose by stating publicly that they have seen a UFO. It takes a lot of courage to do so. And in many cases these reports from trustworthy observers have indeed remained unexplained when they have been investigated in detail.

Hynek has produced good evidence that there are flying objects that cannot be explained by our present-day knowledge of natural events. But this does *not* mean that UFOs are alien spaceships. The unfortunate, widespread belief which equates them with aliens visiting Earth has meant that it is difficult to investigate UFO reports seriously, and pin down what causes them. The most likely explanation is that the objects that give rise to such reports are entirely natural phenomena of the Earth's atmosphere which are so rare that scientists have not previously been able to investigate them.

Ball lightning is a rare phenomenon, which illustrates just how difficult it is for scientists to investigate objects which appear only occasionally, and in circumstances where they cannot be tested. For centuries there have been reports of brightly-glowing balls of gas, about the size of footballs, that float around after a thunderstorm. It is very difficult to explain ball lightning in terms of the ordinary theory of electricity in gases and, until recently, most scientists refused to believe that it actually existed. Now it is accepted—though not yet explained. UFOs may be another type of rare atmospheric phenomenon.

Another explanation is that UFOs are some kind of psychological phenomenon. In the Middle Ages, educated people saw portents in the sky, including bloody swords, golden crosses and dragons. No one sees such things these days. Instead, there are high speed 'discs' seen during daytime and reddish balls of light visible at night. Since several people often see the same object, mass hysteria must be considered as a possible originator of some claims. Hynek claims that some UFOs have left physical marks on the ground, and of course there is a widely-reported ability to stop car engines to be considered. Such evidence does seem to indicate that at least some UFOs are actually physical objects. Whatever they may be, there is no compelling reason for believing them to be alien spaceships. We cannot use such reports to decide we are alone in the Universe.

All the trails that Man has so far followed have led to dead-ends. Radio astronomers have picked up no messages from the stars; biologists have not been able to show that life would inevitably arise on an Earth-like world circling another star; astronomers have not even proved that other stars have planetary systems; and there is no evidence that aliens have ever visited Earth.

After a heady optimism in the 1960s and 1970s that life is common in space, some scientists are now taking this catalogue of failure as indicating that we are indeed alone—some even feel that we should be exalted by the thought of our uniquely-developed attributes, rather than feel wistful for lack of interstellar company. For all that, hope has not been completely abandoned. The search for interstellar messages continues; scientists discuss how we should best communicate with alien intelligences, and what difference interstellar contact might make to human society here. If it does happen, our meeting with alien creatures will be the most important event in the whole of human history.

A group of four UFOs photographed at a US coastguard station at Salem, Massachusetts, on 16 July, 1952. This was a peak year for UFO sightings around the world.

CHAPTER 8

THE STILL-TICKING PULSARS

If they had set out to look for interstellar messages, Cambridge radio astronomers would have felt in 1967 that they had succeeded beyond their wildest dreams. A new, very sensitive radio telescope was picking up regular pulses from somewhere in the constellation Lacerta, the lizard.

But the transmitter was not artificial. It was a new kind of celestial object—quickly dubbed a *pulsar*. And in our understanding of the Universe, the discovery was certainly as important as finding alien life, even if it did not have the same impact on our day-to-day existence. After all, we know of one type of life in the Universe already, the kind represented by cells on Earth. Pulsars are a completely new *type* of object.

A pulsar is a rapidly-spinning neutron star. It is made up not of atoms, nor of the electrons and nuclei that are the broken-up fragments of atoms, but of neutrons, a kind of particle which is normally found only in conjunction with other subatomic particles within an atom's nucleus. Neutrons are very much smaller than atoms. If all the matter which makes up the Sun were suddenly turned to neutrons, our star would shrink from its present 1,400,000 kilometre (865,000 mile) diameter to only 25 kilometres (15 miles) across. Its matter is so condensed that a pinhead of neutron star material contains a million tonnes of matter.

The Russian physicist Lev Landau predicted the existence of neutron stars in 1932—within a year of the discovery of the neutron itself. Astronomers were not certain how to take the idea but two of them, Walter Baade and Fritz Zwicky, did accept Landau's theory and postulated that such stars might be caused by the crushing force at the centre of a supernova explosion compressing its centre as its outer layers exploded. It was a daring prediction that was indeed confirmed 35 years later. At the time, however, it was considered unlikely that neutron stars could ever be seen from Earth. How could you hope to detect a star only the size of Malta, from hundreds or thousands of light years away?

What no one took into account was magnetism. Everything in the Universe has some magnetic field, and if a star collapses to a smaller size it squeezes its magnetic field tighter and increases its strength. Although a star such as the Sun has a magnetic field only as strong as the Earth, compress it to a neutron star and the magnetism will become a million million times stronger. So although the neutron star itself may be invisible, its magnetic field can give it away. If there are *electrons* moving in this magnetic field, they radiate radio waves, rather like the electrons which make up the moving current in a broadcasting transmitter on Earth.

The collapse of a supernova's core does not just intensify its magnetism, it also makes the star spin faster. While our Sun rotates in just under a month, a neutron star spins round about once a second. If the star's magnetic poles are at an angle to its poles of rotation—as with the Earth—then we do not see continuous emission from the neutron star. Astronomers pick up signals only once or twice per rotation, as one or both magnetic poles are presented to us. The result is a regular train of pulses.

At first the Cambridge astronomers did not know what they had detected. The pulses reappeared to their radio telescope every day, at intervals of 23 hours 56 minutes—the time that the stars appear to circle the Earth as our world rotates. The pulsar was definitely a source out among the stars. Team leader Tony Hewish, later to win the Nobel Prize, realized that in this case the time interval between the pulses should change slightly as the Earth swings around the Sun. His accurate timings did indeed reveal this.

Although Hewish's team never really believed that they had picked up radio signals from another civilization, they decided to keep the discovery to themselves until they could check. If the pulsing beacon were indeed on another planet, it too would have been circling about its star, and this would also have made the pulse intervals gradually change. But Hewish found that the observed changes were exactly accounted for by the Earth's own motion. The pulsar 'beacon' was not orbiting another body. And his student Jocelyn Bell (who had picked out the first, unexpected, pulses) later found three more of these pulsars in the records from the new telescope. It was inconceivable that four different civilizations should be beaming radio pulses of the same type at Earth simultaneously.

The news was published early in 1968. Pulsars are amazingly accurate checks, 'ticking' more precisely than even a quartz wrist watch. But they are natural, not artificial, and after a flurry of theoretical activity astronomers reached the conclusion that they must be highly magnetized, rapidly-spinning neutron stars. Although neutron stars had never been detected before, nothing else ever predicted would fit the bill. Later observations have confirmed this interpretation to the hilt.

PULSAR PUZZLES

In one sense, the mystery was very quickly solved: pulsars *are* neutron stars. But, like everything in the sky, the solution of one enigma seems to lead to the discovery of others.

Although radio astronomers are convinced that electrons are broadcasting the radio waves as they whip around the star's intense magnetic field, there is still dispute over *where* this is happening. The electrons may be right at the magnetic poles of the neutron star, where the magnetic field is most intense. They would then shoot out waves along the magnetic field direction, like a crazily-tilted lighthouse lantern. It is a very appealing idea because it is beautifully simple but, unfortunately, detailed investigations have shown that the pulses are not so elementary after all. If you could slow down a pulse and listen to it, it would not be a simple 'bleep' but a crackle. Each 'sound' in the crackle is called a 'subpulse', and the changing behaviour of the subpulses from one pulse to the next is very difficult to explain.

Jodrell Bank radio astronomer Graham Smith has instead proposed that the radiating electrons are far out in the magnetic field. Since the field is tied to the rapidly-spinning star, the far out region is carrying the electrons

The simplest possible form for a radio telescope, Cambridge's Four Acre Array (*right*) is a vast grid of wires strung on vertical posts. With this very sensitive telescope, pulsars were accidentally discovered in 1967.

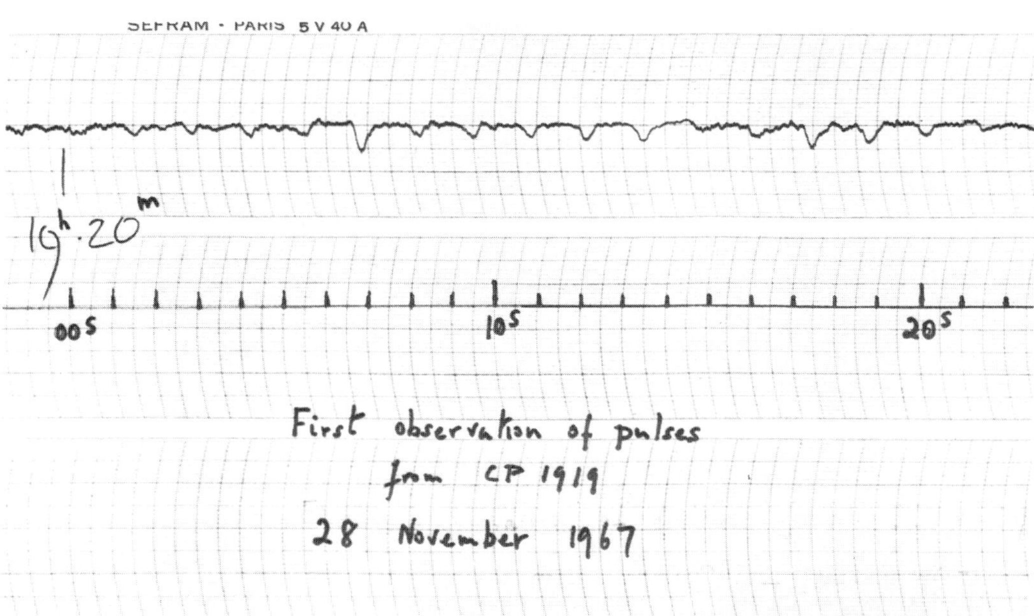

Paper-tape record (*right*) of CP 1919, the first pulsar to be discovered. The trace shows extremely regular pulses repeating every 1.3 seconds.

The neutron star at the centre of the Crab Nebula is a dense ball of matter, only 25 kilometres (15 miles) across—too small to be seen from Earth. We do, however, detect light from electrons trapped in its magnetic field. In one interpretation (shown here) the radiating electrons lie far out in the field, and effects of relativity cause light and other radiations to shine out in a narrow cone. Each time the star turns, we see a 'pulse' of light.

around at a tremendous speed. We have to apply Einstein's theory of relativity to such objects moving at almost the speed of light, and relativity shows that this motion intensifies the radio emission from these electrons hundreds of times over, and also directs it forward in a narrow beam.

Whatever the exact reason for their broadcasting, the interval between pulses tells us incredibly accurately the rate at which the pulsar is spinning. The slowest of the 320 known pulsars turns once in $4\frac{1}{3}$ seconds; the fastest in only one-thirtieth of a second.

When astronomers measure to the limits of accuracy—to millionths of a millionth of a second—they do, in fact, find that the celestial clocks are not quite constant. They are all very, very gradually slowing down. The radiation that a pulsar constantly broadcasts to space robs energy from the small star, slowing down its rate of spin. On the whole then, neutron stars which twirl quickest should also be the youngest. Very soon after the pulsars' discovery, astronomers had proof of this.

In the constellation Taurus lies a twisted cloud of gases, expanding at colossal speed into space. Because of its shape, astronomers call it the Crab Nebula. And we know it is the wreck of a supernova which the Chinese saw explode in the year AD 1054. In 1968, radio astronomers picked up pulses from the heart of the Crab Nebula. The prediction which Baade and Zwicky had made 35 years before was at last vindicated: there was now proof that the cores of old supernovae *do* collapse to become neutron stars, and give themselves away as pulsars.

Its birth in a supernova only 900 years ago means that the Crab Pulsar is the youngest we know. And it is also the fastest-spinning,

turning round once in one-thirtieth of a second. Because it is so young and energetic, the Crab Pulsar flashes not just radio waves but also light—it is visible in photographs as a bluish-coloured star. The second-youngest pulsar, in the constellation Vela, also sends out pulses of light but, unlike its radio emission, a pulsar's light dies away rapidly, and the 10,000-year-old Vela Pulsar was only spotted optically with specially-sensitive electronic light detectors attached to the Anglo-Australian Telescope in New South Wales.

We can measure the slowing of the Crab and Vela Pulsars; the Crab Pulsar in particular is losing energy at a vast rate as it powers the glowing nebula around it. But on 3 March 1969, radio astronomers found that the Vela Pulsar was spinning *faster* than it had been two weeks before when they had last looked at it, although thereafter it continued to slow down in its usual way. It is now known that both the Crab and Vela Pulsars suffer such sudden 'glitches' every few years: and the only reasonable explanation for them is that the star suddenly shrinks very slightly so, like a spinning ice-skater who draws in his arms, the pulsar spins more rapidly.

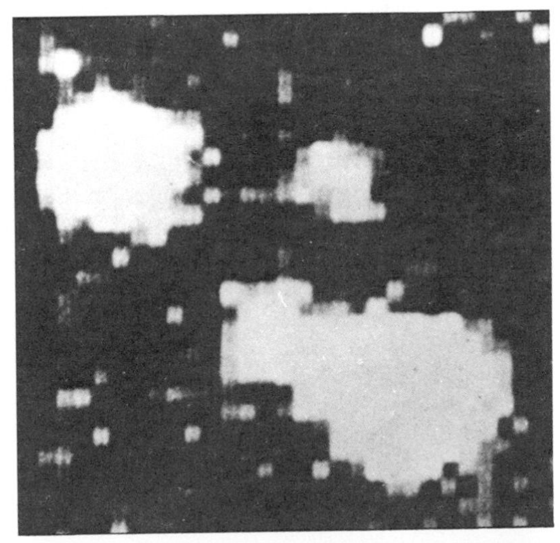

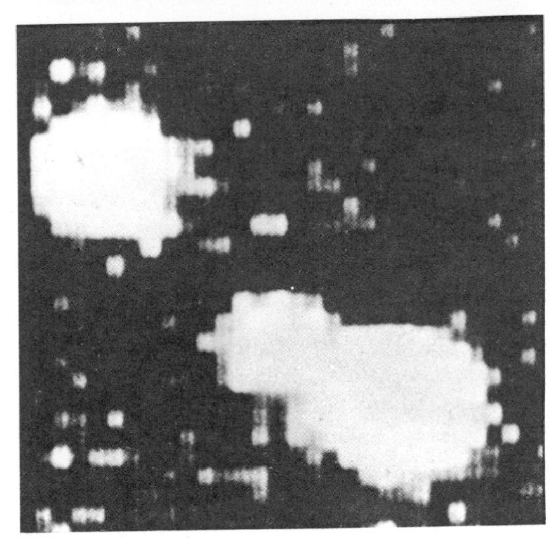

The Vela Pulsar (*right*) is the only pulsar apart from the Crab Pulsar which is known to flash light in addition to radio waves and X-rays. These photographs, taken $\frac{1}{25}$ second apart, show the pulsar 'on' and 'off'. The Vela Pulsar is a thousand times fainter than the Crab Pulsar, and these photographs were taken with an extremely sensitive electronic detector (described on pages 148–9).

The actual shrinkage during one of these glitches is only microscopic—a fraction of a millimetre—but its effect is measurable. From it we can deduce something about the interior of neutron stars. Theory tells us that the star is largely made of neutrons in a liquid state, surmounted by a thin solid crust, which has a very smooth surface: 'mountains' on the star cannot stand more than a few millimetres high, because larger ones would be flattened by their own weight in the immense gravitational pull. Gravity is so strong that an astronaut foolish enough to land there would find that his body weighs some 50,000 million tonnes.

The Crab Pulsar seems to suffer glitches as its surface cracks during the microscopic shrinkage. Astronomers think that the larger glitches in the Vela Pulsar are due to changes at its centre, in a solid core of neutrons within the liquid bulk of the star. Radio telescopes on Earth now reveal to us the microscopic changes deep at the heart of the densest concentrations of matter known, across thousands of light years of space.

Once a pulsar has slowed down to spinning once every few seconds, it ends its radio-emitting life—and finishes as an invisible neutron star in space. The pulses don't just fade gradually, however: the star switches off its pulses entirely for a period of time, then broadcasts for a spell at full strength; but its periods of silence grow longer until it is eventually heard no more.

A radio pulsar's 'life' is about 10 million years. Astronomers reckon there must be a total of around a million radio-emitting pulsars in our Galaxy, calculating from those we have already observed nearby to us. That means that a pulsar must be born every 10 years or so. Supernovae, however, are rarer than this. During each century, two supernovae explode, producing two new pulsars. But, according to the pulsar census, 10 pulsars are born each century.

There is as yet no accepted answer. The Crab and Vela Pulsars tell us that some pulsars at least are born in the cataclysm of a supernova explosion. Taken at face value, the figures imply that the majority—four out of five—are not. Yet we know of no other situation where there would be sufficient force to compress an amount of matter equal to the Sun's mass right down to the tiny size and incredible density of neutron stars. Perhaps the majority are born in violent events in our Galaxy that we have yet to discover.

PULSAR IN ORBIT

Because pulsars are basically such reliable timekeepers, radio astronomers R. A. Hulse and J. H. Taylor were astonished by the behaviour of a pulsar they had discovered in

1974. Named PSR 1913 + 16 after its position in the sky, it produced pulses that were inconstant: sometimes they ran fast, sometimes slow, changing in five minutes as much as an average pulsar would in a year due to its natural slowing down.

They soon realized why: PSR 1913 + 16 is in orbit in a double star system. Just as Hewish had found evidence for the year-long revolution of the Earth about the Sun in pulsar periods, so Hulse and Taylor's timings were affected by the pulsar itself orbiting its companion star. The other star cannot be seen, either by optical or by radio telescopes, but it is most likely to be an old neutron star whose radio emission has switched off. The two neutron stars swing around each other in only 7¾ hours, so the timing changes are easy to spot.

Once they had worked out the pulsar's orbit, Hulse and Taylor could predict exactly when future pulses would occur. Joe Taylor has observed the 'binary pulsar' twice a year since its discovery, and it is so regular that he can predict six months ahead when the pulses will occur. Even though the neutron star pulses 17 times per second, he can calculate the exact number of pulses that have elapsed since he last looked at it, a number that runs to over 200 million in a six-month period.

Astronomers were not the only scientists to be fascinated by the orbiting clock. Physicists, too, were intrigued. In 1915, Albert Einstein had produced a new theory of gravitation: called general relativity, it predicted subtle differences from Newton's Law of Gravitation, then 250 years old. Although Einstein's theory was based on a radically new conception of space and time, the practical differences from Newton's theory only show up when gravitation is very strong and there is no interference from anything else. On Earth, and in the solar system, measurements showed that Einstein's was the better theory; but here gravitation is always weak, and the effects difficult to measure. The binary pulsar is an ideal natural laboratory to test Einstein's theory unambiguously: it is an exceptionally accurate clock, orbiting in the strong gravitational pull that it and its companion exert on each other.

By Newton's law, two bodies in orbit will continue to orbit forever. According to Einstein, they will lose energy very, very gradually and will eventually spiral in towards one another. No one had been able to test Einstein's prediction before the discovery of Taylor's binary pulsar. But if Einstein were correct then the two neutron stars in the new twin system should be moving together fast enough to be detected, so the time they take to orbit each other should be gradually increasing.

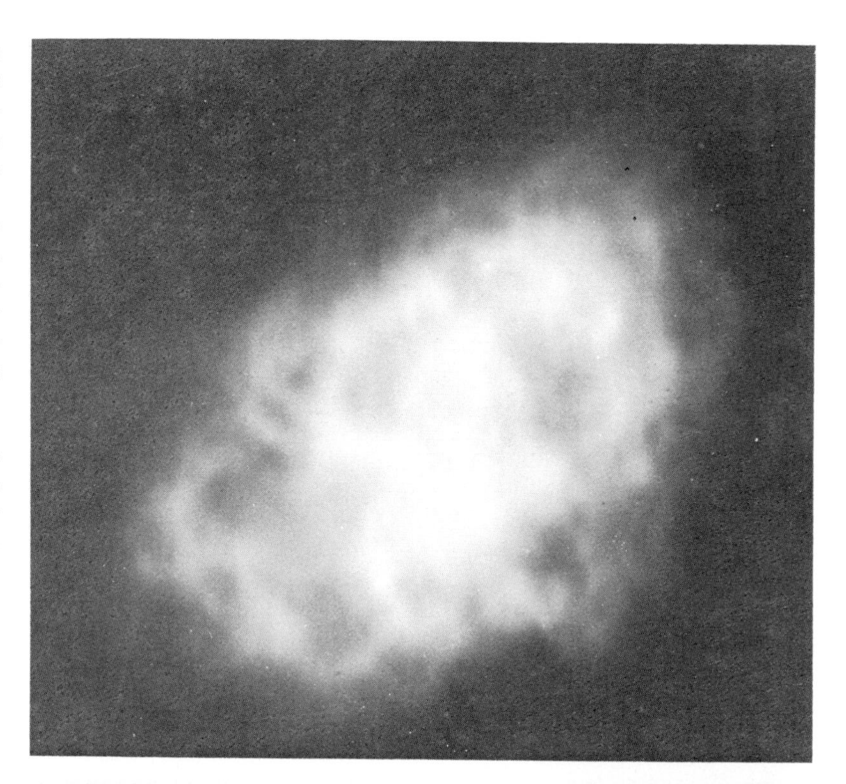

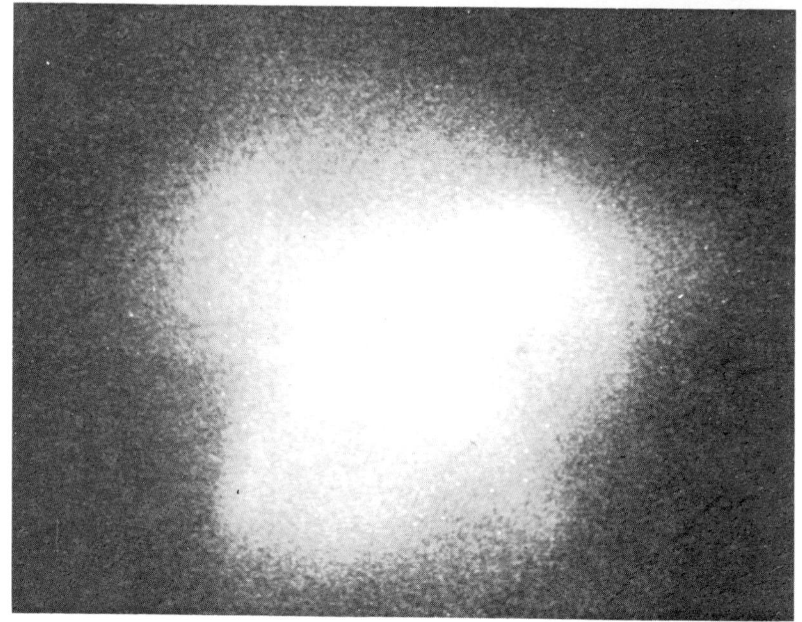

After four years' observations, Taylor announced that the orbital time had indeed shortened significantly. The stars were now closer together and whirling around each other more rapidly. The very regular pulses from the pulsar in the pair enabled him to deduce that the time for one orbit had decreased from 7 hours 45 minutes 6.9818 seconds to 7 hours 45 minutes 6.9817 seconds—a tiny change, but one that the incredible precision of pulsar timings makes it possible to measure. The change fits Einstein's theory precisely—and the confirmation of the theory was appropriately announced in 1979, the centenary of his birth.

Radio telescope image of the Crab Nebula (*left*) computer-processed to look like an optical photograph. The tangled knots are regions of strong magnetic field where spiralling electrons give out powerful radio waves.

The central regions of the Crab Nebula emit strongly at X-ray wavelengths, as seen in this larger-scale photograph (*below left*) from detectors carried into space aboard the orbiting Einstein Observatory. The small blob near the centre is the pulsar itself.

To astronomers, the binary pulsar has opened another door. Now that scientists had no doubts about Einstein's theory, an analysis of the pulsar timings allows them to use another part of the theory to 'weigh' a neutron star for the first time. Both the binary pulsar and its neutron star companion weigh in at 4 percent heavier than the Sun—in nice agreement with the theory of neutron stars, which had predicted that they should be heavier than one-tenth the Sun's mass, but lighter than three Suns.

Other binary pulsars are now known—one may even be a pulsar with a planet—but none revolves more quickly about its companion than PSR 1913 + 16. It is still the only one which can furnish evidence for relativity, and the longer that observations are continued, the more accurately Einstein's theory can be tested. Undoubtedly, astronomers will continue to refer to it familiarly as *the* binary pulsar.

PULSARS REBORN

For every pulsar which is still transmitting its radio 'ticks', there are a thousand too old to broadcast. Most of these are adrift in space, tiny dark dots insignificant and undetectable in the interstellar vastnesses. A few are not alone, though. And if the companion is an ordinary star, like the Sun, rather than another star-corpse, the neutron star may feed off its living companion's matter and return to life.

Although the neutron star's magnetic field is too weak to produce radio waves, its gravitational pull is as strong as ever. And in two kinds of circumstance the tiny star can get its gravitational hands on its companion's gaseous layers.

If the companion is a heavyweight star, its outer atmosphere is continually 'boiling off' into space, like a superpowered version of the solar wind which sweeps outwards through the solar system from the Sun's atmosphere. It is so intense that a supergiant loses as much matter as the Sun contains in only a million years. The neutron star can pull in gas from the wind as it sweeps by. A smaller companion star—similar to the Sun—loses only a little matter in its 'wind'. But eventually it begins to swell as it becomes a red giant, and in doing so its gradually-expanding gases come perilously close to the neutron star. A tug-of-war develops over this gas, and the neutron star gains control of some of it.

Whichever way the compact star steals matter from its companion, the gas does not give in without a struggle. It has orbital energy, so it does not fall straight onto the neutron star's surface. Instead, the gas forms into a disc rotating around the star, very like the accretion disc around the white dwarf star in a star system that suffers nova outbursts (see chapter 6).

These celestial power struggles would mean little to us, if it were not for friction in the gas. As it spirals down towards the neutron star's surface, this friction heats the gas to incredible temperatures—100,000,000° or more. Hot gas on the Sun's surface produces light, like the hot filament in an electric light, but these have temperatures of only a few thousand degrees. Gas hotter than a million degrees shines by producing not light but the far more energetic X-rays. Neutron stars with accretion discs are amazingly powerful natural X-ray tubes: it would take a thousand Suns to produce as much light as a single neutron star's accretion disc generates in X-rays.

Unlike a star, an X-ray neutron star is not a constant source. The neutron star is spinning, and still has a slight magnetic field. The gas from its disc cannot just spiral down to its equator, for the magnetic field channels it up to the star's magnetic poles. From these regions, the X-rays shine out in a beam, and as the star rotates, the beams flash past the Earth to appear as regular X-ray pulses. The pulsar is reborn. This time its 'ticking' is revealed not to radio telescopes but to X-ray telescopes.

X-ray pulsars are not such regular clocks as the younger radio pulsars. The whirling gas hitting the surface gradually changes its rotational rate—and can either spin it up, or slow it down. As a result, X-ray pulsars do not rotate within periods of 30 seconds or so, as you would expect from an old, naturally-slowed radio pulsar. They have either been braked severely by the infalling gases, and turn only once every few *minutes*; or else they have been speeded up again, and rejuvenated to pulse with periods of a few seconds once more.

The most powerful X-ray sources in our galaxy do not quite fit this pattern. These sources lie in the direction of the Galaxy's centre, and one of them is the first-detected cosmic X-ray source, Scorpius X-1. Most astronomers think that they are identical to X-ray pulsars, but are more powerful because they are sweeping up much larger quantities of gas. But these sources do not pulse. Their brightness changes incessantly, with no sign of a regular period to their restless output of X-rays; there is also no real evidence that they are double stars.

Another type of X-ray source also poses problems: X-ray *bursters* are generally rather weak and inconspicuous; but every few hours or days they will spring to life with a sudden outburst of X-rays. For a few seconds they will shine a hundred times brighter, then fade back to insignificance. The extrovert among them is a remarkable source in the constellation Serpens, catalogued MXB 1837+05, but usually known simply as The Rapid Burster.

It lies dormant for several months at a time, then for a few weeks indulges in a celestial fireworks display. Its X-ray outbursts follow hot on one anothers' heels, as The Rapid Burster spits out a thousand a day like a cosmic machine-gun.

The spectacular bursters are probably neutron stars on a diet: they are sweeping up gas from their companion at an unusually slow rate. The neutron star's magnetic field can suspend the infalling gas at the inner edge of the accretion disc, keeping it above the surface for hours or days at a time. Eventually, though, the gas builds up so much that its weight crushes down the magnetic field, and it collapses onto the star surface in a burst of X-rays.

All these X-ray sources seem to result from the rapacious greed of neutron stars for their companions' gas. But the last word may not have been said. There is still discussion about the bursters, as well as Scorpius X–1 and its bright kin, the galactic bulge sources. It may be too simple to try to pin all these sources down to the single common denominator of a hungry neutron star: perhaps something stranger is responsible for some of the stars of the X-ray sky.

For if our eyes were sensitive to X-rays rather than light, we would not think of the night sky as a placid, serene, eternal place. The 'stars' we would see would be incessantly fluctuating: the brightest would flicker the most violently; other stars would pulse with a regularity that would give us a natural feel for the passing of seconds and minutes; while the impression of motion and impermanency is heightened by erratic bursts from the faintest stars. More than light, X-rays from space underline the true drama of the Universe.

GAMMA RAYS FROM SPACE

X-rays are very penetrating and energetic radiation—as we know from hospital photographs where they travel straight through our flesh and soft tissues. But the distinction of being the most energetic radiation of all is reserved for gamma rays. On Earth, we generally come across them only where they shoot out from radioactive elements, emanating from the concentrated energy reserves of the atom's nucleus.

In space, however, regions of unfathomable violence can generate huge amounts of them. Satellites in Earth orbit have detected them coming from sources in space, just as they have pinpointed X-ray sources. Astronomers must use satellite-borne observatories because neither of these penetrating radiations, ironically enough, can get down through Earth's atmosphere to the surface. They are so energetic that they can knock electrons out of

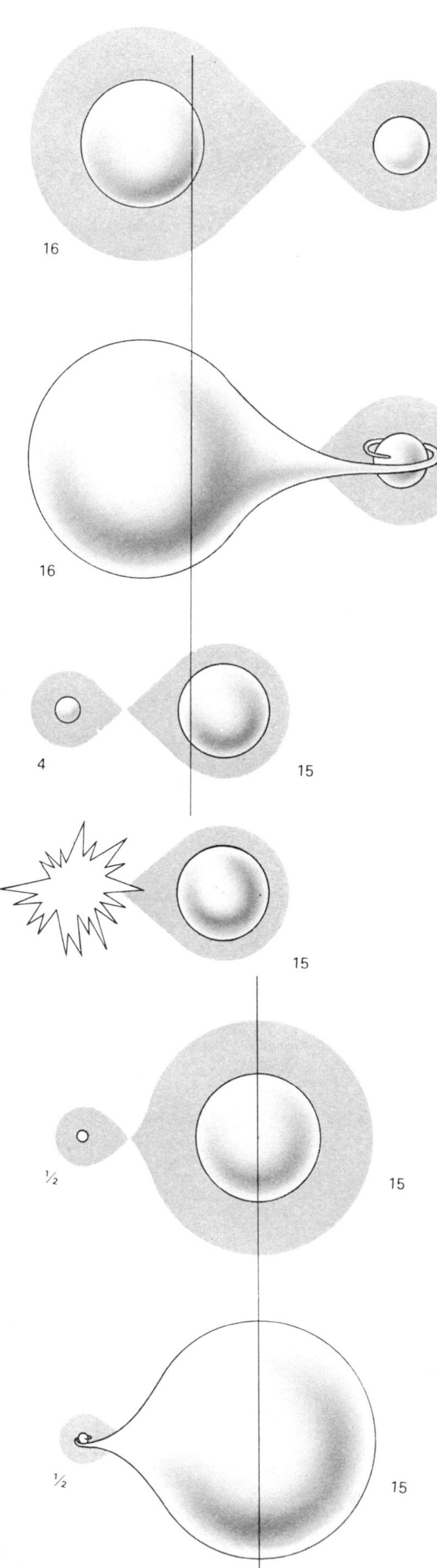

The formation of an X-ray pulsar starts with a pair of stars whose masses are different. The figures give the star masses in terms of the Sun's mass, and the shaded areas are the regions of space where each star's gravity can control gases.

After seven million years, the heavier star expands to become a supergiant. Its outer gases overflow into the gravitational domain of its companion, and spiral down on to it.

The originally-heavier star loses most of its mass to the other star in only 20,000 years, and ends up much the lighter. The period in which they orbit has decreased from 3 days to 1.5 days.

The core of the initially-heavier star explodes as a supernova after a further 1.5 million years.

The remains of the supernova core is a neutron star, only half as heavy as the Sun. The orbital period is now 2.2 days, and the centre of gravity of the system (vertical line) has suddenly jumped sideways as the supernova blows matter away.

Six million years on, the second star begins to expand to supergiant size. Its gases now overflow and spiral down on to the neutron star. Heated by the fall, the gas emits X-rays, and these are directed into pulses by the spinning star's magnetic field: it is an X-ray pulsar.

atoms in the air, and in so doing dissipate all their energy high over our heads.

Most of these infernos have yet to be identified. Some are undoubtedly quasars and other distant violent galaxies, while others lie within our own Galaxy. It is possible, for example, that gamma rays are generated when a supernova explodes, not into virtually empty space but inside a dense cloud of a gas. Its pent-up fury may be released as the most energetic radiation of all.

The most exciting discovery in gamma-ray astronomy so far was not, in fact, made by astronomers. Unknown to the world at large, the American government had launched a series of surveillance satellites into orbit to monitor nuclear explosions that violated the 1963 Partial Test Ban Treaty. These Vela satellites were equipped to detect bursts of neutrons, X-rays and gamma rays from the detonation of a nuclear weapon. They were launched in pairs, and separated in orbit to opposite sides of the Earth; between them a pair of Vela satellites could continuously monitor the entire Earth and the whole of the sky—to check on nuclear tests in space, and even to keep a look-out for radioactive debris shot into space from a nuclear explosion on the Moon's far side.

The stream of data from the Vela satellites was being investigated at the Los Alamos Scientific Laboratories in 1974, when it was realized that they were indeed detecting bursts of gamma rays not from the Earth, nor from anywhere in the solar system, but from the depths of space beyond. The discovery astounded astronomers. They had expected gamma-ray sources to vary—like X-ray sources, to pulse, perhaps occasionally to burst out. But these 'bursts' were different: 'flash' would be a better word. If our eyes were sensitive to gamma rays, it would be like looking up at the dark night sky and suddenly seeing it lit up by the brilliance of a lightning stroke, for just a fleeting fraction of a second.

The reason for these bursts remains a mystery. After the Los Alamos discovery it was found that other satellites had also picked up some of these flashes, but that they had been dismissed earlier as false signals from stray electrons. With several observations from widely-spaced satellites, the sources in the sky could be pinpointed. They did not coincide with any known astronomical objects.

What puzzled astronomers most was that they were fairly evenly spread over the sky. Our Galaxy is a flattened system: because of perspective, most kinds of astronomical objects are commonest near to the line of the Milky Way. Most of the nebulae where stars are born lie in the Milky Way band, as do the remains of old stars, supernova remnants and pulsars, and the X-ray sources and bursters. The only exceptions are the very nearest stars. Stars which are closer than the thickness of our flattened Galaxy appear all around us in the sky. It seems then that the gamma-ray flashers are relatively close to the Sun, less than a few hundred light years away.

Their energy must be immense. In less than a second, one of these objects flashes out more energy than the Sun emits in three days as light and heat, yet because the rays flash so rapidly, we know that they are far smaller than the Sun—smaller than even Earth.

Early suggestions as to their origin included massive flares on otherwise insignificant stars. Astrophysicists Martin Harwit and Ed Salpeter even suggested that they occurred as comets hit neutron stars. Now it is considered more likely that the gamma-ray flashers are, in fact, related to the X-ray bursters discussed earlier in the chapter, where gas held up in a neutron star's magnetic field suddenly collapses onto its surface. In the case of the flashes, this happens a hundred times more quickly, so the flash is instantaneously a hundred times brighter and shines out as the more energetic gamma rays. And so the debate rested until 1979 when a unique event occurred which threw the question open again.

A flash of gamma rays winging its way through space hit our solar system on 5 March, 1979—and it was the most intense that man has ever witnessed. The immense flood of radiation activated gamma-ray detectors in a whole variety of spacecraft spread across our planetary system. It first reached the Russian probe Venera 11, which passed Venus later in 1978 and is now endlessly circling the Sun, then two more Sun-orbiting probes, Venera 12 and the German Helios 2, felt its force. It reached Venus, and activated the American Pioneer Orbiter, busily mapping the planet; and on to the Earth, where it triggered the still-watching Vela satellites, the American International Sun-Earth Explorer 3 which floats 1.5 million kilometres (1 million miles) from Earth where Earth and Sun's gravitation balance out, and finally the Russian Prognoz 7 in high Earth orbit.

Astronomers combined the results from this interplanetary network of probes to pin down the position of the source with unprecedented accuracy. And to their surprise, it coincided not with anything in our Galaxy, but with the gaseous remains of a supernova in our nearest neighbour galaxy, the Large Magellanic Cloud. If it does indeed lie in the Large Magellanic Cloud, then this flasher is 180,000 light years away—and it must be incredibly powerful, some 100,000 times brighter in gamma rays than the typical gamma-ray flashers in our Galaxy. Yet its rapid turn-on indicates that it is, if anything, even smaller in size.

We are faced with three options, all with their drawbacks. The most daring is that the 5 March, 1979 flasher did lie in another galaxy and that it was a typical flasher in its power output. This would mean that the commoner events are *not* due to nearby gamma-ray sources in our Galaxy, but are all in distant galaxies. Distant galaxies are, in fact, evenly spread over the sky, so this suggestion would also explain their regular distribution. But most galaxies are much farther off than the Large Magellanic Cloud, and a typical gamma-ray flasher must therefore be intrinsically even more powerful than the 1979 source, outshining at peak brightness 10,000 whole galaxies. Most astronomers refuse to accept that such huge quantities of energy could be released so rapidly, but it is not possible to rule out the suggestion altogether: the Universe has produced strange surprises before.

Or the source may not lie in the Large Magellanic Cloud at all. There is no independent way to measure its distance, so it could be a nearby flasher in our Galaxy which happens to appear superimposed on our companion galaxy. It is odd, though, that it coincides so well with a supernova remnant in the Large Magellanic Cloud. Old supernovae leave behind neutron star corpses, and all astronomers agree that these are precisely the kind of object that somehow produce gamma-ray flashes.

This leaves the possibility that there are two kinds of flasher. The fainter ones are common and we spot them near to us in our own Galaxy. Powerful flashers are very rare. We have been lucky enough to spot one exploding in the Large Magellanic Cloud; and as the interplanetary spaceprobe patrol continues to watch, it should eventually find a powerful flasher in our own Galaxy—as it inundates the solar system with rays many times more intense than those which swept through in 1979.

Its rapid changes point to the flasher in the Large Magellanic Cloud being the neutron star left behind by the supernova whose exploded gases are still visible, as the nebula catalogued N49. Yet even with its intense gravitational pull, a neutron star would not find it easy to generate the huge power in the 1979 gamma-ray flash. The explanation could be a variant on the neutron star grabbing gas from a companion theme, the all-embracing picture which explains so many X-ray sources. Another possibility is a super-glitch in the neutron star itself, a rearrangement of its interior similar to the Crab Pulsar's glitches, but on a much larger scale. As the neutron star collapses to a smaller size, its magnetic field could convert some of this energy to an intense burst of gamma rays. Theorists are still struggling to interpret precisely what did

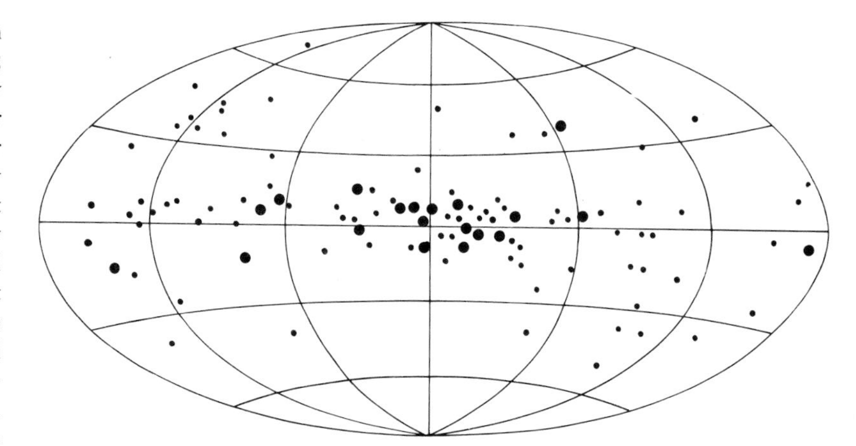

X-ray sources

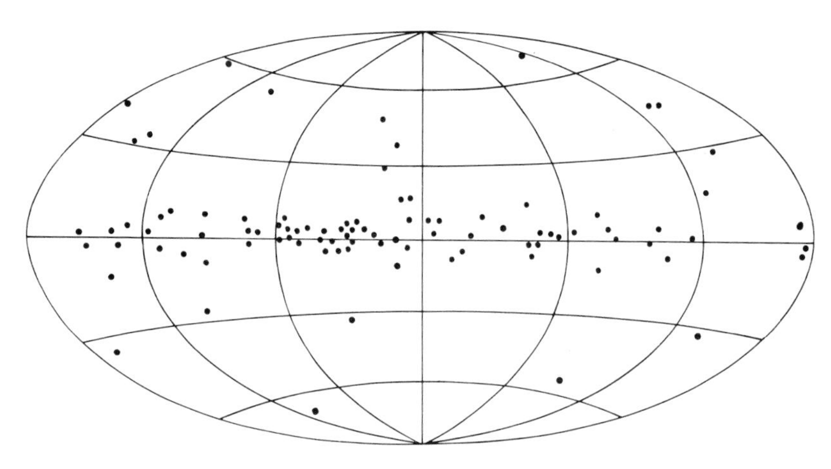

radio pulsars

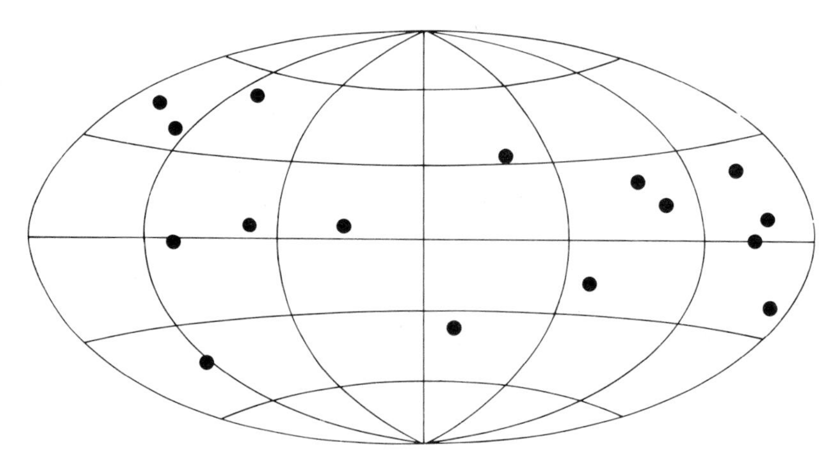

gamma ray 'flashers'

Distribution of gamma-ray flashers in the sky (*bottom*) shows marked differences from that of radio-emitting pulsars (*middle*) and X-ray sources—mostly pulsars—(*top*). Both kinds of pulsar concentrate towards a band on the sky which marks the plane of our Galaxy, showing that they are members of the disc of our system, like the stars which make up the visible band of the Milky Way. The spread of the gamma-ray flashers indicates either that they are so close as to show no concentration towards the plane, or so distant that they lie well outside our Galaxy and have no connection with it.

happen in N49. They are helped by an independent piece of evidence that a neutron star was involved, a clue that brings us back to the opening pages of this chapter. Although the first flash of gamma rays lasted only a fraction of a second, the source continued to shine more faintly for a couple of minutes—and not to shine steadily but to pulse regularly every eight seconds.

Yet again, astronomers have discovered pulsars. The first pulsars to be found were the youngest, with strong magnetic fields broadcasting radio waves. Later, astronomers trying to explain X-ray sources found their key in the X-ray pulsars; since the pulses undoubtedly came from neutron stars, these pulsars confirmed the theory that X-ray sources are neutron stars ripping gas from a companion star. Now gamma-ray astronomers have found pulsars, too, in the afterglow of the dazzling flashes. Again, the pulses implicate neutron stars.

We still cannot say what makes the gamma-ray sources flash so blindingly, or even whether they lie within our Galaxy or far beyond. Pulsars are once more presenting us with new problems at the limits of our knowledge.

Through a red filter sensitive to hot hydrogen gas, the Large Magellanic Cloud is seen to be surrounded by supernova remnants, expanding bubbles of gas which remain after stars have exploded. An intense flash of gamma rays came from one of these.

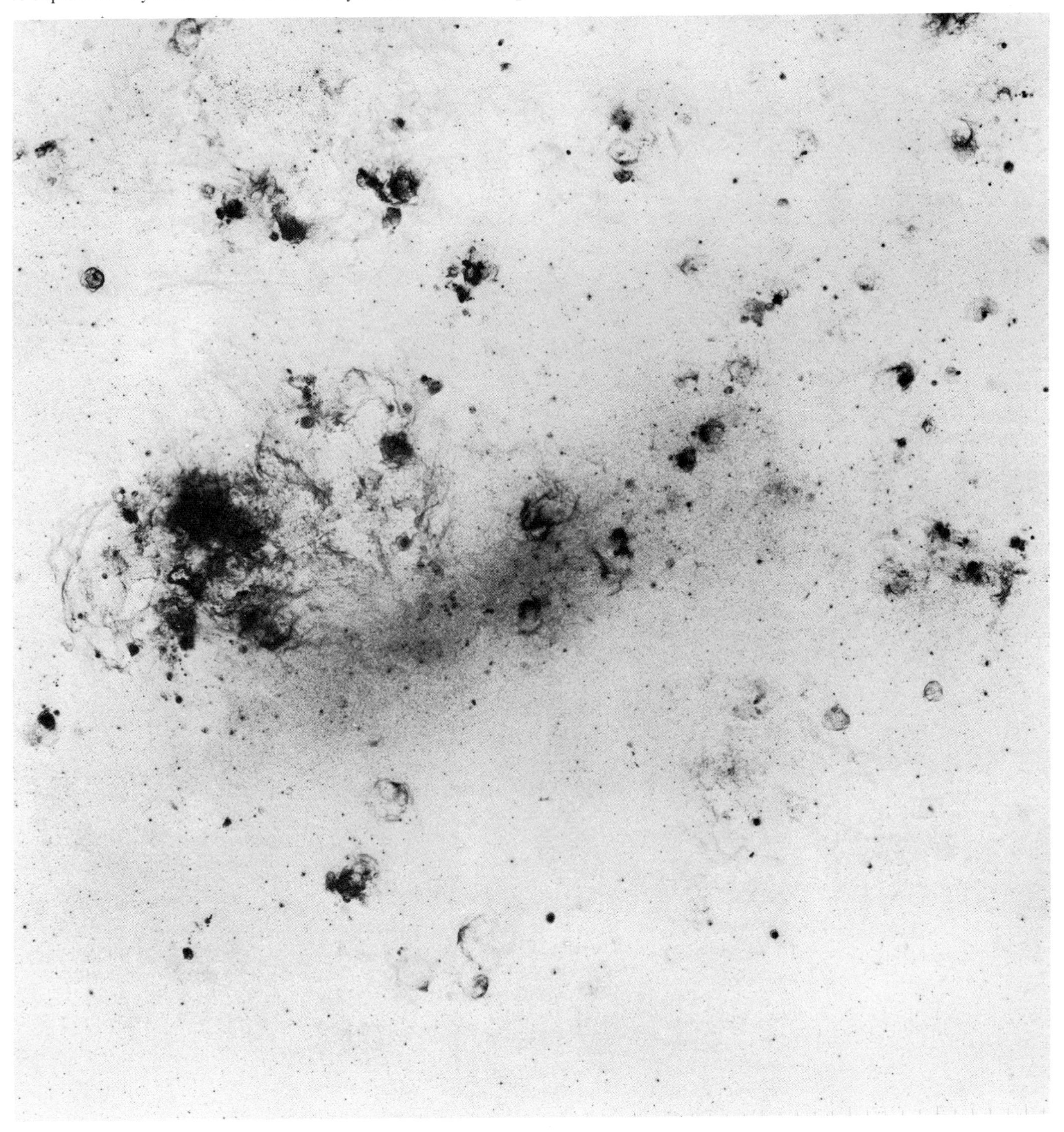

CHAPTER 9

BLACK HOLES

Six thousand light years away, in the constellation Cygnus, a huge star circles an invisible, cannibalistic partner. The star, HDE 226868, is a giant, 20 times larger than our Sun, containing 20 times as much matter and shining 50,000 times brighter. Its ferocious output of heat raises its surface to a searing 27,000°C, an incandescent bluish-white in contrast to the Sun's dimmer yellow-hot surface. Its companion is even more remarkable. Bound to HDE 226868 by the force of gravity, it circles it in just under six days, just as the Earth and the Moon revolve about each other once a month. Here, however, the companion is tiny, only a few kilometres across; yet it contains fully a quarter as much matter as its partner.

If we could approach this star-waltz closely, we would find an unseeable shrunken entity swallowing up the outer layers of the giant star's huge gaseous body, and this gas, falling onto the tiny star, disappears from view too, lost not only to HDE 226868 but also to our Universe. It is the most bizarre of all the mysteries of modern astronomy—a black hole.

'Black hole' is a graphic name, and a true description. It is so voracious that even a beam of light shone into it will be captured. It emits no light of its own; it cannot be illuminated from outside. It is the blackest thing in the Universe. And its gaping maw constitutes the ultimate in holes—anything which falls into this chasm can never return to our Universe, but is bound to perish unseen in a space and time outside our own.

Where does such power come from? It is simply the force of gravity, familiar in holding us to the surface of Earth. All bodies in the Universe exert a gravitational tug on objects near to them, and although Earth's gravity can seem fairly strong, it is very weak in the cosmic gravity league.

One way to measure gravitational pull is by escape velocity, the speed a projectile needs to attain in order to escape from surface gravity. Anything shot upwards at 11 kilometres (7 miles) per second or faster will escape from Earth's gravitation. The Sun's escape velocity is, naturally, much higher—620 kilometres (385 miles) per second.

Escape velocity does not just depend on mass, but also on size; smallness going with a higher escape velocity. Neutron stars are only a little more massive than the Sun, but because they are tiny they have an immense gravitational pull at the surface—and their escape velocity is correspondingly high, around 100,000 kilometres per second, which is almost nature's speed limit, for the velocity of light itself is 300,000 kilometres per second. In principle, there is no reason why the Universe should not contain 'stars' so massive yet so compact that their escape velocity is higher than the speed of light.

Even if such a star shone very brightly, its light would be trapped by its own gravitation and none could leak out to the Universe at large. The star would seem to be completely black, perfectly invisible except as a silhouette against a bright background. And any object which fell into the region where gravitation is so intense that light cannot escape would itself be unable to get away—because nothing can travel faster than light. The region is thus a perfect hole, too, which objects can fall into, but never escape from.

Although the idea of black holes seems too bizarre to be real, astronomers do know of one situation where they should form in space. In a supernova explosion, a star's core is crushed down to a neutron star—a pulsar. But such stars are only stable if they weigh less than three Suns. If a star core starts off heavier than this, it will continue to collapse, smaller and smaller, and will quickly shrink so much that it is surrounded by a region a few kilometres in diameter where the escape velocity is higher than the speed of light. It has become a black hole. The imaginary surface surrounding the black hole is called the *event horizon,* because we can never know of any events occurring within it.

Proving that black holes exist is another matter. Finding an isolated black hole in space is far more difficult than locating the proverbial black cat on a dark night. It is only a few kilometres across, insignificant compared to the distances between the stars, and on its own remains well concealed, shrouded in its own mystery. But astronomers have one sure guide to the whereabouts of some black holes, those in partnership with an ordinary star. The most famous of these is HDE 226868 and its black hole companion, with which this chapter opened. About half a dozen such partnerships have now been located and although the black hole cannot actually be seen in any of them, an astronomical detective hunt leads us to the inevitable conclusion that they are there, companions to the stars that we can see.

Black holes do not inexorably 'suck in' everything which encroaches on the outermost fringes of their gravitational fields. Outside the event horizon, the gravitational pull falls off gradually in exactly the same way as Earth's gravity does. And just as a satellite or the Moon can orbit the Earth without falling to the ground, so a spacecraft or a star can orbit a black hole quite safely as long as it keeps outside the event horizon. So the partnership of a star circling a black hole can last indefinitely.

There is no likely way a star can capture a neutron star, whose existence had been proved by the discovery of the first radio pulsars. But in a few cases, there are clues enough to show that the companion is not a neutron star, and in these the only alternative is a black hole. After the gas has shone out its final blaze of glory as a pyrotechnic X-ray display, it disappears inside the event horizon, lost forever from our Universe.

The chain of reasoning goes as follows. Only if the companion is very compact will the gas ring become hot enough as it spirals in to produce X-rays. Hence an X-ray source in the sky associated with a double star system must contain either a neutron star or a black hole. Even a white dwarf is not small enough, for its gas ring would not heat up enough before it crashed on the star's surface.

Cosmic gravity

	Mass (relative to Sun)	Radius (kilometres)	Gravitational pull (relative to Earth's)	Escape velocity (km per second)
Moon	0.000 000 04	1700	0.2	2.4
Earth	0.000 003	6400	1	11
Jupiter	0.001	71,400	2.6	60
Sun	1	700,000	28	620
White dwarf (Sirius B)	0.6	7000	200,000	5000
Neutron star (pulsar in binary pulsar system)	1.4	12	130,000,000,000	170,000
Black hole (in Cygnus X-1 system)	6	18	250,000,000,000	300,000 (speed of light)
Massive black hole (in centre of galaxy M87)	5,000,000,000	15,000,000,000	300	300,000 (speed of light)

black hole partner previously drifting freely in space. Astronomers believe that these star-black hole partners have grown up together. Originally the pair consisted of two ordinary, but heavyweight, stars; the heavier of the two shone more brilliantly, and reached its death-throes first, throwing off its outer layers in a supernova explosion. Its core collapsed to become a black hole, which ever since has been circling the second star. This star is so hot that gases are continuously boiling off its surface, and streaming away into space in huge clouds. Normally these gases would dissipate into the near vacuum of interstellar space; but when there is an other star on the doorstep its gravity will latch on to some of the gas and draw it in to itself.

The conservative astronomer prefers not to introduce new and strange beasts into the astronomical menagerie if he can help it. In the early days of X-ray astronomy it seemed that all the 'X-ray binary stars' could be explained by invoking the presence of a Sometimes the X-ray source pulses regularly, revealing the culprit to be a neutron star. (See Chapter 8.)

What of the non-pulsing X-ray sources? It is possible that some of these are caused by neutron stars, and strong evidence is needed before astronomers will believe a black hole is responsible. The 'proof' comes from neutron stars' natural weight limit of three Suns.

Conveniently enough, a system of two stars in orbit about each other under their mutual gravitational forces provides an ideal star-balance. The time it takes them to revolve once around each other can be measured even if only one is actually emitting light. The visible partner can be 'weighed' simply by comparison with stars of the same type whose mass is known. Once this is done, the mass of the unseen star partner can be calculated. (See Chapter 6.) If it turns out to be less than three Suns, it is possible to play safe and assume the presence of a non-pulsing neutron star; if it is heavier, it must be a black hole.

THE BLACK HOLE SUSPECTS

The 12 December 1970 was Independence Day in Kenya and a memorable day, too, for man's exploration of black holes. As celebrations erupted over the country, a rocket blasted off from a converted oil rig off the Kenyan coast, carrying one of the most famous of all astronomy satellites into orbit.

The launch was remarkable in many ways. The San Marco platform is run by the University of Rome, but its site, only 2° from the equator, makes it an excellent take-off point for any satellite to be put into an equatorial orbit. America's space agency, NASA, chose San Marco for the first launch of an American satellite outside the US.

The satellite was originally code named Explorer 42, and was the latest in a distinguished series: Explorer 1, the first American satellite, had been beaten to the position of first man-made object in orbit by Russia's Sputnik 1, launched four months earlier in October 1957, but it had recovered American pride by its discovery of the Van Allen radiation belts surrounding the Earth. Explorer 42 was to be the first of three 'Small Astronomy Satellites' designed to investigate short wavelength X-rays and gamma rays from space, radiation inaccessible to ground-based astronomers for they are absorbed in Earth's atmosphere. From this came its alternative name of SAS-A. It was the date of the launch which suggested the name by which this phenomenally successful satellite has been popularly known since, however—*Uhuru*, Swahili for freedom.

Before *Uhuru*, X-rays from space had only been detected by brief rocket flights. Some 30 sources were known in 1970, and *Uhuru* was to discover over a hundred more. But even more important was its study of individual sources in detail. *Uhuru's* remote-controlled X-ray telescopes homed in on the brighter sources, probing them to try to unravel the mystery which had baffled astronomers in the 1960s: What can produce such huge amounts of X-rays that they are detectable even thousands of light years away?

Uhuru pinpointed the position of X-ray sources with unprecedented accuracy. Optical astronomers were then able to study that part of the sky, and they often found there a star evidently in orbit about an unseen companion—frequently with indications that it was losing streams of gas. In this way *Uhuru* provided proof that most of the X-ray sources in our Galaxy are intensely hot streams of gas flowing from a giant star onto a small companion. And it proved, too, that X-rays often came in pulses, so confirming that the companion star (an X-ray pulsar) was a rotating neutron star.

But one source was different. It lay in the constellation Cygnus, which astronomers of old had seen as a flying swan. A rocket flight in 1967 had found a source of X-rays in the region of the swan's outstretched neck. It was called Cygnus X-1, the first X-ray source known in the constellation; its position was not accurately located by this flight, and at that time there was nothing to distinguish it particularly from a dozen other bright sources in the X-ray sky.

Uhuru refined Cygnus X-1's position, and studied its behaviour in meticulous detail. Although the satellite's telescope could not pin the source to any particular visible star—Cygnus is in a region of sky where stars are abundant—it narrowed the field considerably. And in March 1971 radio astronomers located a radio source in the neighbourhood of the supergiant star HDE 226868. Subsequent observations at optical, radio and X-ray wavelengths left no doubt that the X-rays also originated near this star.

Uhuru also found that Cygnus X-1 flickered rapidly in its X-ray output, was never constant for more than a fraction of a second and, in addition, never pulsed regularly. A forensic exploration of the double star system showed that the compact companion was so heavy (at least 6 Suns and probably 10) that it could only have been a black hole.

Cygnus X-1 turned out to be the gas ring surrounding the black hole, which was the companion to the visible star HDE 226868. The X-ray flickered as swirls and eddies in the gas disc momentarily brightened and faded. The radio waves probably come from stray gas splurged out around the rotating pair.

Many astronomers were not happy with this scenario, however, for black holes had not yet been taken seriously by the conventional astronomical fraternity. Perhaps the accumulated data could be interpreted in other ways? The attempts to avoid the complications of introducing the mysterious black hole into the Universe merely to explain one X-ray source, led to some fairly improbable theories being propounded.

One suggestion was that the compact companion 'star' was actually two neutron stars in very close orbit, together circling the giant just as the Earth and Moon together orbit the Sun. Each neutron star could then be just under the critical weight limit of three Suns. Later research, which indicated the companion weighed nearer 10 than 6 Suns, effectively scotched this proposal. Perhaps there were three or even four neutron stars? Such a bizarre barn dance was, in fact, not feasible for it would have been impossible for more than two stars to orbit together in the tight group necessary to act as counterbalance to HDE 226868's huge mass.

Astronomers resort to indirect methods to track down the invisible black holes. A false-colour photograph of galaxy M87 (*right*) shows a dense clumping of stars at the centre (coded blue) as well as the famous 'jet'. The dense star accumulation must result from the gravity of a very massive black hole at the galaxy's heart. X-ray observatories, like Uhuru (whose launch is shown below) can pick up radiation from hot gas streams swirling around black holes.

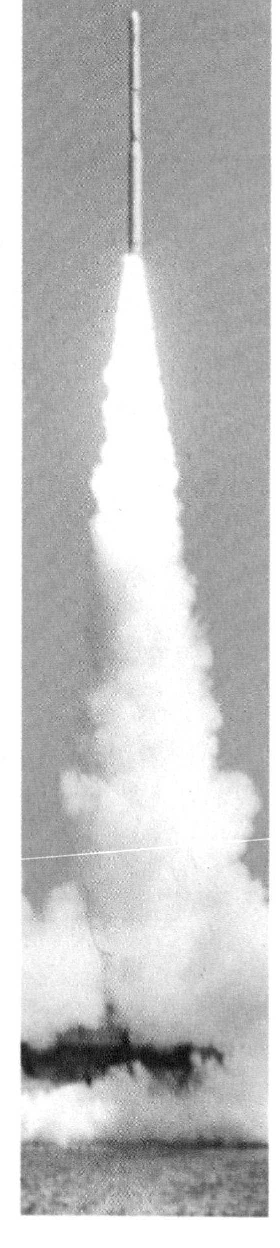

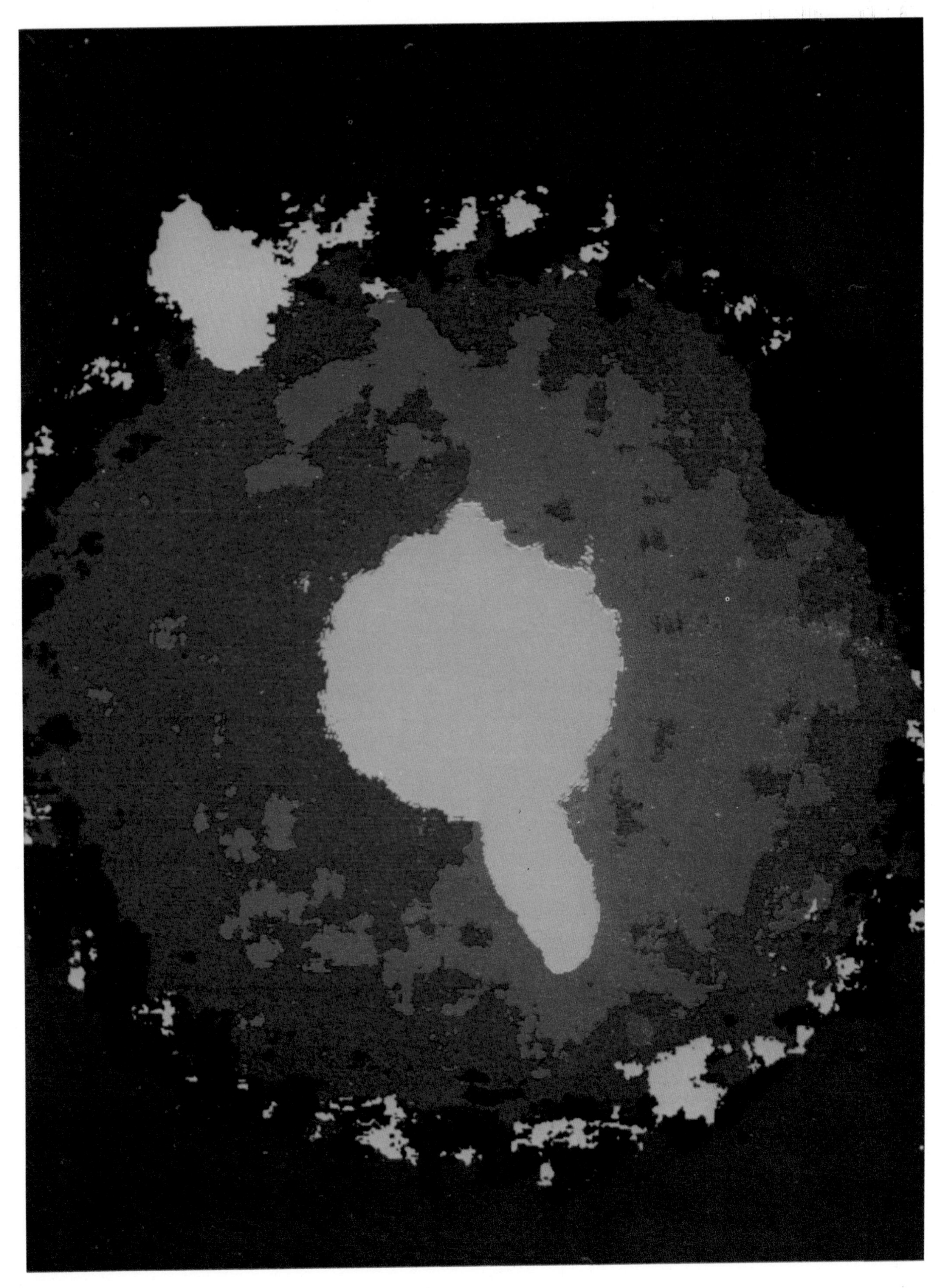

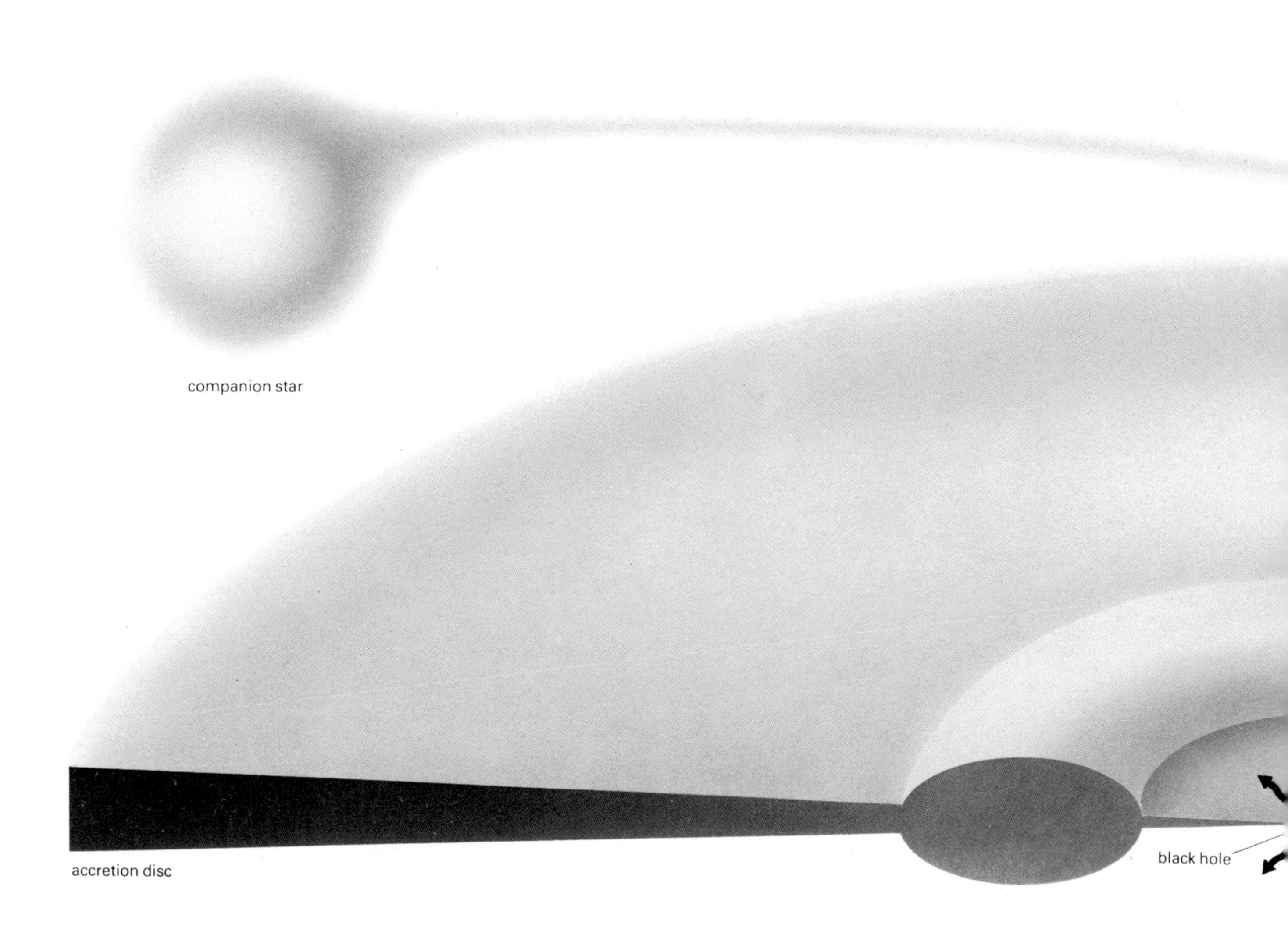

The gas disc surrounding a black hole in a double star system, like Cygnus X-1, should have a distinctive profile (*above*). The black hole's gravity pulls a gas stream from the companion star (see in background). The gas settles into a disc whose matter is continuously spiralling in towards the hole. In the outer parts, the disc's thickness decreases inwards in proportion to its radius, producing a wedge-shaped cross-section.

Other astronomers suggested that HDE 226868 was not the giant star it appeared. If it were intrinsically less bright, they argued, it would lie closer to Earth than the indicated 6000 light years, and it would be correspondingly less massive. Its companion would then be lighter in weight, too, and could fit the bill for a neutron star. All subsequent evidence, however, including a study of its ultraviolet light by the unmanned International Ultraviolet Explorer satellite, shows that it is in every way a typical giant star. Anyone now taking this stance is reduced to trying to produce some kind of subluminous star mimicking in every way a more distant giant. It is far more reasonable to believe in black holes!

A final question was often put to those who, in the first instance, supported the notion that the Cygnus X-1 system contained a black hole: If there is a black hole orbiting HDE 226868, presumably there are others orbiting other stars—and where are they? The search was on. And in the intervening years, others have begun to come to light.

Some are not too convincing to the sceptic. An X-ray source in the constellation of the compasses, Circinus X-1, flickers like Cygnus X-1 and has the same unusual ratio of high energy to low energy X-rays: it seems to be a similar system. Later and much improved X-ray satellites, such as the Einstein Observatory, have detected more such sources, which to the converted bear all the signs of a black hole at work.

In April 1978, however, the X-ray telescope on the remote-controlled Copernicus satellite pinpointed a much stronger candidate. This source coincides with a known supergiant star, variable in brightness and catalogued V861 Scorpii. Lying in the constellation of the Scorpion, near the top of its long tail, this star is situated 6000 light years from the Sun. After the protracted arguments over Cygnus X-1, optical astronomers decided to clamp down quickly on this new suspect. Observing with the large telescope in the clear air of Mauna Kea volcano in Hawaii, they could deduce that the supergiant V861 Scorpii is 50 times heavier than the Sun. The mass of the compact companion star? Between 10 and 15 Suns. It cannot be anything but a black hole.

So the tide of belief has turned. HDE 226868 (Cygnus X-1) and V861 Scorpii, and

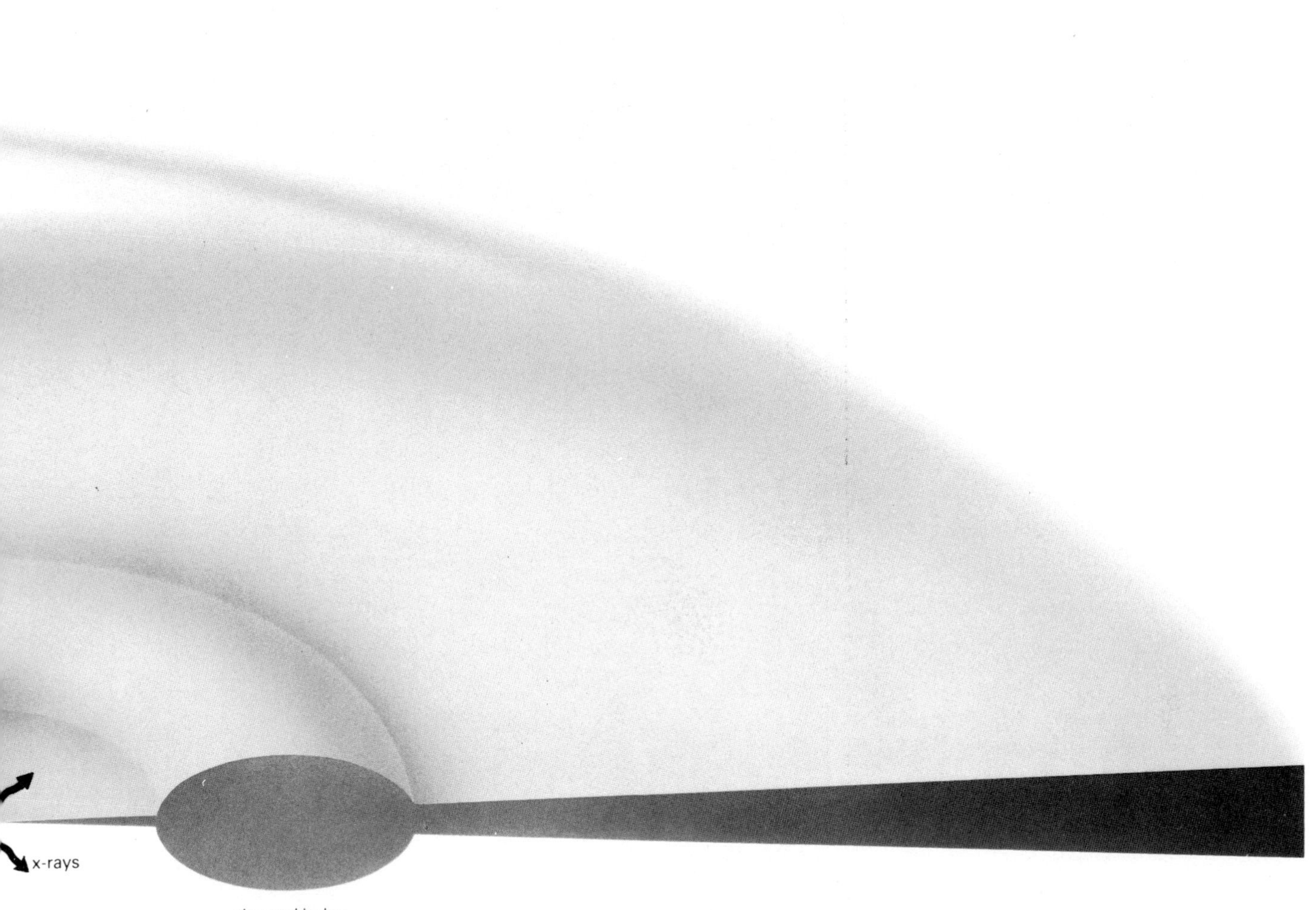

other stars like Circinus X-1, are evidently just the most conspicuous of a legion of cannibalistic star-partners in our Galaxy. It is just too difficult to explain them any other way. The mysterious X-ray sources have revealed the even deeper mystery of black holes in our Universe.

THE OSCILLATIONS OF SS433

Astronomers delight in finding unexpected objects in the sky. The accidental discovery of the pulsars is one classic story of this kind. The years 1978 and 1979 saw another, less publicized—and as yet unexplained—astronomical find. Its name is SS433.

The discovery depended on the thriving collaboration between researchers observing the sky at different wavelengths of radiation. Back in the 1960s, American optical astronomers C. Stephenson and N. Sanduleak had noted a star with an unusual spectrum in the constellation of Aquila. It duly took its place as number 433 in the Stephenson–Sanduleak catalogue of peculiar stars.

The year 1978 saw several astronomers, including the British Astronomer Royal Sir Martin Ryle, take a personal interest in SS433. It was found to be emitting both X-rays and radio waves in copious amounts, making it a most unusual 'star' indeed, and it was discovered to lie right in the centre of the expanding gaseous remains of a star which exploded 40,000 years ago, situated 10,000 light years from the Sun. It is evidently the collapsed core of that star. But what kind of object is it now?

It should be either a pulsar (neutron star) or a black hole. Yet a pulsar this old should not be visible in ordinary light and, in addition, SS433 does not pulse. And a black hole should be invisible, unless there is a companion star in orbit with it, and first investigations of the light from SS433 revealed no sign of a normal star.

It was the light from SS433, in fact, which confirmed its oddness. A team of seven astronomers led by Bruce Margon of the University of California, Los Angeles, investigated its light output in late 1978. At their disposal were the four largest telescopes in the United States. And they analyzed the light with the most sophisticated modern electronic devices available.

The gas heats up as it moves inwards, and eventually its heat energy is sufficient to make the disc thicken up in a characteristic 'bulge'. Closer in still, the black hole's gravitation is strong enough to flatten the disc again. It is this small region of the gas disc (exaggerated in size above) that produces the X-rays in a source such as Cygnus X-1. Eventually, the gas disappears down the black hole itself.

The secrets of starlight are cracked by splitting it into a spectrum of colours. Each chemical element shines at only a few particular colours—wavelengths of light—which appear as lines crossing the spectrum. So the pattern of wavelengths present as spectral lines in a star's light reveals which elements it is made of. But SS433 was different. Although it had some lines at the wavelengths of the commonest gases in the Universe, there were also three lines at wavelengths corresponding to no known element, and these wavelengths changed from night to night. In no other object in the sky had lines been observed marching inexorably along the spectrum. There was consternation in the astronomical world. These lines must come from hot gas clouds—but what had altered their wavelengths, and in such a way that they could change so drastically from night to night? The answer came soon enough, but only heightened the mystery.

As fascinated astronomers all over the world watched this inconspicuous speck of light in Aquila, they saw that the shifting spectral lines did not continue their motion. They stopped and turned back; then stopped and resumed their original motion. In the star's spectrum of light, these lines were oscillating backwards and forwards, and the midpoint of the swing was, in each case, marked by an ordinary line of hydrogen or helium. The light in the shifting lines evidently came simply from hydrogen and helium gases—but something was moving the wavelengths back and forth, back and forth, swinging through a cycle once every five months. And the shifting lines came in pairs, one either side of the fixed line, and moving in and out like mirror images.

Sophisticated apparatus was again used to rule out some of the hurriedly proposed explanations, such as strong oscillating magnetic fields. The answer had to be the Doppler effect. When a cloud of gas is moving rapidly, the spectral lines have their wavelengths changed—in SS433, there must be twin clouds of gas, each alternately approaching and receding from us, along with stationary gas. But where exactly are these clouds? What would SS433 look like close-up? Even astronomers were confused and puzzled, and it is little surprise that the headline writer in one daily paper was reduced to describing SS433 as the 'star that's coming and going at the same time'!

The evidence now suggests that the central part of SS433 is squirting out jets of hot gas in opposite directions, while it rotates. The jets always move outwards at the same speed, but the changing orientation means that the part of their motion towards or away from us continuously changes: each jet is alternately

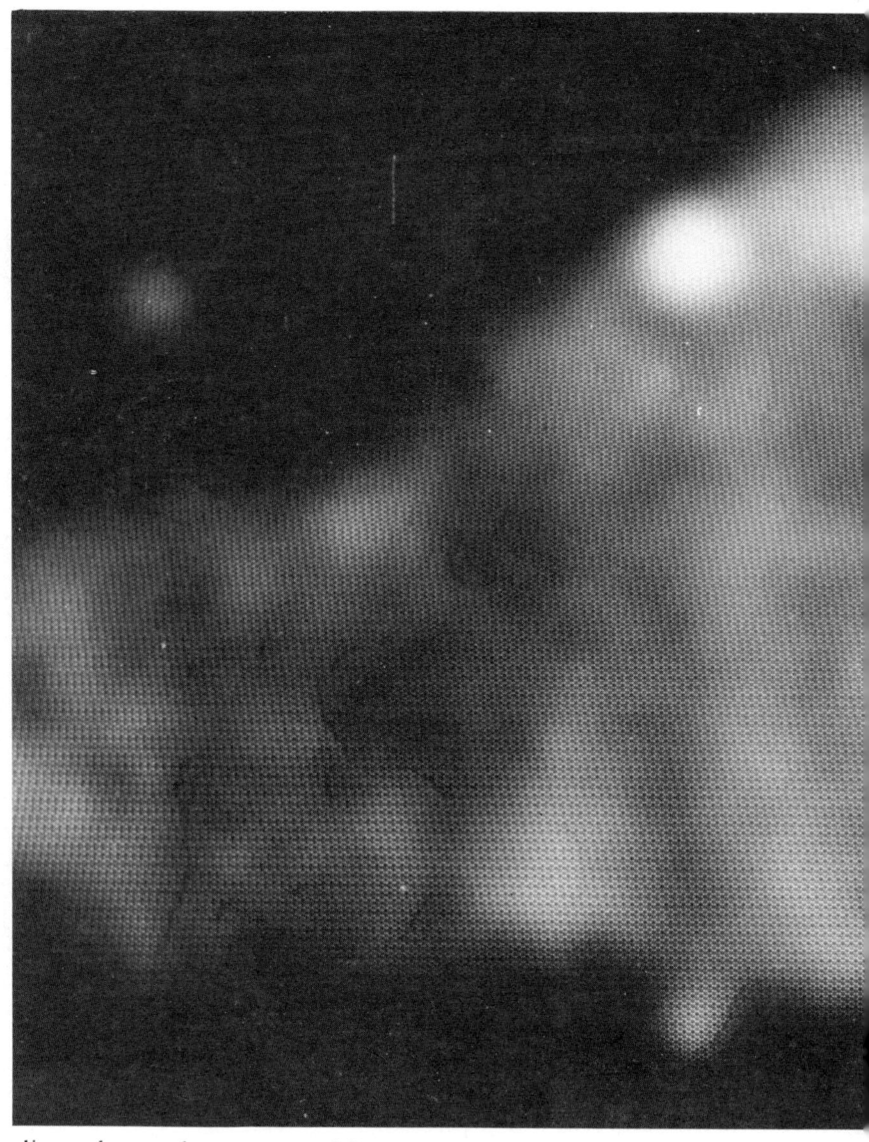

directed towards or away as SS433 turns, and the other jet is orientated in the opposite direction, giving rise to the mirror image motion. Astronomers are still amazed by the speed of the gas ejected in these powerful jets—at 80,000 kilometres per second it is over a quarter the speed of light! What can be forcing gas out at such a rate? And where is the supply coming from?

We return to the black hole to answer at least the first question. It is quite likely there would be a black hole in the heart of SS433 anyway, left over from the original supernova explosion 40,000 years ago. And matter falling into a black hole can produce copious amounts of energy before it finally tumbles through the event horizon and disappears. In Cygnus X-1 and its brethren this energy in the gas ring shines out as X-rays, but perhaps in other circumstances it could propel superfast jets of gas. Certainly it seems at the moment that gas falling into a black hole would be virtually the only way to provide enough power to propel the enormously fast jets of hydrogen and

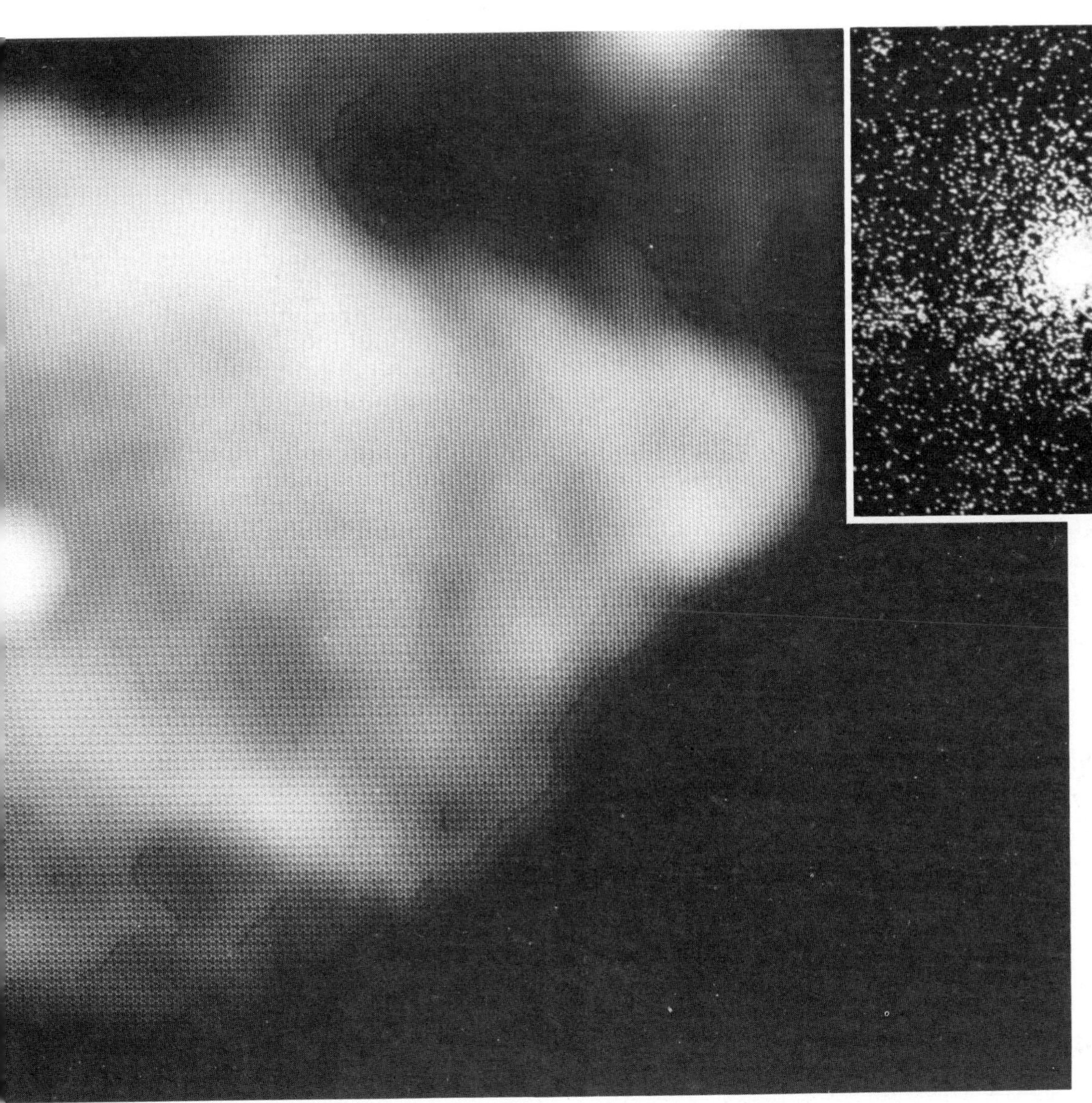

A radio-photograph of the region of SS433*(left)* shows its appearance as 'seen' by the 100-metre radio telescope near Bonn, West Germany. SS433 itself is the bright source at the centre (the other small radio source is probably unrelated, lying much further away). The large shell of radio-emitting gases is the remains of a supernova which must have exploded some 40,000 years ago, and whose imploding core became the strange object SS433. Observations at higher resolution show that jets of gas shoot out from SS433 at high speeds, towards the left and right in this orientation. The jets have clearly pushed out 'ears' in the normally circular shell of gases from a supernova explosion, giving the gas cloud an elongated shape.
The inset is an X-ray photograph from the Einstein Observatory. The jets can be seen as diffuse bands stretching left and right across the picture from the SS433 itself (*centre*).

helium emanating from the core of SS433. The two opposed jets could be caused by gas boiling off either side of the thin gas ring around the hole.

The gas is probably coming from a companion star. As in Cygnus X-1, the black hole's gravity pulls some of its companion's gas into the accretion disc, and some of this is ejected in the jets. The hydrogen and helium lines originally described as 'stationary' do, in fact, change slightly in wavelength over the period of 13 days, and this light probably comes from gas near the companion star, circling the black hole every 13 days.

It is likely that the mystery of SS433, like the earlier case of Cygnus X-1 and its giant companion HDE 226868, will not be satisfactorily solved until astronomers have identified more examples of this exotic kind of star. Already they have found a few more objects in the gaseous remains of other old supernovae, star-like points of light emitting radio waves and sometimes X-rays. None so far has shown the strange spectral line shifts with which SS433 drew the world's gaze. But perhaps SS433 is giving away the secrets of a whole new kind of celestial object—weird star-corpses whose existence in our Universe has been unsuspected until now.

THE EINSTEIN REVOLUTION

We now have impressive indirect evidence that black holes do exist. For the space traveller of the future they will be a hazard; probably from a distance a pleasurable change of scene from endless stars; and perhaps the gateway which will usher him into another universe, another space and time.

Science fiction writers have recently promoted journeys through black holes as the way around nature's speed limit, to enable Man to travel instantly from one part of space to another, from one universe to another, and to travel through time—although there are good reasons for believing such feats to be impossible. But black holes have achieved this reputation because Einstein's theory of gravi-

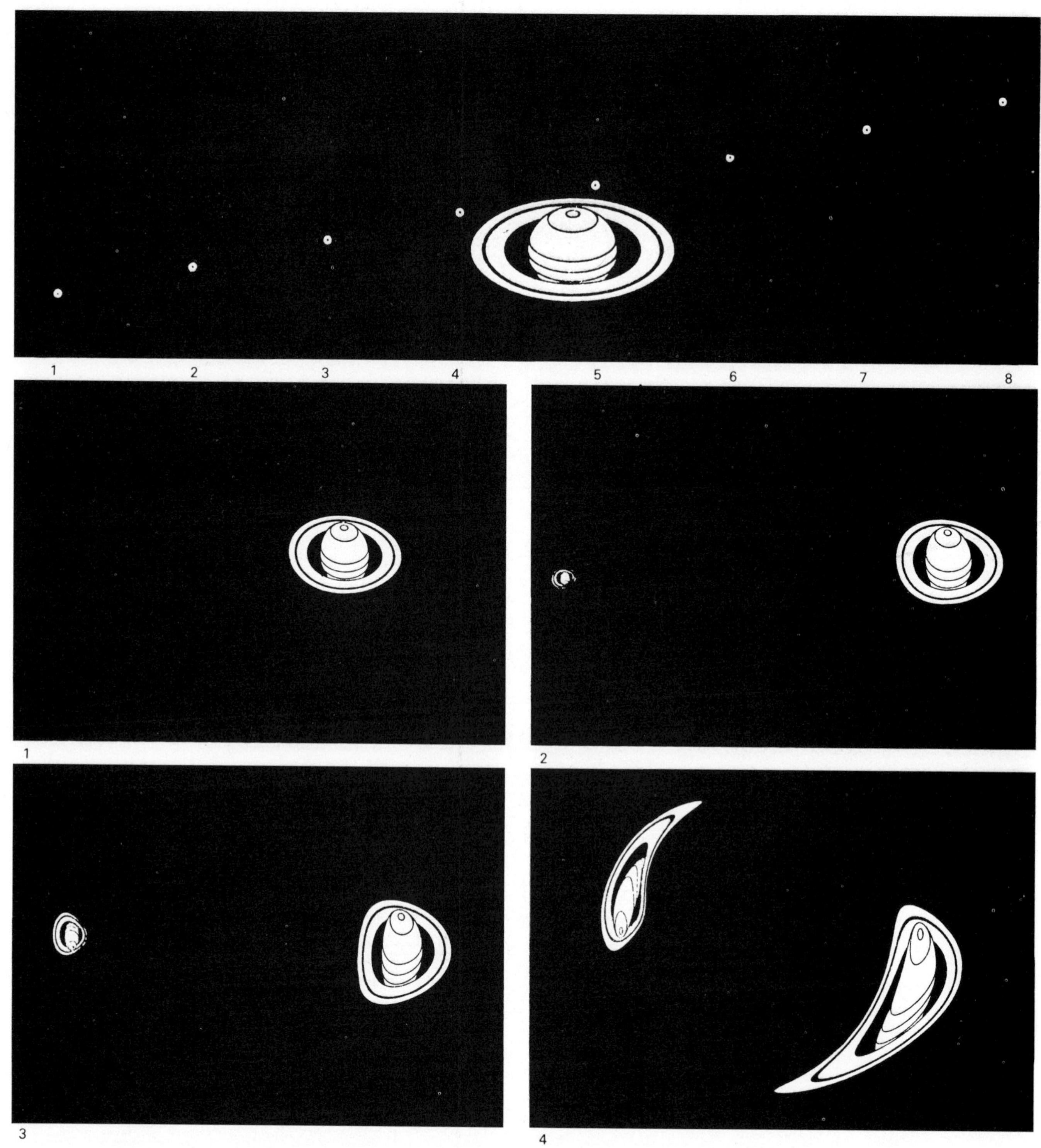

tation does indeed seem to predict such strange happenings.

Einstein saw gravitation, not as a force like the tug of a rope, but as a distortion of space and time. In fact, space and time form an interlinked canvas of 'space-time' against which the Universe is set. This has measurable effects, even if gravitation is not involved. When two spaceships pass at high speed, each sees the other distorted in both space and time: the length of both seems less, while clocks on board seem to be running slow. In Einstein's view, these two effects represent a different 'perspective' in space-time, a change in 'space' being offset by a change in 'time'. It is similar to the change in perspective as you walk around a building, to take an everyday analogy. The front wall appears to shorten, and the side wall correspondingly to lengthen as you turn the corner.

In Einstein's theory of gravitation, general relativity, a body's mass distorts the fabric of space-time around it. While a spaceship out in space drifts at a constant speed, the instant it

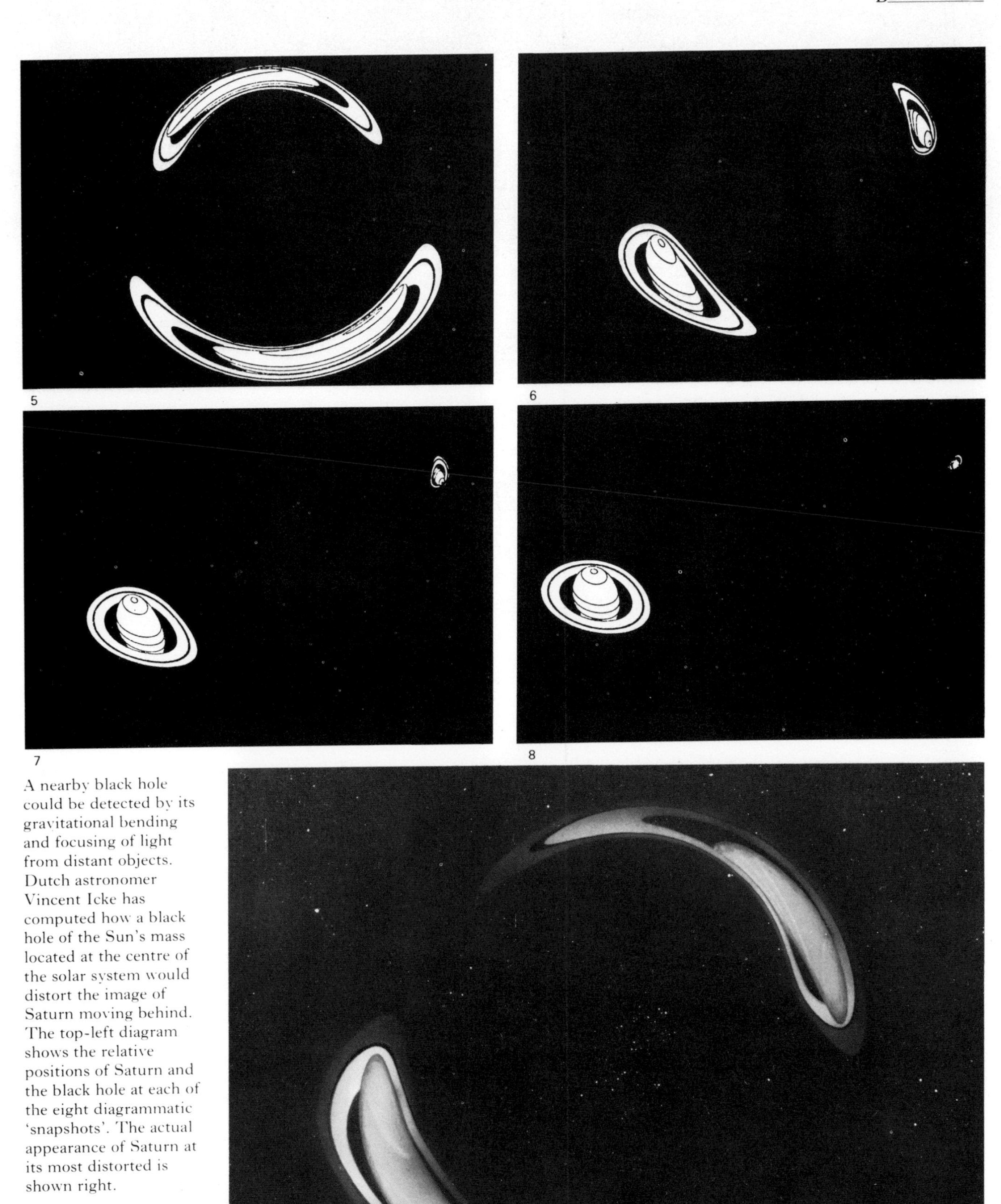

A nearby black hole could be detected by its gravitational bending and focusing of light from distant objects. Dutch astronomer Vincent Icke has computed how a black hole of the Sun's mass located at the centre of the solar system would distort the image of Saturn moving behind. The top-left diagram shows the relative positions of Saturn and the black hole at each of the eight diagrammatic 'snapshots'. The actual appearance of Saturn at its most distorted is shown right.

comes near to a planet it finds that space itself is warped. Following this 'bend', the spaceship's path is distorted. In everyday terms, we say that the planet's gravitation pulls the spaceship towards it. But note that this is not really a 'force' acting at a distance and 'drawing-in' the craft. The spaceship is responding to a local bending in space-time caused by the presence of the planet. The bending manifests itself as a 'grain' in space-time which makes motion towards the planet the natural course open to it.

This grain in space and the ship's forward momentum continually compromise, to swing the spacecraft around in a tight curve behind the planet and back out into space. America's Voyager robot spacecraft swung around Jupiter and Saturn in exactly this way in 1979.

At a distance, a black hole's gravity is no different. A future spacecraft coming across an isolated, invisible black hole would most likely only trespass on the outer fringes of its gravitational domain. And the crew would simply find their craft flung around in space and sent off in another direction, as if batted by an unseen hand.

Since the black holes likely to be lurking in space are only a few kilometres across, the chances are remote that a spaceship would accidentally score a bulls-eye on the black hole itself. The odds are considerably worse than hitting a pinhead-sized target on Earth with a dart thrown randomly from space.

A spaceship will only end up in a black hole if its pilot intends to explore it. And even at the present time, we can use Einstein's theory to say what he is likely to find, without the disadvantage of cutting ourselves off inexorably from our Universe.

At the event horizon, the grain of space becomes so pronounced that it is impossible to cut across it. Space itself is directed purely inwards, towards the collapsed star, relativity's way of describing the unconquerable inpull of gravity at the black hole's surface. At first, it may seem no better a description than simply saying that here the escape velocity is equal to the speed of light; but Einstein's theory has far more depth. For it is not just space, but space-time, which is contorted; and in these extreme conditions the alteration in the flow of time is no less remarkable. Only Einstein's theory can describe this, and the even stranger world within the black hole.

From a safe vantage point outside the black hole let us—in imagination at least—drop a regularly flashing beacon. As it falls, we see it speed away from us, travelling faster and faster towards the hole. The light waves from the speeding beacon are stretched; the initially white flashes appear redder and redder, until we need a device sensitive to the longer wavelength infrared radiation to see it at all.

It is natural to expect that the beacon will instantly disappear from sight when it reaches the event horizon, lost from our Universe along the irresistible inward-directed grain of space. But it won't. It is affected by the tremendous distortion in the passage of time at the event horizon.

Although the beacon initially sped away from us faster and faster, we now see it slowing again as it approaches the event horizon. In fact, its speed decreases more and more, in such a way that however long we keep watch, we never see it reach the event horizon at all.

In practice, the shift in wavelength of the beacon's light means that it quickly becomes invisible anyway. This shift towards longer wavelengths—infrared, then the even longer radio waves—is not due just to the beacon's speed but to the strong gravity of the hole. Light fighting its way up against gravity loses energy, and this loss of energy also means that its wavelength increases. In reality, as the beacon appears to slow, its radiation very quickly increases in wavelength through infrared to radio waves, and the radio wavelengths become so long that it is impossible, even in theory, to measure them. Whatever detectors we have at our disposal, the beacon will disappear in a fraction of a second from when it starts to slow in its inevitable approach to the hole; but at that time it has not yet reached the event horizon, and to an outside observer it never will.

The decreasing speed of the beacon is not an effect of *space* distortion; it occurs because of the changed flow of *time* near the black hole. Einstein's theory shows that the distortion of space-time which we call gravity has only a small effect on time when it is as weak as the Earth's (although even on Earth the effect can be measured with sensitive atomic clocks), but that in the vicinity of a black hole's event horizon it is as important as the irresistible distortion of the grain of space.

To an outside observer, time seems to flow slower and slower at points nearer and nearer to the event horizon. If it were possible to position clocks on a strong cable hanging down towards the event horizon, then the clocks successively further down would appear (to someone at the top of the cable) to be running more and more slowly. This is not a mechanical effect of gravity on the clock mechanism. Each clock is truly recording the passage of time at its own locality.

If the clocks were replaced by manned space colonies, the people in the outermost colony would see their brethren in the inner ones age more slowly; by their reckoning the colonists nearest the black hole would live well beyond a normal life-span. This Methuselah factor cannot be achieved with an ordinary black hole; as we shall see later there are forces which would tear the space colonies apart. And anyway the inhabitants of the lower colonies would not themselves actually feel the time to be passing more slowly; everything, including their thoughts, would be happening 'in slow motion', and their life-spans would seem to them no different.

The reason for the beacon's decreasing speed, then, is simply that time runs more slowly near the black hole. A distance we would expect it to traverse in a second will take it far longer, simply because time runs more slowly where the beacon is. The choice of a regularly flashing beacon as a black hole probe is useful, for it will show up directly time's seemingly inconstant flow.

If the beacon is constructed to send out regular flashes, then we will see successive flashes arrive later and later than expected. The interval will become longer without limit, and the rate will slow down in such a way that only a certain, finite, number of flashes will ever reach us, no matter how long we keep up our vigil.

From the very definition of a black hole—nothing inside can get out, not even radiation—it is obvious that we cannot expect a probe to send us messages from inside the event horizon to tell us what is there. The only way to find out is to go and see. And an astronaut venturing into a large black hole will satisfy only his own curiosity, because there is no way he can let the Universe outside know what he has found.

INSIDE THE BLACK HOLE

An astronaut setting out on a black hole exploration trip has one initial problem even before he has to worry about the effects caused by relativity near the event horizon. Even well outside the hole, he will discover that his feet (assuming he is attempting a foot-first entry)

Hypothetical view from a starship porthole of another craft about to explore the black hole in the Cygnus X-1 system. In reality, any astronauts aboard would be dead long before reaching the black hole. They would first suffer from the intense X-radiation generated by the hot gas disc, and then from 'spaghettification'—stretching out by the black hole's ever-increasing gravitational pull in the space around.

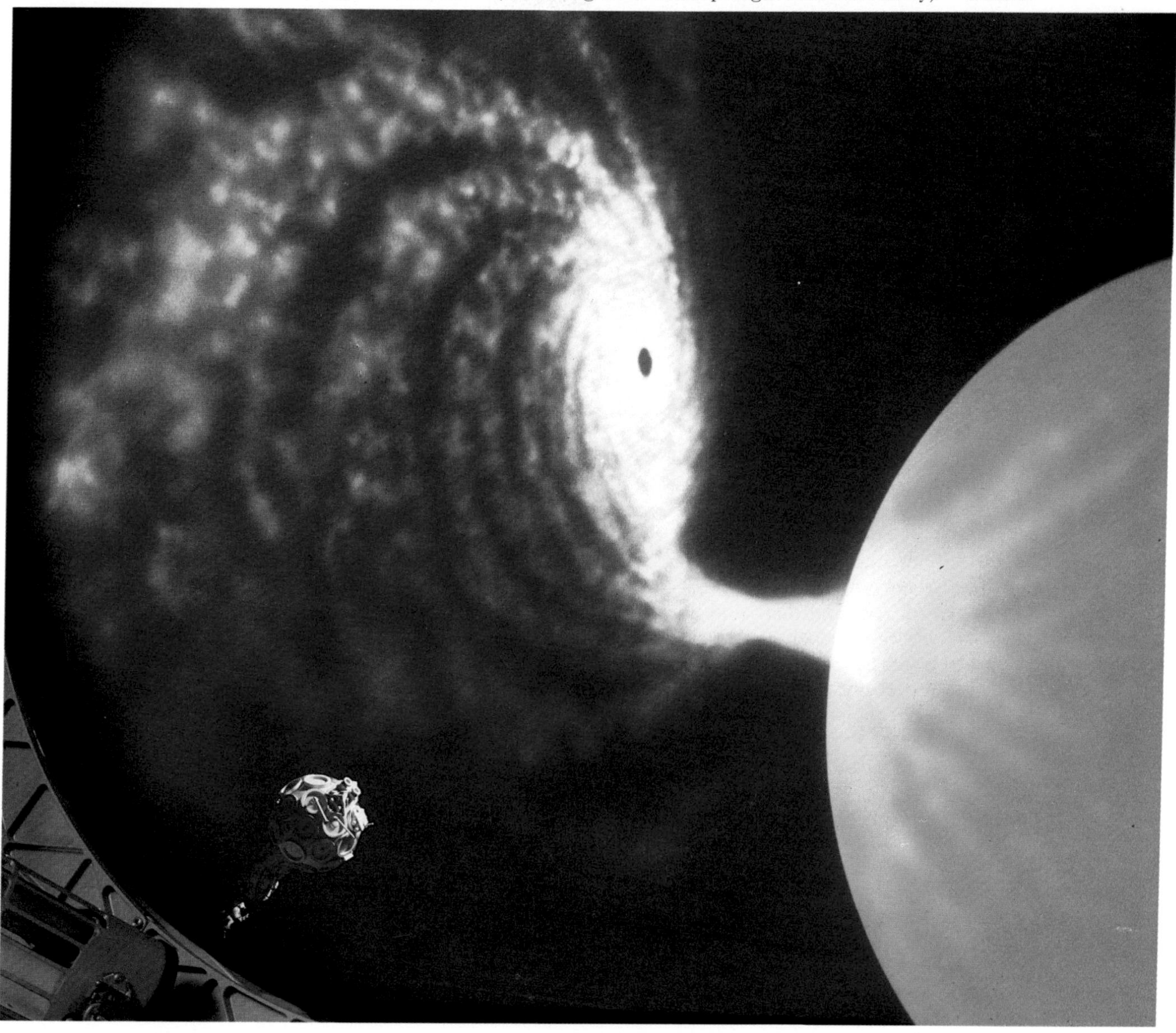

Exploding mini black holes

In 1974, Cambridge physicist Stephen Hawking announced that black holes do not last forever: they slowly 'evaporate'. A black hole gradually loses mass in the form of an outward-flowing stream of electrons and other subatomic particles. As a result, it becomes lighter in weight, and will eventually dissipate altogether.

Although nothing can escape from a black hole's gravitational inpull when it is inside the hole, Hawking's particles are created by the gravitational field *outside* the hole. As a result, these particles can escape to space beyond. The mathematics of the process is complex, but Hawking explains it this way. As the gravitational field's energy creates the evaporating particles, it must produce an equal amount of 'negative mass' to compensate. The negative mass falls into the hole, where it annihilates the matter at the centre. Since it is this matter which produces the hole's gravitational effect, the black hole becomes weaker and weaker. The evaporation, in fact, speeds up as the hole shrinks, until it culminates in an explosion.

The rate of evaporation from the black holes created by supernovae—like that in the Cygnus X-1 system—is so slow that it makes no practical difference. But the Hawking process is more important for smaller, lightweight black holes. A black hole weighing as much as a mountain (very light on the cosmic scale) would be about one-thousandth the size of an atom, and would last about 15,000 million years—the present age of the Universe. If the Big Bang produced a range of small black holes, the holes with this mass should be exploding right now.

Astronomers have searched for the bursts of radiation from such exploding 'mini black holes', but so far without success. Hawking's theory is not in doubt; it appears that the Big Bang did not, in fact, produce such small black holes.

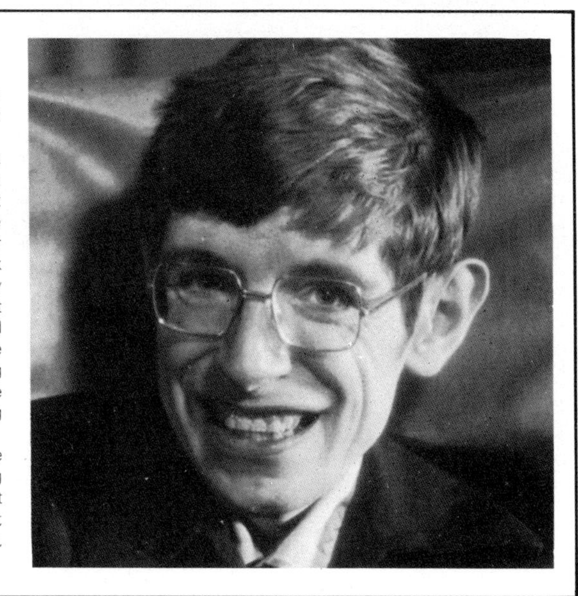

Stephen Hawking (*above*) has been hailed as producing the greatest advance in our understanding of gravity since Einstein's work early this century. His achievement is the more remarkable because he is partially paralyzed, and has to perform all the calculations in his head.

feel a stronger gravitational pull than his head, which is further from the hole. This is simply because gravity decreases steadily away from the gravitating object; it happens to us in Earth's gravity. But in the approach to a black hole the effect is far stronger: the astronaut's body is stretched out on a gravitational rack. Before he comes near the event horizon, his body will be pulled out beyond the limits that flesh and blood can stand and he will suffer the excruciating death of 'spaghettification' long before he is in any position to unravel the secrets of the black hole.

The strength of the spaghettification effect (usually known more prosaically as 'tidal stretch' for it is related to the way the Moon raises tides on Earth) depends on the mass of the rack-inducing black hole. It is not so bad for larger holes, oddly enough, so a spacecraft and pilot could indeed venture safely into one of the huge black holes believed to lie at the centres of some galaxies.

But is it not a contradiction to talk of an astronaut entering a black hole, when we have already seen that a probe would take an infinitely long time to reach the event horizon from outside? In fact not, for one of the apparent contradictions in relativity comes into play here. In Einstein's theory, space can appear on a different scale, and time can run at a different rate, for people either moving relative to one another or simply being in different parts of a gravitational field. And what one person feels to be perfectly normal may appear completely bizarre to someone looking at him from outside—like the flow of time in the very hypothetical space colonies strung up outside a black hole.

So it is with the infall to the event horizon. Time appears to pass no more slowly to the astronaut himself as he approaches the black hole's boundary. His crewmates outside will see his accompanying beacon flash more and more slowly; but to him it will be as regular as ever. Say it flashes once per second and his friends see 60 flashes altogether (the last few being calculated as the ones they would have seen if they could have waited for ever on watch), then the explorer will find that only a minute has passed (to him) before he finds himself at the event horizon, about to pass through.

During this brief infall, the black hole will loom steadily larger, blocking out the sky in front as a jet-black disc ringed with distorted patterns of stars as its gravitation bends the path of light from the stars beyond. Looking back, life among his friends on the spaceship seems much speeded up, just as they see him slowed down.

And then it is through the event horizon into the hole. Inside the black hole only relativity can describe what we find; and the answer is overwhelming. Under this strong gravity, space and time have virtually changed roles. In ordinary life, we can move freely in space, but there is no choice about our direction in time: we can only progress forwards. It is not only poets and philosophers who regret that we cannot put the clock back—we all feel it at some time. But inside a black hole, there is only one direction to space: inwards. It is as impossible to travel outwards—or even to remain stationary—within a black hole, whatever power you have available, as it is to travel backwards in time in the ordinary Universe.

Inexorably, then, our traveller is dragged towards the black hole's centre, where the matter of the original star's core is now concentrated. And just as we on Earth cannot see forward in time, so the black hole explorer cannot see forward in space to ascertain what awaits him at the heart. He can certainly look ahead of him, but his view is simply of an area of blackness. Light is falling into the black hole from stars outside, however, so the traveller can still see the Universe outside. His heroic feat can tell him nothing about the prime enigma of the black hole: what resides at its very centre. He will quickly hit it, and be crushed by its infinite gravity into a small point in space.

NO HAIR

Real black holes would not be this simple to explore. For a start, there would generally be a ring of gas to circumnavigate, whether the hole were in a double star system or at the centre of a galaxy. But even more interesting, real black holes undoubtedly spin.

The known remains of dead stars, white dwarfs and neutron stars (pulsars), certainly rotate on their axes. They inherit the spin from their predecessor stars, and the tiny pulsars are born spinning round once in less than a second. The even more compact and shrunken black holes should spin far more rapidly.

But how does a hole in space spin? And how could it appear different from a non-rotating black hole? The answers lie with Einstein's theory again. Since the gravity outside the black hole is a distortion of space-time, a spinning black hole bends the outside structure of space-time in the direction of its rotation. It drags outside space around with it, the strength of the drag decreasing with distance from the hole.

So if we were to drop a beacon into a real black hole, it would not just fall straight in. Even if it were aimed dead-centre at the hole, as it fell closer it would feel the grain of space directed not simply inwards but angled in the direction of the hole's rotation. We would see it, apparently unaccountably, veer sideways as it feels the effect of space itself carrying it into a decreasing spiral path in towards the event horizon.

Black holes are evidently a little more complicated than they seem on first acquaintance. But only a little—from the outside at least. Einstein's equations show that its effects on the outside Universe are related to only three things: its mass, its spin and its electric charge. The first two affect the distortion of space-time around the hole—its 'gravitational field'. The last is not likely to be important in real black holes for most stars are not highly electrically charged.

Otherwise, there is absolutely nothing we can know about the contents of a black hole. Make one from a dead star, or from an equal weight of old beer cans, and there will be no discernible difference. Even stranger to the scientist is the fact that a hole made from antimatter is exactly the same as a hole made from ordinary matter. The proof of this 'theorem' on the outside properties of black holes took many years of mathematical calculation to achieve: it is usually known affectionately as 'a black hole has no hair'.

Much of the credit for the 'no hair' theorem rests with Oxford mathematician Roger Penrose. Penrose has explored (mathematically that is) some of the odd things that can happen near a rotating black hole. Such a hole still has a spherical event horizon, where the inpull of gravity is inexorable, but surrounding this is a region where the rotation of space is all-important. This *ergosphere* is a part of space where an infalling spacecraft can escape again into the rest of the Universe, for it lies outside the event horizon; but while the craft is within the ergosphere it cannot remain stationary or move counter to the hole's rotation. In the ergosphere space is twisted so strongly in the direction of the hole's spin that no amount of power will suffice to either hold a craft still or to send it in the opposite direction to the grain of space.

Penrose sees possible ways of extracting energy from a spinning black hole. One idea is to build a large frame all around the hole: being symmetrical, it would not be pulled in and the rotation of space would drag it around. Perhaps its turning could be harnessed to drive a superpowerstation of the far future. Or a two-part unmanned craft could be sent into the ergosphere in such a way that one part falls through the event horizon and is lost, while the other shoots out of the ergosphere with a much higher speed than it started with. Again, human ingenuity could no doubt provide some way of tapping the speed of a stream of these projectiles to produce power.

In both schemes—and others which have been proposed—the energy is drawn from the hole's rotation. Eventually it will be braked to a standstill by the continuing energy drain on it, and settle down as a simple nonspinning gravitational trap. But even if these futuristic powerstation scenarios do come to fruition there will be no need to worry about the black hole's energy reserves. Drawing on a typical spinning black hole (with the same amount of rotational energy as a pulsar) for our present level of power consumption would affect it hardly at all. Even a continual energy demand at this level for the entire present age of the Universe (15 billion years) would reduce its spin by less than a billionth part.

KINKS IN SPACE

Whether a hole is rotating or not, its ultimate mystery must be: Where does the matter of the collapsed star go? If the hole does not

1
2
11
10

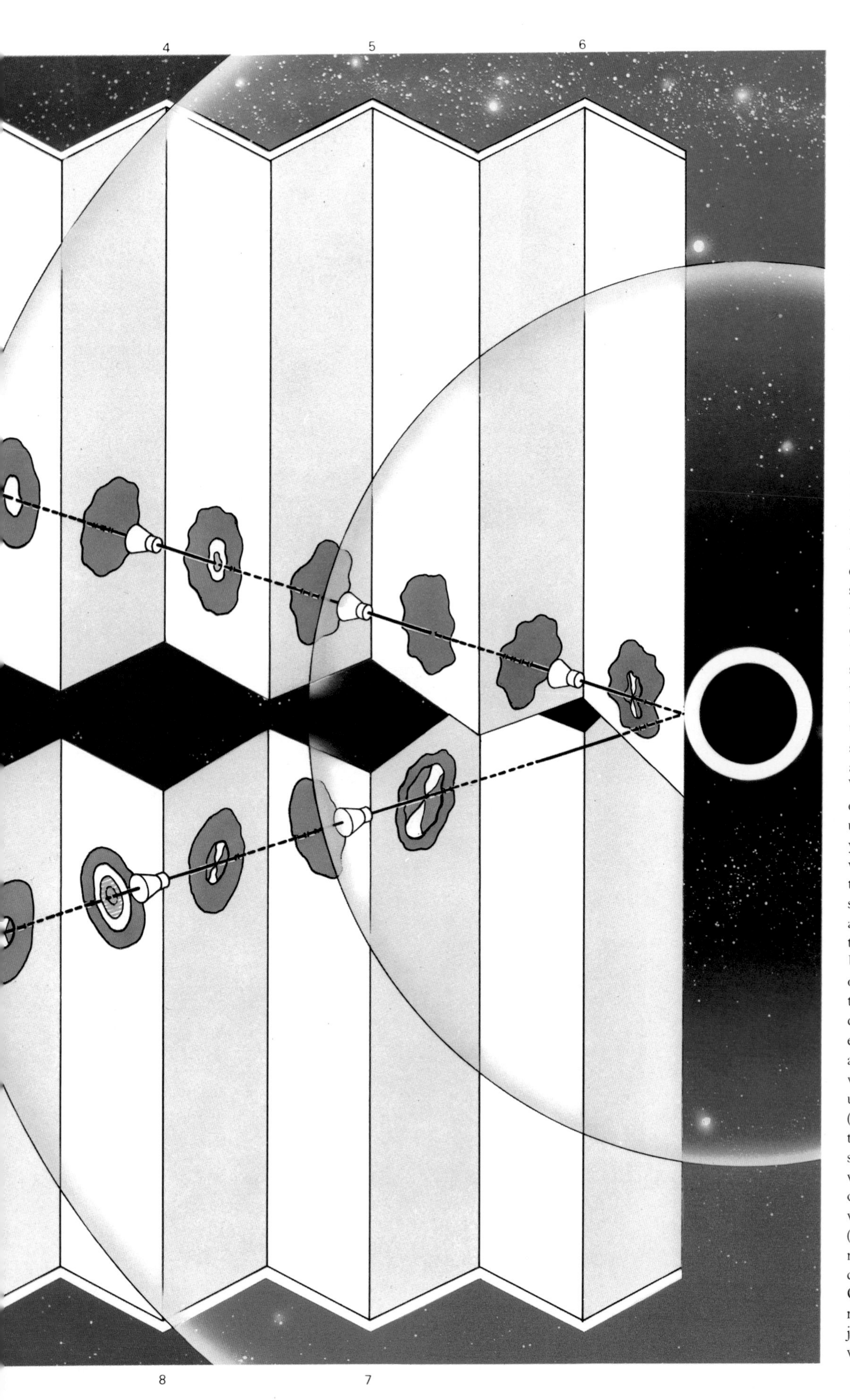

Journey through a rotating black hole into another universe. The circles show the views from the front and back of the spacecraft at successive positions. Several other universes come into view, each indicated here by a different colour. Far from the hole (1) we are surrounded by our Universe (blue) and see two universes within the outline of the black hole. These shrink until only one is visible, within a halo of 'trapped' light (2). At the outer event horizon (3), only the trapped light is seen, while further in (4) a fourth universe becomes visible. At the inner event horizon (5), we see our Universe in front again. The climax of the journey is a 'bounce' off the ring-shaped singularity at the hole's centre. Just before this, yet another universe (the sixth) is seen (6), and grows after the bounce.
We are now heading out of the hole, towards a universe which is not yet visible. The 'rear' view (now looking in the direction of the spacecraft's motion after the bounce) begins to show not just our Universe (8) but two other universes, and trapped light which dominates at the outer event horizon (9). Only after emerging from the white hole does the final universe become visible (10). Well away from the hole, this universe is seen all around: the white hole appears to contain our Universe wrapped round another (11). These purely mathematical calculations by Cunningham, would not apply to an actual journey: in reality, it would prove fatal.

rotate, the answer is fairly simple. It just collapses into a single point at the black hole's centre. Here, at the *singularity,* all the matter from the star-core is crushed by its own irresistible gravity into literally zero volume, into a mathematical point in space having no length, breadth or height. It is hard to comprehend, or even believe.

In fact, even scientists have some difficulty swallowing the concept of a pointlike singularity. Within Einstein's theory are built in signs that it will not work when the distortion of space-time becomes infinite, as it does at the singularity. So there is the unhappy situation that relativity predicts a situation with which it is not able to cope. This in no way affects the accuracy of relativity elsewhere; both inside and outside the event horizon it is undoubtedly our one true guide. The singularity, though, is a mystery still to be solved by the mathematicians, and its solution will without doubt provide us with more clues as to what those seemingly innocuous things, space and time, really are.

If the singularities do exist, as infinitely compressed points of matter, then at least they are safely tucked away behind event horizons. According to mathematics, singularities are 'pathological'—a word as fraught to physicists as to physicians—in the way they affect space and time, but as long as they are not visible in the real Universe there is no great cause for alarm. But if a black hole is rotating, its singularity may be visible from outside. Scientists still argue over this; the complex calculations continue; agreement has yet to be reached. On one side are those who believe that singularities in the real Universe are always modestly veiled by a surrounding black hole and event horizon. But they have yet to prove this theorem, the principle of 'cosmic censorship' (that nature always hides naked singularities) to everyone's satisfaction. The opposition says that unless cosmic censorship can be proved, we must accept that, in some rare cases at least, singularities may be visible. Our space-time may contain pathological kinks. And if they exist, they could be of more than academic interest. Some scientists have argued that the proven existence of singularities would upset the whole chain of 'cause' happening before 'effect', and as a result open the way to time travel. The anticipation is as exciting as it is somewhat unnerving.

WHITE HOLES

Matter disappears into a black hole, and is lost from our Universe. As it flows into the central singularity, it produces an infinitely distorted kink in space-time. And one of the most exciting proposals in black hole science is the possibility that the matter may re-emerge, not back through the event horizon, but into another universe.

We can know only one Universe—with its three dimensions of space stretching out to infinity, and the fourth dimension of time—but there may be other universes we cannot know, set in different dimensions. Einstein's theory is powerful enough to describe other universes mathematically, as easily as it can our own and, using relativity, matter disappearing into a black hole in our Universe could travel through an 'interuniverse tunnel' to emerge in another. There it would reappear in a brilliant fountain of matter and energy in empty space, apparently from nowhere.

If we accept the existence of such tunnels, then it is just as likely that matter reaches our Universe through the black holes of others. We should see these stardregs from another universe burst forth in a blaze of glory, from the aptly-named *white holes.*

Astronomers first called on white holes to explain the *quasar* mystery. Latest research, however, shows that quasars seem to be related to huge black holes, as the next chapter will discuss, although that in itself does not rule out the possibility that white holes exist—there must be many things in the Universe that astronomers have yet to find.

Unfortunately, scientists who have looked deeper into the theory of white holes, have come up with disappointing news. For a start, white holes cannot come into being spontaneously. Even though a black hole can form in a region of ordinary space when a star collapses, a white hole just does not happen out of the blue. Any white holes there are in our Universe must have formed right at its very beginning, during the Big Bang. But there are problems here, too. In order to survive, a white hole must continuously expand; otherwise it would collapse back into a black hole. A white hole expanding since the beginning of the Universe would now be far too big to account for the compact quasars, still the only places known where white holes could conceivably be spewing out matter and energy in quantity.

And the white hole's problems do not end there. Even if some have always existed, and are still around, they will not be the bright energetic fountains that astronomers thought at first. Recent calculations have revealed that light and other radiations from a white hole will have their wavelengths stretched—much like the light from the beacon falling into the black hole. And this stretch not only makes the light redder, but weakens its intensity. So instead of a marvellous, dazzling blaze of glory, the white hole would simply be dim and red; in close up appearing as a ring of light, but from a distance as an undistinguished reddish point of light, seemingly no different from millions of ordinary stars in the sky. Or, more likely, it may be too faint to be seen at all.

White holes currently have the thumbs-down from scientists. Although there is a chance there are some in our Universe, they will not be particularly exciting to find. Despite their initial glamorous reputation, it turns out that they would be fairly uninteresting, far tamer than their black brothers which underlie the excitement and fury of the X-ray stars and quasars

Into another universe

Tunnels between black holes in one universe and white holes in another are still a possibility. Some of the most interesting theories suggested by present day scientists envisage space explorers of the future exploiting these tunnels to travel from our Universe to others.

A simple nonspinning black hole would be no good for interuniverse travel. We have already seen the fate of the unfortunate explorer of such a hole, crushed inexorably by the one-way flow of space within, into the unseen central singularity. The entrance to the interuniverse tunnel must be a rotating, or an electrically charged, black hole, large enough for the intrepid explorer not to be spaghettified. The large black hole at the centre of a galaxy should be ideal.

Falling into and through the event horizon of a charged black hole, an interuniversal explorer would experience much the same sights as his unfortunate colleague who was unwise enough to venture into the nonspinning hole. The crew on his homeship (to which he could never return again) would seem to be living faster and faster, light coming in from outside would have its wavelengths shortened, so that everything outside appeared bluer. If he watched the homeship leave for Earth, it would, to his view, speed up precipitously fast. The scene outside would take place in faster and faster time, like a film being inexorably speeded up. Stars would die as he watched, and he would see new stars suddenly come to life. As in an astronomer's dreamworld, he could see stars of different masses born, live their different lifetimes, and die in their own ways—as gently expanding planetary nebulae or as explosive supernovae. Faster and faster, more rapidly behind him than in front, the Universe's future would unfold.

But what would he see out of his front window? Straightforward mathematical calculations suggest a bewildering variety of other universes, perhaps as many as six, in rings of various sizes interspersed with circles of brilliant light, light trapped by the singularity within the hole. Like a circular kaleidoscope, the images of the other universes would expand and contract from the point dead-ahead, in an astounding succession of colourful and distorted appearances and disappearances. And as he emerges from the far end of the 'tunnel', his new universe would expand to surround him, while our Universe would shrink behind, until it is only a small image lying within the white hole through which he has emerged.

And what he sees would never be reported back to Earth. For no one can tell where an interuniverse tunnel may lead. If our universe traveller entered a black hole in his new universe, there would be no guarantee it would connect with a white hole in ours. Among the infinite number of possible universes, it is almost certain he would end up somewhere else, fated endlessly to journey from universe to universe, seeing marvels vouchsafed to him alone.

If he did by chance end up in our Universe again, calculations indicate that he could return at any time in our Universe's history, not necessarily at the time he left. And this raises all the conundrums and paradoxes of time travel. For this reason alone, very many scientists are sceptical of interuniversal travel.

There are more straightforward objections to interuniverse tunnels, in fact, which have emerged recently. Where the traveller would see the far future of our Universe appear behind him—the point in the tunnel known as the inner horizon—strange things would happen. For a start, radiation from the Universe outside would be so shifted in wavelength that it would have enormously enhanced energy. The spaceship would be in such a concentrated region of energy that it would vaporize instantly. And quantum theory (which describes matter on the very small scale) predicts even worse conditions here: the structure of space breaks down on the microscopic scale.

At the inner horizon, the traveller would be destroyed by the jumble of space-time. In this confused mishmash, every atom in his body would be given a different direction and speed in space, a different sense of time: both his body and his spacecraft would be reduced to an utter chaos, broken down not just to atoms but to the ultimate constituents of which atoms themselves are made.

This bleak picture is the scene painted at present. Perhaps in the far future, though, man may learn to survive the jungle of the black hole's interior and perform his spacefaring exploits. The 'laws of Nature' are not immutable, after all, but merely summaries of our present knowledge. Even if black holes do prove to be impenetrable, there may be undreamed-of portals to other universes.

CHAPTER 10

QUASARS

Our Sun, and the stars we see in the night sky, are all part of the Milky Way, a huge galaxy consisting of 100,000 million stars, along with the gas from which new stars will be born. Although it is 100,000 light years across, the Milky Way inhabits only a tiny corner of the Universe. Beyond it stretch the huge empty reaches of intergalactic space.

Our nearest important neighbour is the Andromeda Galaxy, two million light years away, a flattened spiral disc of stars and gas, very like the Milky Way. Andromeda and the Milky Way are the two major galaxies in our Local Group: in proportion, they are spaced like two dinner plates at opposite sides of a room, while around each one, rather like house flies, swarm some dozens of smaller galaxies.

Farther out into space, there are more and more galaxies, mounting up into tens of millions. Some are spirals, like the Milky Way and Andromeda; others are irregular-shaped relatives of the dwarf galaxies that swarm near us. As numerous as the spirals are oval-shaped *elliptical galaxies,* which cover a whole range of sizes. They are distinguished by their plainness, having neither beautiful spiral arms nor reserves of gas between the stars to make the glowing nebulae where stars are born. Farther out in space, too, are more exciting, more energetic objects, seemingly very different from ordinary galaxies. The most mysterious and frightening of these objects are quasars.

The quasar mystery began in the early 1960s. Radio astronomers surveying the sky for natural broadcasters, had asked optical astronomers to photograph the sky with large telescopes wherever a source lay, in the hope of finding out just what was producing the radio waves. Already they had identified some sources within our Galaxy, including nebulae where stars are born and the clouds of gas thrown out by dying stars. To their surprise, some of these sources turned out to be distant galaxies, emitting a tremendous quantity of waves: these *radio galaxies* will reappear later in this chapter, for they hold an important clue in the quasar story.

No one had expected ordinary stars to emit measurable amounts of radio waves, so it came as something of a shock in 1962 when a star was found to coincide with a powerful radio source. Australian astronomers pinpointed a source called 3C 273 by timing precisely when the Moon passed in front and cut off its waves. Although 3C 273 looked like a star in photographs, it had a short 'jet' of light extending from it, something never seen near a star before.

Surprise turned to puzzlement when the light from 3C 273 was examined in detail. When starlight is spread out into a spectrum—the rainbow of colours corresponding to different wavelengths of light—there is usually a characteristic pattern. The spectrum is bright all along, from blue to red, except for particular individual wavelengths where the bright band is crossed by narrow dark lines, caused by atoms in the star which absorb specific wavelengths. Each element produces its own unique set of lines at particular wavelengths, like a fingerprint, so by investigating spectral lines it is possible to say what elements are present in the star—and by refined analysis, estimate its temperature and other important properties.

Astronomers expected the spectrum of 3C 273 to reveal what kind of star it was. The first surprise was that the lines stood out brighter than the rest of the spectrum. This was not unheard of, though, since very hot stars do have bright-line spectra. But the wavelengths of the lines did not correspond to those of any known element. This *was* baffling: All stars are made of the same elements that we know on Earth, so what was causing these strange wavelenths of emission?

Early in 1963, Dutch-American astronomer Maarten Schmidt hit on the answer. The spectrum was the fingerprint of hydrogen, but all the lines were systematically shifted along the spectrum towards the red (long-wavelength) end. Schmidt found that each hydrogen line had been moved in wavelength by 16 percent. By convention, astronomers express this wavelength change in decimal form, so that 3C 273 'has a redshift of 0.16'. This shift had made the hydrogen spectrum unrecognizable at first sight, and it was now realized that three other 'stars' previously identified as radio sources were similar to 3C 273—their spectra, too, had defied interpretation, because their lines were, if anything, even more shifted.

Explaining the spectra only made these 'stars' more enigmatic. Although the shift differs from one to another, they are always towards the red end of the spectrum, never towards the blue (short-wavelength) end. And it is difficult to explain why the lines should have shifted so much. One straightforward interpretation is that the 'stars' are travelling away from us at very high speeds. When radiation comes to us from any source that is moving away, the wavelengths are stretched out, so that the radiation we pick up has all the wavelengths increased in proportion. With light, the effect shows as a redshift. But the speeds implied are tremendous: 3C 273 is shooting away at almost 50,000 kilometres per *second*; in everyday terms, 170,000,000 kilometres per hour (100,000,000 miles per hour). And 3C 273 is among the slowest; the object known as OQ 172 has a redshift of 3.53, which means that it is moving at 95 percent the speed of light. The waves from OQ 172 are stretched by 353 percent so that the light we receive at the Earth actually started off as short-wavelength—and invisible—ultraviolet radiation.

There are two ways that a 'star' could have such a high velocity. Our Galaxy could have shot out unusual stars at tremendous speeds. Although this could explain the first few, over 1500 of these objects have now been catalogued, and it is estimated that an extensive search by large telescopes would turn up over a million. Our Galaxy just could not have the power to fling out so many stars with speeds approaching that of light. Besides we live in a very ordinary galaxy; according to this explanation, we should see similar objects spraying out from other nearby galaxies, and some of these would be coming towards us, and so have their wavelengths *shortened*—a

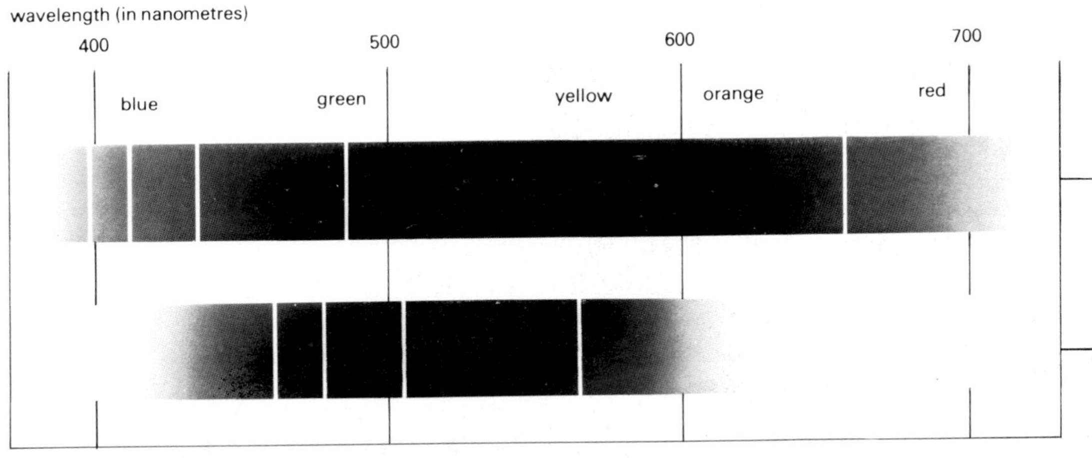

The spectrum of the quasar 3C 273 (*left*, lower diagram) has a set of bright emission lines. These do not fall at any expected wavelengths, but they form the pattern characteristic of four of the lines of the hydrogen spectrum (upper diagram). The 3C 273 spectrum is clearly a hydrogen spectrum with all the wavelengths increased by 16 percent: this was confirmed when the usual red hydrogen was found in the infrared region of the quasar's spectrum.

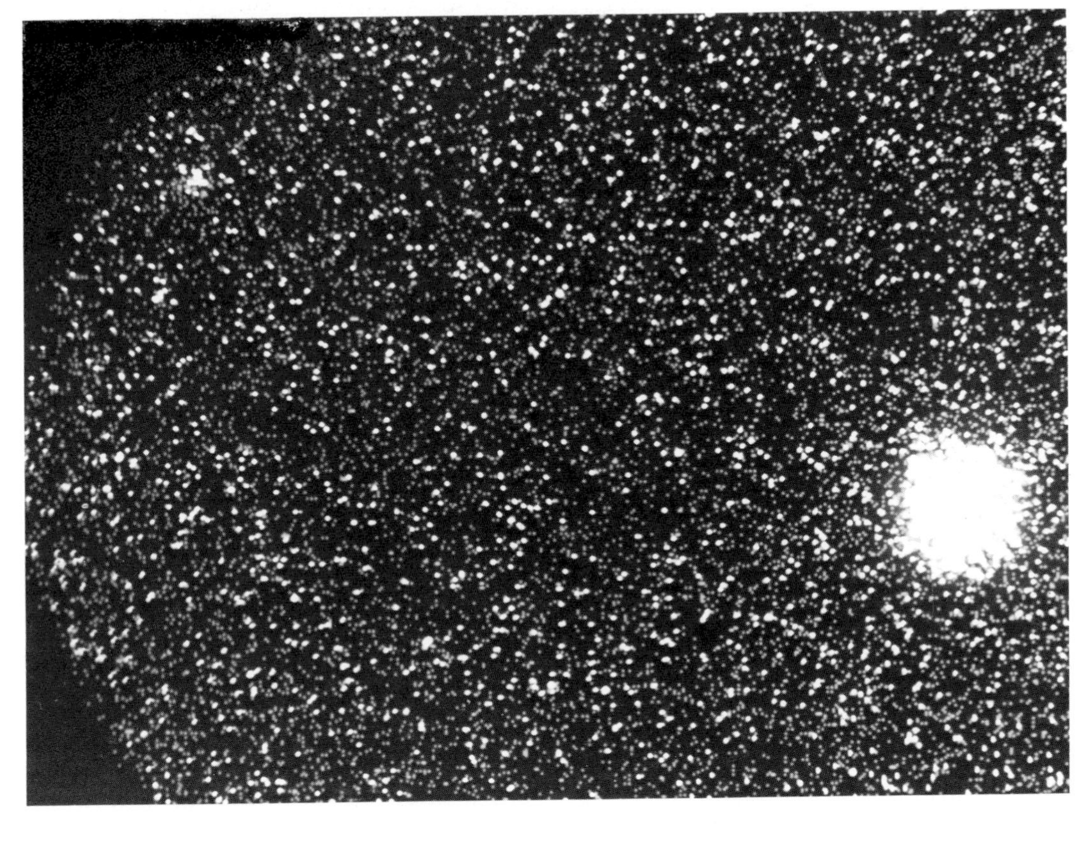

The quasar 3C 273 is a strong source of X-rays as well as light and radio waves. In the long exposure 'photograph' from the Einstein X-ray observatory (*left*) the image of 3C 273 is overexposed. The faint group of dots (near the top left edge) is X-ray emission from a very distant, previously unknown quasar.

Modern electronic light detectors can analyze the light from even very faint quasars into a spectrum. One of the most sensitive is the Image Photon Counting System (IPCS) devised by Alec Boksenberg. The IPCS can take the very dim light from the second most distant quasar known, OH 471 (*right*, arrowed), and produce a remarkably detailed spectrum (across bottom of page). The height of this trace indicates the number of particles of light (photons) at each wavelength. The large peak at the right is Lyman-alpha light from hydrogen: although this is normally an ultraviolet spectral line, the high redshift of OH 471 has stretched the radiation to a green-yellow wavelength.

blueshifted spectrum. None has ever been seen.

But there is a more 'conventional' explanation. The entire Universe is expanding, as we will see in the next chapter. The farther away a galaxy lies from us, the faster it is carried away by the expansion of space. So all galaxies (except the very nearest) have their spectra redshifted, and the redshift increases in proportion to distance. The simplest possibility is that these 'stars' lie far, far beyond our Galaxy, and that their tremendous speeds occur just because they are riding on the expansion of the Universe. In that case, their redshifts indicate how far away they are. And the distances are astounding. 3C 273 is 2000 million years from the Milky Way, while the high-redshift OQ 172 is 12,000 million light years away.

Such enormous distances mean that these objects must be intrinsically, incredibly luminous. Take even a giant galaxy, 10 times brighter than Andromeda, to the distance of OQ 172 and it would be so dim that the world's largest telescopes could not detect it. Yet OQ 172 is easily seen. Objects like it must be a hundred or more times brighter than a galaxy, although they are far smaller. They

The first stage of the IPCS is a sensitive television camera (*right*) which is fitted to a large telescope (not shown). The telescope gathers light from the quasar, and a spectroscope spreads the light into a spectrum. This is too faint to be photographed in the usual way. Instead, it is scrutinized by the IPCS camera.

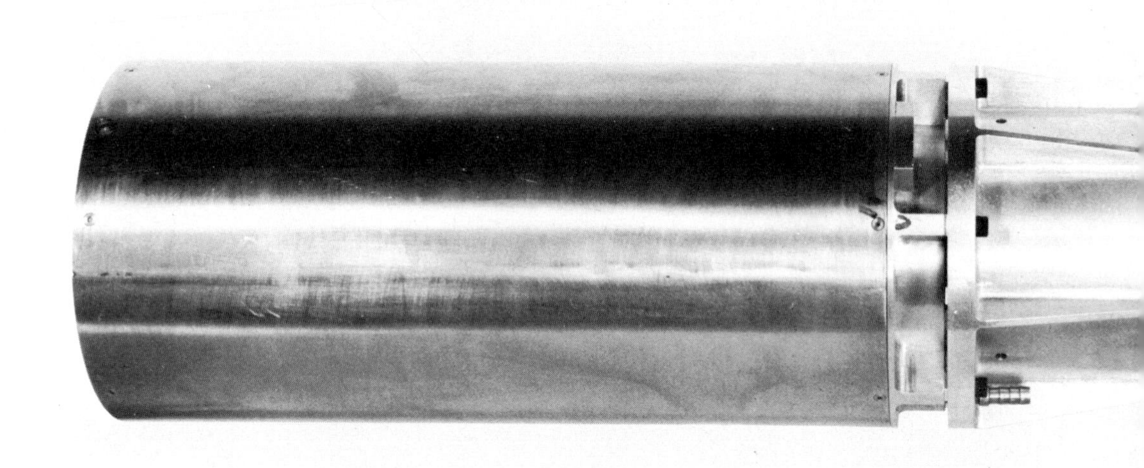

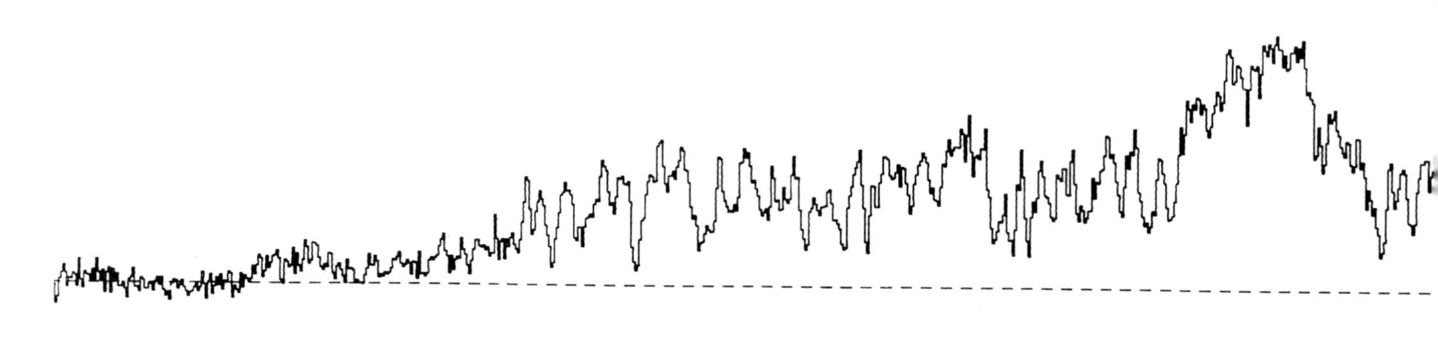

The raw output from the IPCS camera is an array of 'blips' (*above*). Some of these are caused by photons from the quasar, as spread into the spectrum, but many are 'noise', caused by stray electrons or gas atoms in the camera. A computer (*right*) looks at each 'blip' and rejects those whose shape indicates they are 'noise'.

The computer analyzes the positions of the remaining 'blips'— those caused by light in the quasar spectrum. It counts the number of photons at each point along the spectrum, and plots these as a graph (*below*) which is not only detailed but also gives the exact intensity at each point.

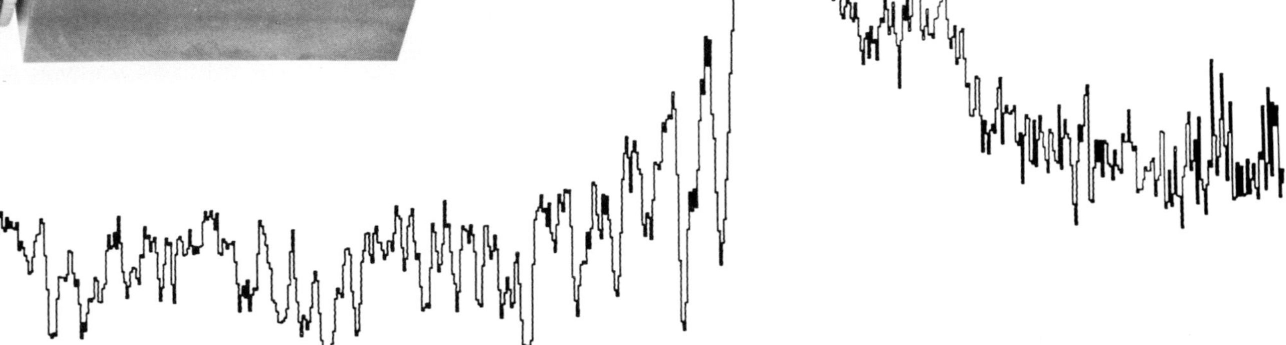

show as 'star-like' points of light and, as we shall see below, we have indirect reasons to believe that they are considerably less than a light-year across, less than a millionth the width of a large galaxy. But we are talking of the light of *10 million million* Suns, coming from a region little larger than the solar system!

In the 1960s, many astronomers baulked at such a proposition. Could there be an alternative explanation for the redshift? A controversy sprang up that was to dominate astronomy for a decade, and which still lingers today. At the same time, an agreed name was settled on for these phenomena. Originally they were *quasi-stellar radio sources*—'quasi-stellar' meaning star-like in their appearance. But Schmidt's colleague Allan Sandage soon found that for every one of these redshifted 'stars' that is detectable by a radio telescope, there are hundreds which do not emit radio waves, so the name *quasi-stellar object* took over, commonly shortened to QSO or quasar.

THE REDSHIFT CONTROVERSY

There have always been problems measuring distances directly. A long line of reasoning and precise determination of positions and spectral characteristics has led to reliable distances being estimated for the stars in our Galaxy, and for ordinary nearby galaxies where individual stars can be seen. By comparison with these neighbours, farther galaxies can be plumbed. For those distant galaxies, astronomers have found the law relating distance to redshift in the expanding Universe, so that they now can use the law simply to convert from a galaxy's redshift to its distance. There is no *direct* way, however, to measure the distance to quasars. Most astronomers have now accepted that the expansion of the Universe is the only known way to produce high redshifts, that the quasars must be far off, and we can therefore find their distances from the redshift exactly as we do for distant galaxies.

There are, however, some dissenters. Quasars are strange by any standard, so it could be argued that some process is occurring inside a quasar to shift all the light to longer wavelengths. In this case, they could lie comparatively near to us in space, among the general run of galaxies. What we need, to prove the case one way or the other, is an independent way of measuring quasar distances. American astronomer Halton 'Chip' Arp claims that he has done this by looking for quasars associated with galaxies whose distances we know. And he believes he has found that quasars are indeed nearer to us than their redshifts would have us believe.

Arp has for long been studying peculiar galaxies and small groups of galaxies, and in some groups he has discovered one galaxy with a redshift very different from the others. Since the galaxies lie in a group, they are presumably all at the same distance from us, and should have the same redshift. Arp concludes that some unknown process gives the odd galaxy an additional redshift.

A galaxy's redshift is measured from the light of all its stars, however, and it is very difficult to accept Arp's assertion that each individual star in a particular galaxy has its spectrum redshifted by some totally unknown process, while the stars in a closely-neighbouring galaxy are completely unaffected. And Arp can offer no explanation within the known laws of science for a process that shifts wavelengths equally for all the stars in a galaxy, spread over thousands of light years of space.

Conventional astronomers explain these results in two ways. In some groups, the gravitational tug of the other galaxies has fortuitously coincided to give one member a hefty kick, so its velocity—and hence its redshift—differs from the rest. In others, the odd galaxy is not a member of the group at all, but lies either much nearer or much farther away, and appears superimposed on the group. It has a different redshift simply because it is at a different distance from us.

Chip Arp is still convinced, although he stands almost alone in believing, that galaxies can have 'anomalous' redshifts. When it comes to quasars, however, he has company. He has marshalled some impressive photographs showing a high-redshift quasar apparently in close association with a galaxy whose low redshift shows that it is not far away from us. In some cases, the quasar appears to lie in a galaxy's spiral arms, or to be joined to the galaxy by a 'filament' of gas or stars; in others, several quasars may lie in a line straddling the nearby galaxy. If these associations are real, then these quasars at least must be nearer than their redshifts indicate. Something within the compact object is lengthening the wavelengths of the light it emits.

British astrophysicist Geoffrey Burbidge has produced statistics which back up Arp's claim. He has looked at the location of quasars in the sky, and found that an unexpectedly large number of them seem to lie near galaxies of low-redshift, rather than being scattered at random. He also cites a study by radio astronomers who have looked for close pairs of quasars. From previous studies of the number of quasars scattered over the sky, they expected to find only one case, if any, of a nearby quasar (low redshift) and a distant (high redshift) quasar lined up by chance to look like a close pair in the sky. In fact, they

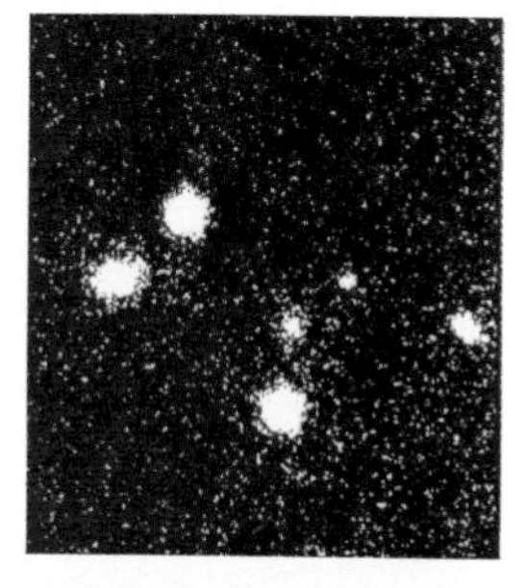

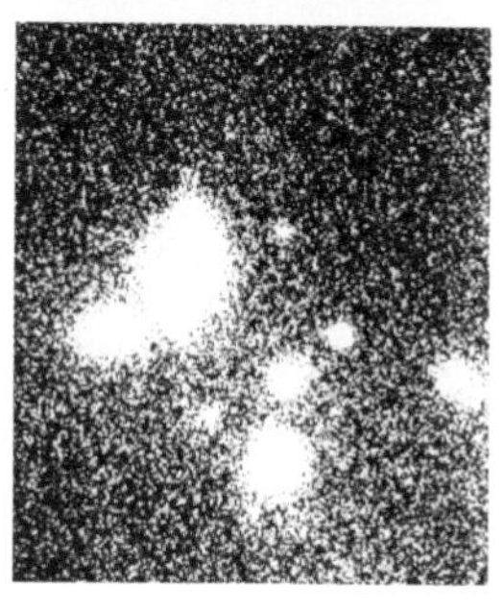

On a short exposure photograph (*above*) a quasar appears as a point of light; but a longer exposure (*below*) shows the larger image of the surrounding galaxy.

Quasars are almost certainly the violent centres of galaxies, more extreme versions of radio galaxies like Centaurus A (*below*). The milder explosions here have not produced a superluminous centre, but have ejected a rapidly expanding belt of gas and dust.

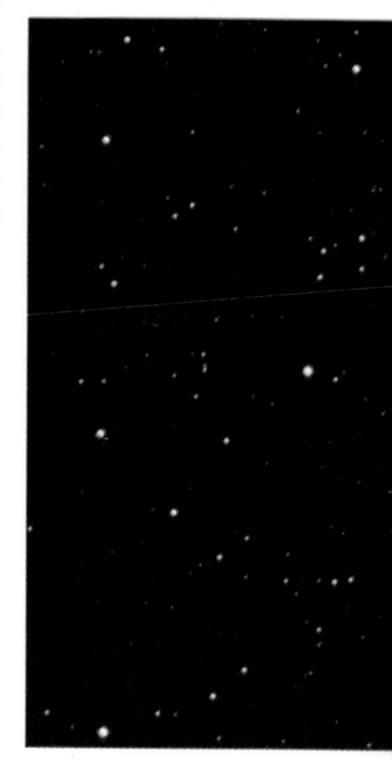

found five pairs. The odds against this number occurring by chance are about 1 in 10,000, so Burbidge concludes that most of these cannot be chance linings-up; they must be real pairs of quasars, lying at the same distance from us, but having different redshifts. Accepting this reasoning, one must accept that one, or both, of them has a redshift which does *not* indicate its distance.

In 1980, Arp and his colleague Cyril Hazard found the even odder case of two triplets of quasars, close to each other in the sky. In each triplet, the central quasar has a comparatively low redshift (0.5), while the flanking quasars have a much larger one, around 2. On the conventional view, it is an almost unbelievable coincidence that four distinct quasars should happen to lie in pairs behind two nearby quasars.

Unlike Arp's galaxies with discrepant redshifts, an anomalous quasar redshift *can* be explained within our current knowledge of the Universe. By anyone's reckoning, quasars are very small, and it is generally agreed that they have strong gravitational fields at the centre. As light tries to escape from a region of strong gravitation, it is robbed of energy, and as a result its wavelength is increased: it is redshifted. Arp's quasars could thus be very heavy black holes which stray near a galaxy. Out in space they were invisible; now they draw in gas from the galaxy's outskirts, and the gas shines briefly just before it falls into the hole. The light is redshifted as it fights its way up through the outer part of the hole's gravitational field.

Attractive though the idea is, the spectrum of light one would expect from such gas is very different from a quasar spectrum. (We will see below that conventional astronomers do now invoke black holes in their explanation of quasars, but in this scenario the gas is much further out and its light is not redshifted by the hole's gravitation.) And Arp's conclusions are weakened by recent observations that show that some quasars *do* lie at distances indicated by their redshift. The astronomers who follow Arp are therefore forced to conclude that there are two kinds of quasar: those lying close to nearby galaxies, which produce their own redshift; and those which are very distant, and whose redshift is due to the expanding Universe.

But there are no legitimate reasons for dividing quasars into two classes. Arp's quasars and the others all have the same kind of spectrum (apart from the redshift), and the same properties when observed with radio telescopes. And more and more quasars have turned out to be associated with galaxies *of the same redshift*. Since even Arp believes that a galaxy's redshift is, in general, a good indicator of its distance, these cases show that a quasar's redshift, too, is a good distance indicator. Virtually all quasars had seemed isolated when discovered, simply because they are relatively rare in space. They are widely spaced apart, and even the nearest ones are billions of light years away. They show up because they are so bright, but associated galaxies would be at the limits of present-day telescopes. During the 1970s, though, electronic image-intensifiers increased the sensitivity of telescopes to faint objects.

Astronomers can now look for galaxies around the nearest quasars—and they are regularly spotting them. These galaxies do, indeed, have the same redshift as the quasar lying near them. The original quasar, 3C 273, for example, has a galaxy lying nearby with a redshift exactly the same as the quasar's.

These techniques also show that a typical quasar is surrounded by a galaxy of stars. Because it lies so far off, the galaxy itself does not generally show on photographs, which reveal only the overpowering light from the central quasar. So a quasar is not an isolated object in space, but the central nucleus of what is otherwise an ordinary galaxy. As such, it is not quite as exotic as it once seemed. We know of many types of active galactic nuclei: galaxy centres emitting huge amounts of radiation—radio waves, X-rays and light. These galaxies are undoubtedly at the distances their redshifts indicate; and the quasars fit in quite naturally at the most violent extreme of these active galaxies.

The redshift controversy was *the* biggest astronomical puzzle of the late 1960s and early 1970s, but it is now generally accepted that quasars are very distant, and very luminous. But Arp and Burbidge maintain their unconventional views. Burbidge's statistics are certainly impressive, and Arp keeps on

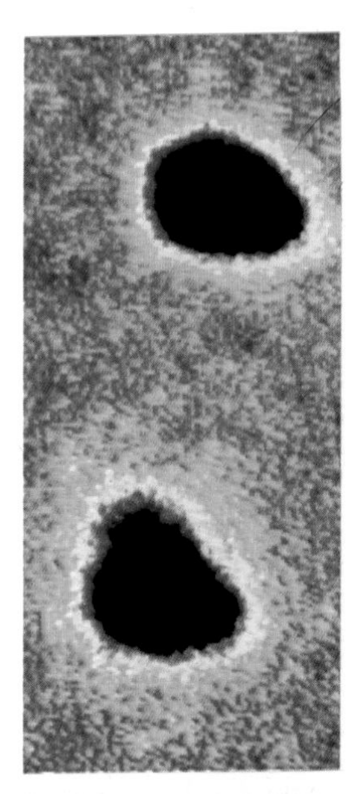

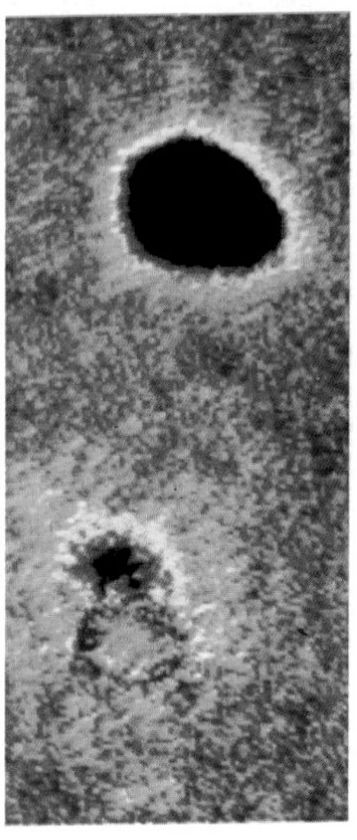

The 'double quasar', seen here as a false-colour negative (*top*), is, in fact, two images of the same quasar. Its light is focused into two separate images by the gravitation of a galaxy which lies between the quasar and us (see diagram overleaf). In the second picture (*above*), the lower quasar image has been removed by 'substracting' an image identical to the upper quasar image. The image which now appears just above the lower quasar position is the galaxy which produces the focusing.

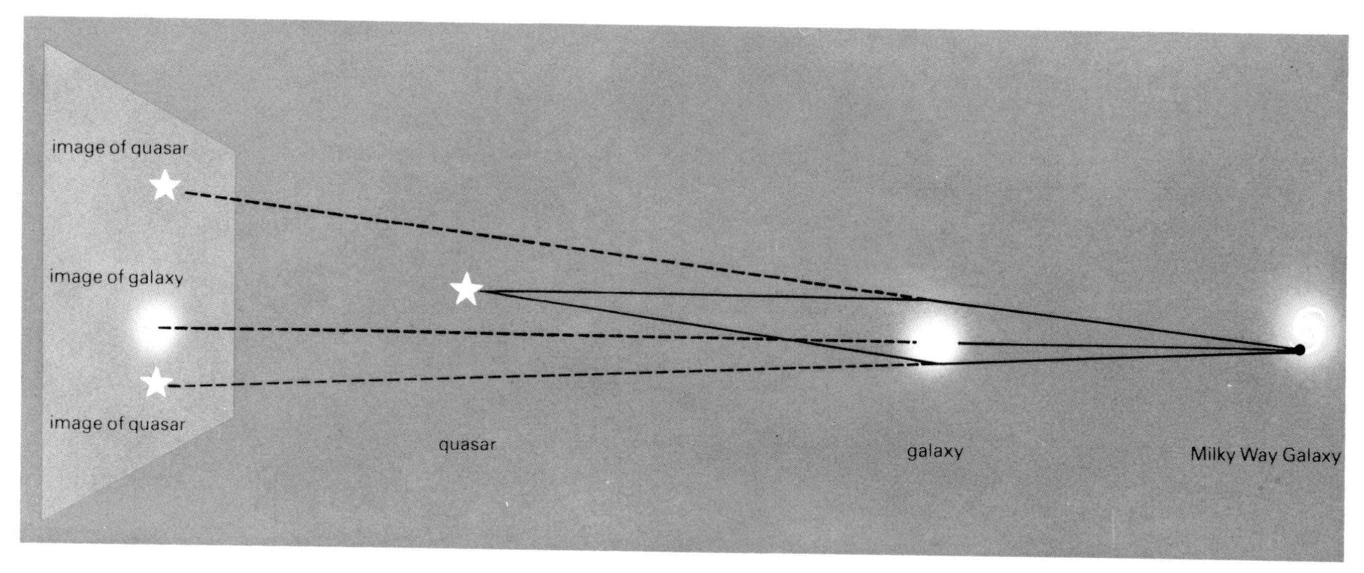

The double quasar effect is caused by the focusing effect of a galaxy lying almost in line with the quasar. Even though the quasar is the more distant object, it is so much brighter than the galaxy that it is easier to see the two quasar images than that of the galaxy. The double nature of a gravitational image also appeared in the calculation of the distortion of Saturn's image by a black hole (pages 136–7): strictly speaking, an extended object like a galaxy should produce three or more images, and astronomers believe the gravitational pull of a surrounding cluster of galaxies must be involved in reducing the number of images to two.

finding disturbing cases of high-redshift quasars appearing to lie near low-redshift galaxies or quasars. So for some quasars at least, the last word may yet to be said. If even one of Arp's 'discrepant redshift pairs' *is* genuine, then a whole new range of theories will be needed, theories that could possibly contravene what we presently understand to be the laws of nature.

THE DOUBLE QUASAR

In 1979, a remarkable quasar pair turned up; remarkable not because their redshifts were different, but because they were exactly the same. Dennis Walsh and his colleagues at Jodrell Bank had for several years been identifying radio sources. One, known as 0957 + 561, turned out to be not just a point of light like most quasars, but two 'stars' very close together, separated by less than one-three hundredth the apparent diameter of the Moon.

The pair could have been a distant quasar and a star in our own Galaxy which happened by chance to lie close to each other in the sky; or they could have been two quasars at different distances, again in a chance alignment. Walsh and his colleagues took the logical next step of dissecting the light from each into a spectrum, using the 2.1 metre (84 inch) telescope at Kitt Peak in Arizona.

They were startled at the result. Each 'star' was a quasar; but remarkably enough the two spectra were practically identical—as if they had looked twice at the same object. Because each had the same redshift (1.41), the obvious interpretation was that they were two quasars at the same distance, orbiting around each other. But the redshifts were *too* similar even for this. If the quasars were in orbit, their orbital speed would have affected their redshifts, increasing one slightly and decreasing the other in proportion. Besides, they were similar in other ways. Quasar spectra cover a wide range of types, and it was too much of a coincidence that two quasars keeping company should have exactly the same spectral features. Another coincidence was that the two were well-matched in brightness, and quasars come in an enormous range of intrinsic brightness. When the fine detail of 0957 + 561 was probed, it was found that the quasars were producing similar amounts of radio waves. The 'double quasar' must, in fact, be two images of the same object.

There is only one reasonable explanation: Between this distant quasar and us lies some object with a powerful gravitational field. Gravity can bend light (and radio waves) and, unlike an ordinary glass lens, a 'gravitational lens' focuses light into two or more separate images. Albert Einstein had predicted the focusing effect of gravitation in his general theory of relativity, and it was only appropriate that the double quasar was found in his centenary year.

The intervening object acting as the gravitational lens was most likely to be a galaxy. Remembering how luminous quasars are, it is quite plausible that a galaxy could lie closer to us than the quasar, and yet be outshone by the quasar's light. And so it turned out. Astronomers using an electronic light detector on the huge 5 metre (200 inch) telescope on Palomar Mountain in California discovered the fuzzy image of a galaxy lying between the two quasar images. This is exactly what the gravitational lens theory had predicted, but was an almost impossible coincidence otherwise. The galaxy has a redshift of 0.39, so it lies 3000 million light years away, about one-third the distance to the quasar.

The galaxy must be far more massive than our own, for it is the brightest in a whole cluster. It is the gravitational pull of both the galaxy itself and the surrounding cluster that focuses the quasar's light into the two images.

The 'double quasar' is thus in itself a pretty undistinguished beast, no different in type from a thousand other quasars. Its light just happens to pass close to a heavy galaxy in a cluster of galaxies on its way to the Milky Way, so making it, for us, such a very unusual sight.

RADIO GALAXIES

Quasars baffled optical astronomers at first acquaintance, because they were so incredibly bright. Radio astronomers took the news more calmly. Ten years earlier they had had a similar shock when they had found that some galaxies poured out huge amounts of radio waves. These *radio galaxies* are now seen as first cousins to quasars: in both cases, the central nucleus produces vast amounts of energy. And the radio galaxies undoubtedly hold clues to how the quasars work.

Quite apart from their sheer power (their radio output equals the light from a million million stars) radio galaxies are remarkable because their broadcasts do not come from the centre of the galaxy, but are produced in two huge regions, one either side, way out in space beyond the point where the stars of the galaxy thin out to nothing. Out there are enormous 'bags' of magnetic field, thousands of light years across, entrapping the fast-moving electrons that produce the radio waves. The amount of energy stored is immense: between them, the magnetic field and electrons contain as much energy as a quasar would produce over a period of a million years.

This energy cannot originate in empty space. It must come from the galaxy's centre, and virtually all astronomers now agree that it is shot out in two directly opposing 'beams' of energy from the nucleus right out into the near-vacuum beyond the galaxy's confines. The beams can travel enormous, unimaginable distances. The radio galaxy 3C 236 is 14 million light years across, measured from one end of its radio-emitting region to the other, across the galaxy itself.

The existence of such tremendous energy in radio galaxies has long been accepted and it is entirely plausible that quasars are the energetic phase of the nucleus's life when it pumps energy out into its magnetic sidepacks. The link looks strong because many quasars, too, have twin regions of radio emission way out in space on either side.

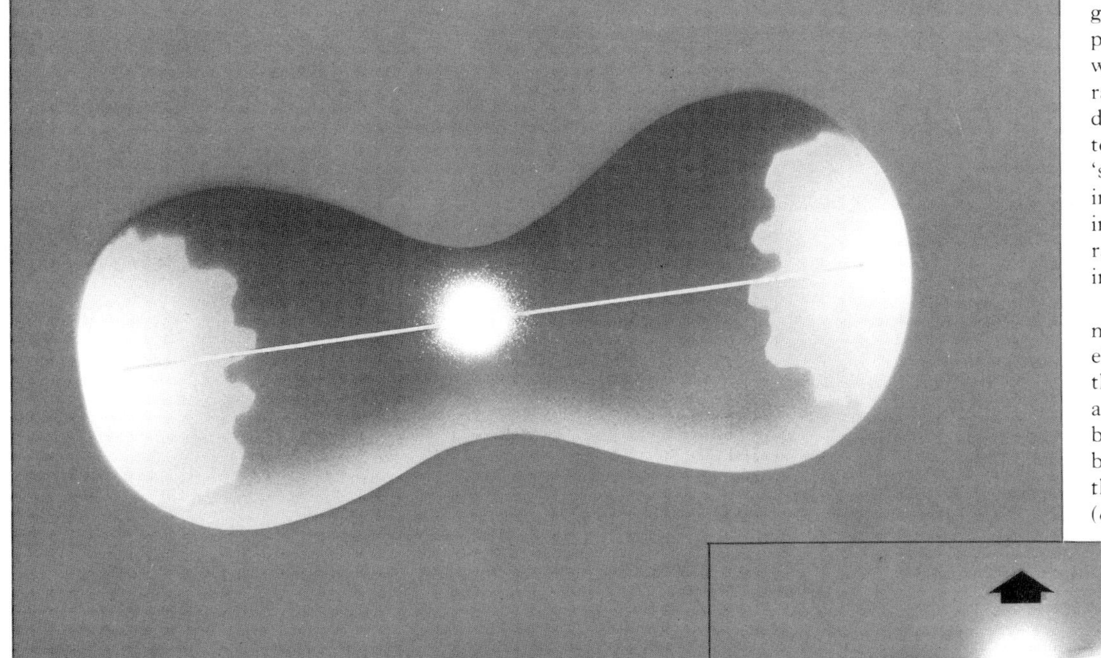

A radio galaxy has two huge 'lobes' which emit radio waves, and which must somehow be powered by the galaxy's active core. The favourite interpretation (*left*) is that the core shoots out two 'beams' of high speed electrons in opposite directions. These are almost invisible to both optical and radio telescopes until they hit the gas which surrounds the galaxy. Here they produce a 'hot spot' which is very bright in radio waves, and shines dimly in visible light, too. The electrons 'splash back', and inflate an emptier cavity in the intergalactic gas, rather like a double hole in a Swiss cheese.

If the galaxy is moving (*below*), the electron beam may hit the side of the cavity at a glancing angle, and become visible as a bright straight 'jet' of the kind seen in M87 (*overleaf*).

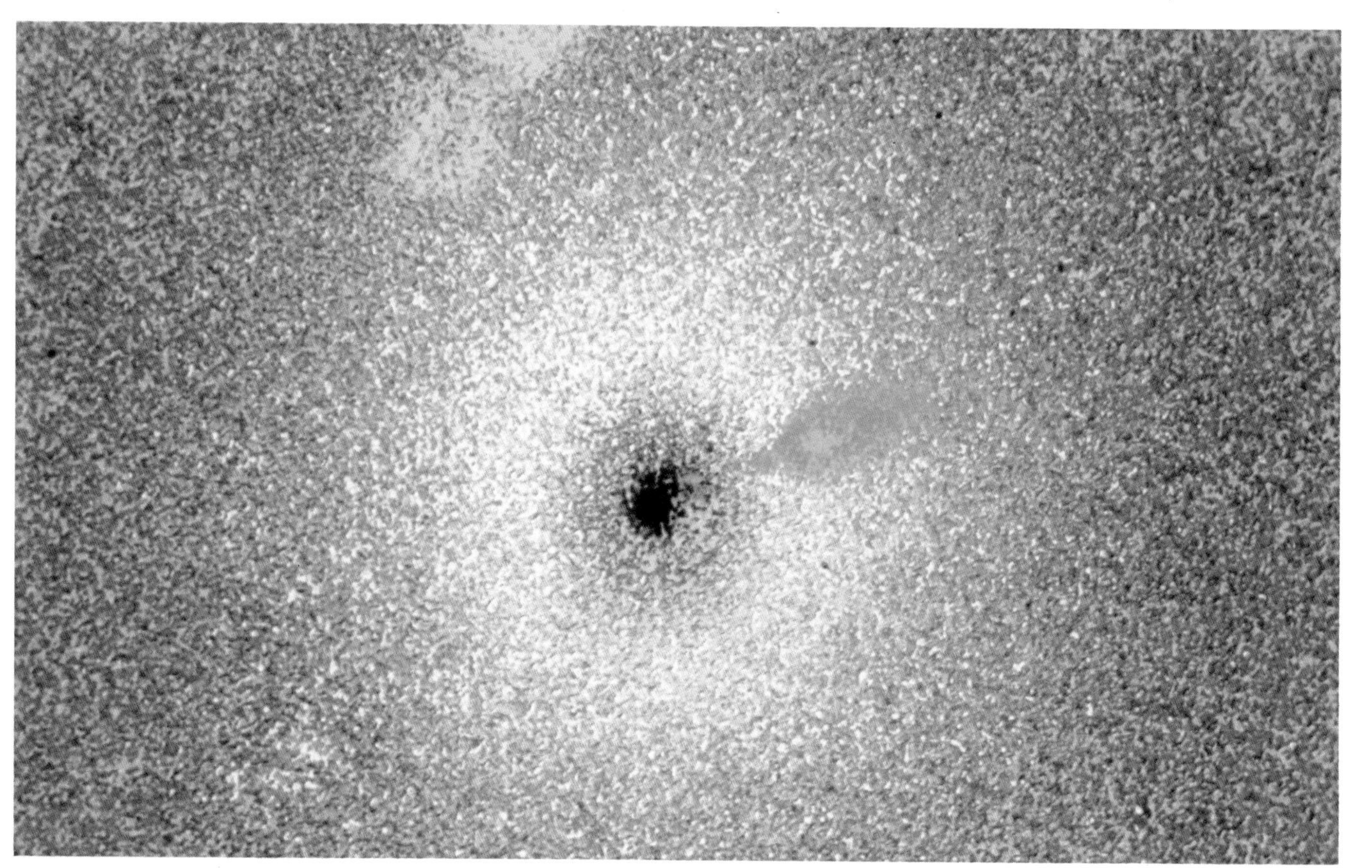

In a false-colour picture which codes different brightness levels in different colours, the jet in the galaxy M87 (*above*) stands out as an elongated blue streak reaching out from the centre. It is probably a beam of high-speed electrons shot out from the galaxy's active core.

Quasar 3C 345 may, in fact, be even larger than the radio galaxy 3C 236. Measurements with the world's largest steerable radio telescope, at Effelsberg in West Germany, suggest it is 65 million light years across, making it by far the largest single object yet found in the Universe.

There are many radio galaxies lying nearer to us than the quasars, though, and they can tell us much that we cannot see in the distant quasars. For example, sensitive radio observations often show up a straight 'jet' linking the nucleus to the outer magnetic bag. This is undoubtedly the beam of energy travelling outwards, probably made up of fast electrons. Where this jet of gas hits the thin gas which fills even the near-vacuum of intergalactic space, it is stopped dead, and splashes back to inflate the magnetic bag in space. The loose electrons in this bag produce the radio waves that our radio telescopes detect.

Radio astronomers have found that this central jet, on occasion, appears to shoot outwards faster than the speed of light—an impossibility according to the theory of relativity. There is no agreed explanation yet, but it is probably only an illusion, perhaps caused by our viewing the jet at an unusual angle.

One of the nearest radio galaxies lies 50 million light years away, in the heart of a huge cluster of galaxies called the Virgo Cluster. The galaxy itself is known as M87, and is a giant, one of the most massive known, containing an estimated million million stars. A powerful radio-emitting jet emerging from

Active galaxies

Type	Is a result of activity in	Optical brightness of galaxy core (in billions of Suns)	can brighten up by:	over a period of
Seyfert	Spiral galaxy	1–1000	½	months/years
Radio galaxy	elliptical galaxy	no optical core	—	—
Quasar	giant elliptical galaxy?	100–100,000	10 times	months
BL Lac object	giant elliptical galaxy	100–100,000	100 times	days/week

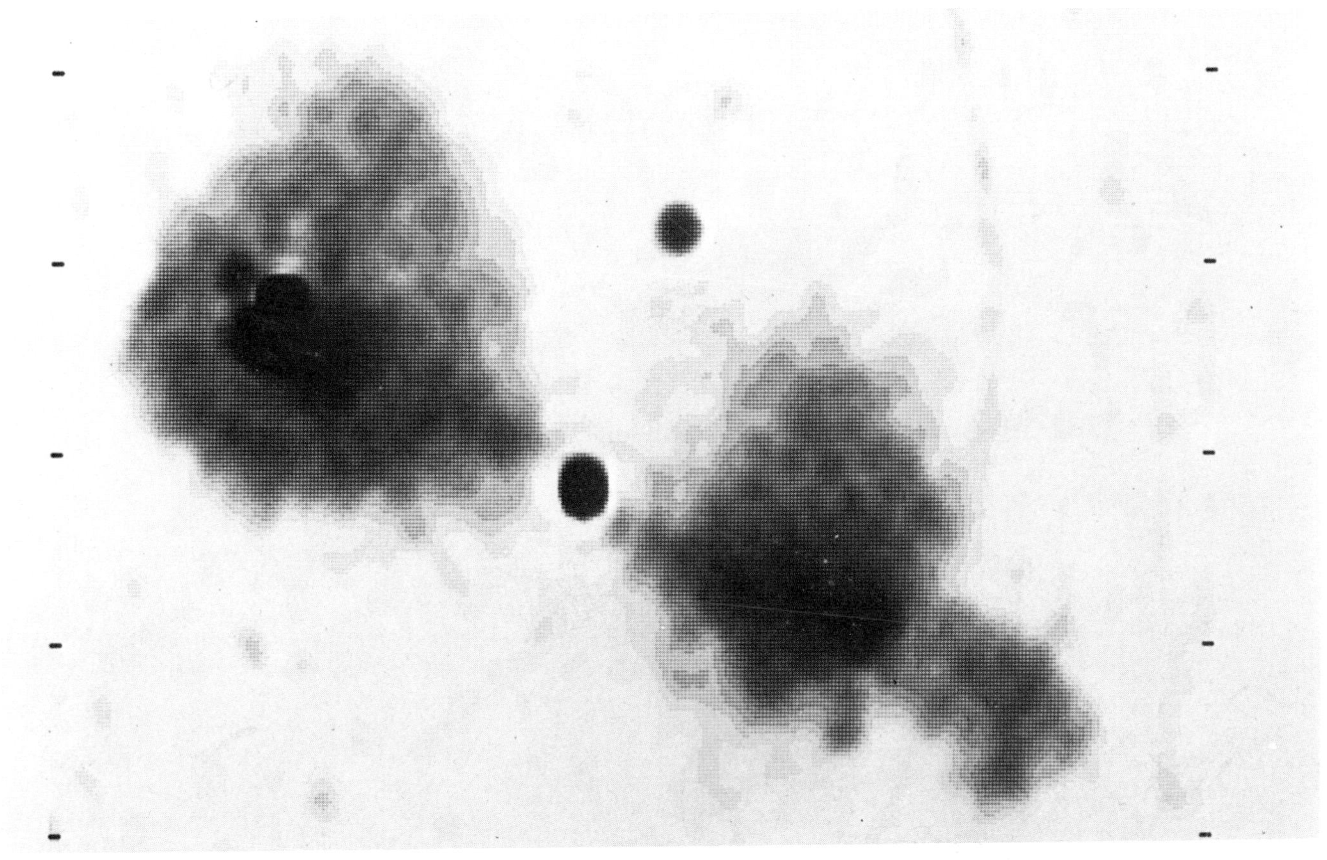

Giant radio galaxy DA 240, shown here in a radio photograph (*above*), is one of the largest structures known. Its radio-emitting lobes stretch across five million light years of space—more than twice the distance between us and the Andromeda Galaxy.

the nucleus has been identified. Back in 1917, in fact, the American astronomer Heber Curtis had photographed M87, and noticed a jet of light, which no one at the time could explain. We now know it coincides with the radio jet: M87 is one of the few cases where the electron beams from a galaxy nucleus generate light as well as radio waves.

The first quasar to be identified, 3C 273, also has a jet which shows up in ordinary photographs. Enigmatic at the time, it also coincides with a radio jet where 3C 273 is beaming fast electrons into space. It underlies the essential similarity between radio galaxies and quasars, the only major difference being that the centres of the latter produce a tremendous amount of light, while the former do not.

The radio galaxies provide a vital clue to the nuclei of active galaxies and quasars: they prove that the nuclei have 'memories'. The outer radio-emitting 'bags' in 3C 236 are the result of jets that left the galaxy's centre in an outburst almost a billion years ago. This outburst seems to have ceased and the magnetic bags with their electron contents are gradually fading away. But a new outburst has taken place within the past million years, and the jets from this are pointing in exactly the same direction as they did in the previous burst of activity. This is not the only instance. Most radio galaxies that have suffered several outbursts 'remember' the direction of previous eruptions. Whatever lurks at a galaxy's heart to power such eruptions has some memory for direction in space.

Optical spectrum of galaxy core	Optical polarization of galaxy core	Radio emission		
		from core	from extended lobes outside galaxy	
broad emission lines	weak	weak	none	
—	—	weak	very strong	
broad emission lines	weak	none strong	none strong or weak	('radio quiet quasars'—99%) ('radio loud quasars'—1%)
no lines	strong: varies in strength and direction over hours/days	strong	weak	

SEYFERT GALAXIES AND BL LAC OBJECTS

Two other kinds of astronomical object belong in the quasar family, and both were discovered before radio galaxies, although their true nature was not immediately suspected.

In 1943, American astronomer Carl Seyfert investigated half a dozen spiral galaxies which were unusual in having an intensely bright point of light at the centre. This small central nucleus was in some cases producing as much light as all the stars in the galaxy itself. Seyfert found that the spectrum of light from the nucleus did not have the characteristics of starlight, and that their most striking feature was the presence of broad spectral lines, which must have originated in clouds of gas moving at very high speeds.

Such activity at a galaxy's heart was so unexpected at the time that no one realized the importance of Seyfert's find. After the discovery of quasars, it was realized that the nucleus of a *Seyfert galaxy* has a very similar spectrum to a typical quasar, the main difference being that Seyfert nuclei are about a hundred times fainter. They are also more common, so there are Seyfert galaxies close enough for us to see the surrounding spiral galaxy as well as the bright nucleus.

The other relative of the quasars was first catalogued even earlier, and its true nature recognized later still. The unwitting discoverer was the German astronomer Cuno Hoffmeister, a world expert on variable stars early this century. Among the thousands he catalogued was an undistinguished star in the constellation Lacerta, which he discovered to be variable in 1929. He named it BL Lacertae—BL Lac for short.

Forty years after Hoffmeister's discovery, radio astronomers found that the 'star' BL Lac emits radio waves. They immediately suspected it was a quasar. But when they looked at the spectrum of its light, optical astronomers realized that they had netted a new kind of object. For BL Lac shows no lines in its spectrum: the light is uniformly bright all along, from blue wavelengths to red.

Without spectral lines, the redshift of BL Lac could not be measured, and therefore its distance could not be determined. But photographs had shown that it was surrounded by a faint 'fuzz', which looked suspiciously like a surrounding galaxy, with BL Lac as its bright core. Beverly Oke and Jim Gunn used the Palomar telescope to look at light from the fuzz—not an easy measurement because they had to block out the light from the brilliant BL Lac itself. The fuzz did have the spectrum of a galaxy, and a galaxy at a redshift of 0.07, corresponding to a distance of 700 million light years from us.

Some 50 other objects similar to BL Lac have now been found. Although each has its own name, a convenient name for the class as a whole has yet to be agreed: proposals include 'lacertids' and 'blazars', but most astronomers still stick to 'BL Lac objects'.

BL Lac objects differ from quasars in several ways, in addition to their obvious lack of spectral lines. They change in brightness very rapidly, sometimes altering during a single night. Quasars do vary in brightness, but take weeks or months to do so. BL Lac objects are generally redder in colour, and their radio emission differs in its wavelength distribution, too. Their strangest property is that their light is polarized: most of the light waves vibrate in the same plane, rather than being jumbled up. And in BL Lac objects, the plane of polarization alters even more rapidly than the brightness changes.

The peculiar object BL Lac appears no different from a star in a short exposure photograph (*below*). It is, in fact, a type of quasar, the bright core of a distant galaxy. Unlike a normal quasar, BL Lac has no strong lines in its spectrum, and its brightness varies rapidly and erratically. Its light undoubtedly comes from closer to the actual central 'powerhouse' than does the light of a normal quasar, and the study of objects like BL Lac could prove the key to understanding quasars of all types.

JOURNEY TO THE CENTRE OF THE QUASAR

Astronomers believe that what is happening at the centre of the modest Seyferts, the violently fluctuating BL Lac objects and the jet-emitting radio galaxies is basically the same as the central mysterious processes in quasars themselves. They have used all these

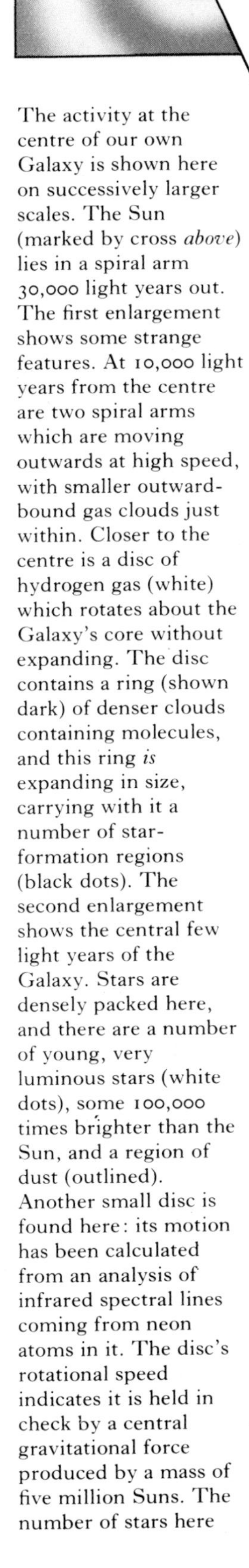

The activity at the centre of our own Galaxy is shown here on successively larger scales. The Sun (marked by cross *above*) lies in a spiral arm 30,000 light years out. The first enlargement shows some strange features. At 10,000 light years from the centre are two spiral arms which are moving outwards at high speed, with smaller outward-bound gas clouds just within. Closer to the centre is a disc of hydrogen gas (white) which rotates about the Galaxy's core without expanding. The disc contains a ring (shown dark) of denser clouds containing molecules, and this ring *is* expanding in size, carrying with it a number of star-formation regions (black dots). The second enlargement shows the central few light years of the Galaxy. Stars are densely packed here, and there are a number of young, very luminous stars (white dots), some 100,000 times brighter than the Sun, and a region of dust (outlined). Another small disc is found here: its motion has been calculated from an analysis of infrared spectral lines coming from neon atoms in it. The disc's rotational speed indicates it is held in check by a central gravitational force produced by a mass of five million Suns. The number of stars here should, however, amount to only one million Suns. Some astronomers have taken the discrepancy to mean there is a black hole of several million solar masses at the Galaxy's very centre.

lines of evidence to build up a picture of what we would find if we were audacious enough to travel to the heart of a quasar.

First we would thread our way through the stars of the surrounding galaxy. Seyfert cores occur in spiral galaxies, while most BL Lac objects lie in giant ellipticals, but we are not yet sure what kind of galaxy surrounds the majority of quasars. Towards the centre of the enveloping galaxy, we would come across wraith-like, almost invisible, clouds of gas shooting past, outwards at high speed into space behind. We know of these because the clouds absorb some of the quasar's light to make narrow dark lines in the spectrum. In some quasars, we would also see the two bright jets of fast electrons beamed out from the centre: it would be wise to steer clear of these concentrated beams of energy.

By the time we are only 10 light years from the centre, we are amid a whole swarm of hotter, brightly glowing clouds, which produce the prominent bright lines in a quasar or Seyfert spectrum, the lines which Maarten Schmidt identified in the first quasar, 3C 273. Detailed study of these has shown that the gas clouds are whipping about at tremendous speeds—millions of kilometres *per second*. But no one is sure whether they are just milling about or being thrown out in some great explosion. Explosion or not, what is pushing them about, and making them shine, is a very tiny, very bright object lying right at the quasar centre.

This object embodies the frightening power of the quasar: its prodigious light output, equal amounts of high-energy X-rays and gamma rays, and the high-speed beams of electrons. The nature of this powerful 'engine' is the ultimate quasar mystery.

The changes in a quasar's light output are an important clue, for they tell us how small the engine is. A gas cloud one light year in size cannot vary in light output over a time shorter than a year: even if all the gas 'switches off' at once, light from the rear part of the cloud is still reaching us a year after the front part has gone dark. Quasars can change their light over a month or two, so they must be smaller than a light year in size. The rapidly varying BL Lac objects can be only a few light-*hours* in diameter, not much larger than the solar system. X-rays from the quasar OX 169 have changed in less than two hours—so it could be contained within the orbit of Saturn.

The difference between quasars and BL Lac objects is probably that in the latter we are seeing regions closer to the centre, where the energy originates. Here the light fluctuations are more violent, and the raw energy emerges without losing its polarization. It is not clear *why* we should see deeper into BL Lac objects. It may just be a matter of our angle-of-view; seen from a different direction, a BL Lac object's central source would be partially hidden, and we would see it as a quasar. Or it may be that BL Lac objects occur in relatively gas-free galaxies. This would explain why they do not show lines of gas in their spectra, and there would be less gas in the way to obstruct our view of the central engine.

But what is this engine? In past years, astronomers have thought it could be a close group of stars going supernova in rapid succession; or a single 'supermassive star' as heavy as a million Suns and spinning rapidly. But both of these interpretations have problems. And observations of the radio galaxy M87 have now convinced most astronomers that the answer is, in fact, a heavy black hole at the galaxy's heart.

In 1977, a team of British and Californian astronomers studied the light from the central regions of M87 in great detail. They found that the stars were unusually closely packed there; a strong gravitational pull seemed to be holding them close to the centre. And they were moving unusually fast, indicating that there must be a strong gravitational tug to balance their motions. Peter Young, Wallace Sargent and their colleagues concluded that an invisible mass of almost 5000 million Suns must reside at the centre of M87: the only possible explanation, is the presence of a black hole containing this tremendous mass.

The memory of radio galaxies such as M87 also suggests a black hole at their centres. To keep a constant reference direction, navigators on Earth or in space use a gyroscope. The axis of any spinning body maintains a constant direction in space unless a force acts on it—in fact, there is nothing else that can act as an invariable direction indicator. What can be spinning at a radio galaxy's heart? The most reasonable answer is a rotating black hole, which carries space around with it. Although M87 is not a quasar, it is a powerful radio galaxy, and its core was probably a quasar or BL Lac object in the past, when it summoned up the tremendous energies to power its electron jets.

It is now generally accepted that the secret heart of the quasars is a huge, massive black hole, fed by gas from broken-up stars which stray too close. The gas spirals inwards, heating up all the time and forms into a disc which rotates around the hole, dragged by the hole's own rotation. This hot, turbulent disc is the quasar engine. It is what we would see, blinding bright in front of us, if we were foolish enough to venture into the blazing inferno near the heart.

The light and other radiations from this super-hot disc would outshine a hundred galaxies like our own, although it is no larger than the solar system. At the same time, it whirls up electrons to enormous speeds, and these are beamed off the flat faces of the disc and head off into space to create the outer radio-emitting 'bags' millions of light years off.

Bizarre though this picture seems, it is taken very seriously today. Indeed, it is difficult to think of any other way of generating the huge amounts of energy from such a small region of space. To fuel the quasar engine of its disc, a black hole need only swallow stars at the quite reasonable rate of one per year. Whole stars are no use. They would simply fall in without letting out energy to the Universe at large. A quasar black hole must be a messy eater, and must gorge itself off loose streams of gas that can form into a rotating gas disc around the hole. In the terrifying central region there are many ways that ordinary stars, like our Sun, can be reduced to a chaotic whirl of gases. A star which strays too close to the hole will be torn apart by the hole's gravitation: the 'spaghettification' effect mentioned in Chapter 9. And the central stars are held in a tight cluster, close to the hole, and here will occasionally collide with one another and spew their gases out into space. The powerful radiation from the quasar engine itself can disrupt stars, boiling off their outer layers of gas. Such gas may, in fact, trail away from the parent stars like the tails of comets.

According to statistical studies of quasars, their cores are active for a million years or so at a time. During this period, the black hole swallows a million stars—a tremendous amount of matter, but only a tiny part of an entire galaxy. The hole consumes only one star for every 500,000 in the galaxy, a very light toll.

And just as radio galaxies repeat their explosions, so undoubtedly do quasars and BL Lac objects. Many completely innocuous-looking galaxies around us must be harbouring dormant quasars at their cores, heavy black holes lurking unseen, but only requiring a

The nearby galaxy M82 seems to be a spiral, seen edge-on, but it is clearly disrupted at its centre, as shown in the colour-enhanced photograph (*below*). These central regions are also a source of radio waves and infrared radiation. M82 is not just a scaled-down quasar, however. Its activity is spread out over several light years of its core, not concentrated at a small point, and may simply

be the result of an intense period of star-formation there. Such outbursts of activity may occur from time to time in all galaxies—perhaps such an event could explain the activity at the centre of the Milky Way without needing to invoke the presence of a black hole.

renewed supply of gas to burst forth once more.

Many ordinary spiral galaxies must have been Seyferts in the past, and still contain dormant mini-quasars. Even the Milky Way shows signs of a Seyfert past. Ironically enough, we know less about our Galaxy's centre than we do of distant quasars, because dense clouds of dust within our Galaxy block off the light from its central regions. Although it is certainly not as spectacular as a quasar core, the Milky Way's core has mysteries to be unravelled. Radio waves indicate a very small, but intense radio source right at the centre, and infrared radiation has shown up peculiar motions in the gas near to it. These motions indicate that there is the equivalent gravitational pull of five million Suns at the Galaxy's heart, although it is estimated that the core contains only one million stars. The missing matter may be in the form of a black hole several million times heavier than the Sun. A disc of gas around it could power the peculiar small radio source that marks the Galaxy's heart.

Such a hole is not heavy enough to be a dormant quasar of the most violent kind—but when it was supplied with bountiful amounts of gas, it would have shone out as a mini-quasar of the kind we see in the Seyfert galaxies. And if our Galaxy was once a Seyfert, then it is only a matter of time before gas once more accumulates near the hole, and our core flares up again.

CHAPTER 11

THE ULTIMATE MYSTERY

Our Universe itself is the ultimate mystery. Has it always existed? If not, when did it begin? And why? Men have always been intrigued by the problems of understanding the entire Universe, estimating its size, its age, its fate, its purpose. Now, many of the veils have been stripped away so that, roughly speaking, the 'when' and 'how' of the Universe can be answered, though such answers can still contain surprises and uncertainties. But the 'why' remains as elusive as ever.

To answer the largest questions of all, astronomers must look on the largest scales, shrinking whole galaxies until they are little more than markers in space—floats adrift in the cosmic ocean. Albert Einstein taught us how to view space on this scale: his general theory of relativity describes 'empty space' as something almost real, and time as another dimension of space. As we saw in chapter 9, space-time together is a real entity, a background fabric against which the events of our Universe unfold. Gravitation is a distortion of space-time which, in the extreme case of a black hole, rips a gap in the fabric of the Universe. Over the wide expanses of the Universe as a whole, the gravitation of all the individual galaxies affects space-time, too, distorting it in more subtle ways, but in ways that will determine the ultimate fate of the Universe. Before looking at the past and future of the Universe, it is important to be sure that galaxies are fully understood. And they are not: many mysteries remain.

GALAXY PUZZLES

Half the galaxies in the sky are spirals, graceful whorls of gas and stars. Although astronomers have known of spiral galaxies for over a century, and realized that the Milky Way was a spiral some 50 years ago, it is still not clear *why* spirals have adopted such beautiful patterns. In all spiral galaxies, the stars and gas lie in a thin disc, its proportions very like those of a gramophone record. The stars revolve in orbits about the galaxy's centre. It is rather like a huge version of the solar system—except that, while the planets are held in orbit by a single object (the Sun), the stars in a galaxy are not being attracted by just one central object, but rather orbit under the combined gravitational pull of all the other stars in the galaxy, which concentrate most towards the centre.

Because of this, star orbits are not simple. They bunch up to make the spiral pattern, zones in the disc where stars are more concentrated. Individual stars do not stay in the spiral arms for ever, though. They spend some time gradually moving through an arm, leave it and travel quickly through the space between the arms, then slow down as they traverse the next arm. As a result, they build up in the arm regions, and are comparatively rare in the parts of the disc which lie between. It works in exactly the same way that a speed restriction on a motorway leads to the cars becoming more closely bunched as they slow down. The speed restriction zone of a spiral arm is, in fact, caused by the gravitation of the stars forming the arms at the time. As a result, the pattern remains, even though the stars making it up are continually changing. In the simple analogy, a bunch-up on the motorway will persist even after the speed restriction is lifted, because cars coming up from behind are slowed by it, even though the front cars are accelerating away. The newly-slowed cars then block those catching up later. The foul-up can thus perpetuate itself for quite a long time—much to everyone's annoyance!—even after the initial restriction is lifted. Spiral arms are a cosmic example of such a self-perpetuating hold-up.

In broad terms at least, that is how the origin of these beautiful traceries is understood. Unfortunately, there are problems. Just as a motorway bunch-up will eventually iron itself out and disperse, so should a galaxy's spiral arms—unless something puts in energy all the time. And it is not always obvious what that 'something' is.

In some, the gravitational pull of a companion galaxy, whether in perpetual orbit about the spiral or just passing by, can drive the spiral pattern. The spiral arms of the Milky Way Galaxy may be due to our two small neighbours, the Magellanic Clouds. In others, supernovae within the galaxy itself may supply the necessary co-ordinated energy. New stars are born in the spiral arms, out of the denser gas-clouds between the bunched-up stars of the disc. The heaviest are short-lived, and explode even before they

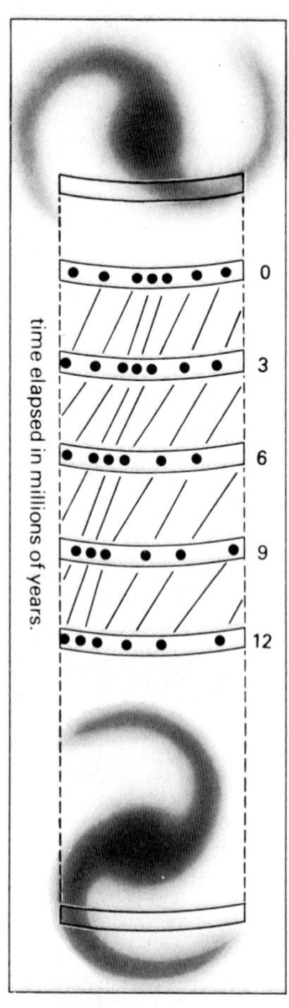

The spiral arms in the discs of spiral galaxies are regions of higher star density. The stars in an arm at any particular time eventually move out of it, and their place is taken by others (*above*). The current favourite theory for the persistence of spiral arms invokes *density waves*: the accumulation of stars in the arm has a gravitational effect which alters star motions (shown by dashed lines above) so that the 'bunching up' in the arm is perpetuated by the stars which are moving into the arm region.

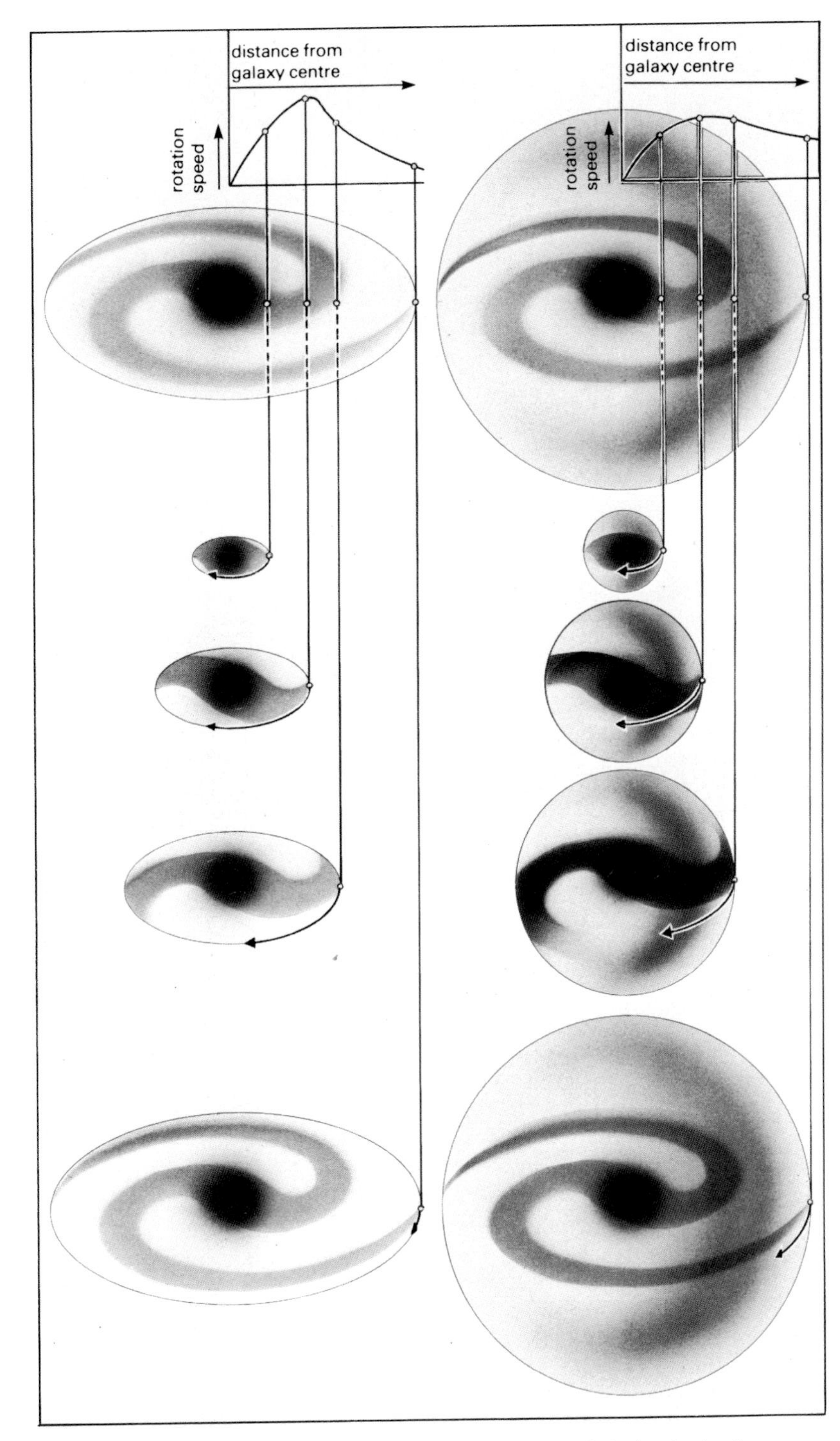

The rotation speed of stars at different distances from the centre of a spiral galaxy depends on the distribution of stars throughout the galaxy. A star at any particular distance feels only the gravitational pull of stars within that radius of the centre (*above*). If the galaxy's matter lies only in the visible disc (*above left*) then the rotation speed drops off at large distances. In a galaxy whose matter resides mainly in a large halo (*above right*), the rotation speeds stay high. Real galaxies often display the latter behaviour, indicating they have massive, invisible haloes. Computer calculations show that massive haloes may be needed to prevent most

Spiral galaxies from developing central bar-like configurations of stars, similar to the bar displayed by NGC 7741 (*right*).

have time to drift out of the spiral arm region. As they explode as supernovae, such stars, sited all along the arms, could provide the energy necessary to keep the pattern stable. But this would give fairly 'patchy', broad and tatty arms. Supernovae cannot be responsible for maintaining the graceful, narrow spirals sported by some galaxies.

The central hub of the galaxy—a large cluster of old stars—may also sometimes be responsible. In half of the spiral galaxies, this is not a simple spherical bulge but an elongated bar, lying in the disc. The stars in it perform complicated orbits, but the net result is that the bar rotates as if it were solid, preserving its shape as it turns. The bar's gravitational pull on a star farther out in the galaxy obviously depends on its orientation at the time for, as the bar turns, its changing gravitation can drive a spiral pattern in the outer parts of the disc. Many of the barred spiral galaxies do indeed have the most magnificent pair of arms, sprouting from either end of the central bar.

The resolution of the mystery of the central bar is not quite what common sense would suggest. Detailed calculations—backed up by computer simulations of galaxies—show that the stars in a galaxy's disc will naturally affect one another's orbits in such a way that they will all end up as part of a rotating bar. The question therefore is not so much, 'Why are some galaxies barred?'; but rather 'Why do some spirals *lack* central bars?'.

The question is not easy to answer. Astronomers have come up with only one possibility—and that opens a Pandora's box of its own. All spiral galaxies have a halo of stars spread thinly above and below the disc, stars which make up only a fraction of its matter. Computer simulations of spirals show that we

An artist's impression (*far right*) of our Local Group of Galaxies. Our nearest neighbours in space, the Large and Small Magellanic Clouds, are 170,000 and 205,000 light years away from the Milky Way respectively.

Radio contours of the Whirlpool galaxy superimposed on an optical photograph (*right*). The radio waves are generated by fast-moving electrons trapped in magnetic fields.

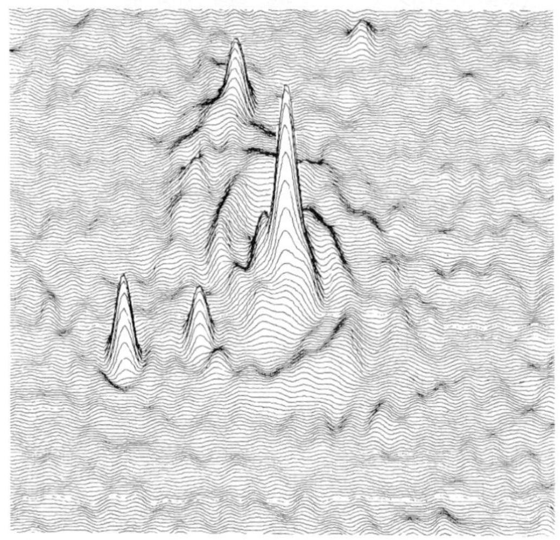

Section through the radio contours (*right*) reveals that the emission is greatest towards the centre of the galaxy. Away from the centre, the bunching of the contours show clearly that there is strong radio-emission from the galaxy's spiral arms where magnetic fields, as well as stars, are concentrated.

can stop a bar from forming if there is a lot more matter in the halo than we think. Its gravitational effect would be more powerful than that of the disc, so the stars of the disc would circle in the uniform gravitation of the halo with relatively little of the gravitational pull from other disc stars which leads to a bar forming. The simulations also show that a massive halo makes the spiral pattern more stable.

But this 'massive halo' could not be in the form of stars alone—or we would see far more halo stars than we do. It must be made mainly of dark objects of some kind, that cannot be detected with optical, radio or X-ray telescopes. Although the idea seems preposterous, support comes from studies of the motions of the stars and gas in the outermost parts of the discs of some nearby spirals. They are generally circling their galaxy's centre faster than we would expect if their orbits were controlled only by the visible stars of the galaxy: it looks as though they are feeling the pull of extra matter around the galaxy.

There is still some dispute as to whether spiral galaxies possess massive haloes. The evidence, taken at face value, suggests that these invisible, surrounding regions could be far more massive than the galaxies that we see in photographs. The same goes for our own Galaxy. Astronomers estimate that the stars and gas in the Milky Way amount to some 100,000 million Suns, but while the dispute about massive haloes continues, no one is sure how heavy our Galaxy actually is.

GALAXIES IN CROWDS

Most galaxies are not alone in space. The Milky Way is part of a small Local Group of a few dozen galaxies, but many live in huge clusters, families comprising thousands held together in a swarm by one another's gravitation.

Some clusters are simple to understand, being groups of spiral, elliptical and irregular galaxies, like our Local Group, but more populous and larger in proportion. Completely different are the so-called regular clusters, crowds of galaxies arranged like a spherical swarm in space, most tightly packed

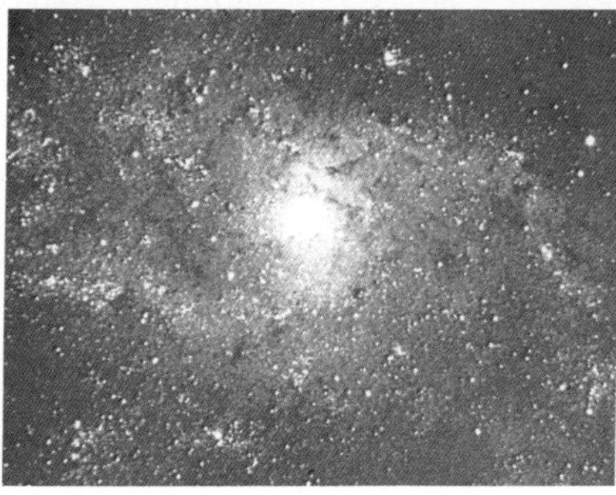

Elliptical galaxies like M87 (*far left*). range in size from sparse collections of a million stars to supergiants with more than a million million members.

M33 (*left*) is a spiral galaxy of about 10,000 million stars. The arms are rich in dust and gas from which new stars continuously form.

towards the centre. Almost all the galaxies in regular clusters are ellipticals. In the centre, there often lurks a monster galaxy (called cD), huge and misshapen, as large as 10 ordinary galaxies together; cDs are the largest and brightest galaxies in the Universe—and are outshone only by quasars.

The cD galaxies may, indeed, be cosmic traffic accidents. Galaxies moving about the cluster are most likely to collide near the centre and, if they do so, they will amalgamate into a larger galaxy. Because they have destroyed each other's momentum, the newly-formed merger will stay near the cluster centre, right in the way of other galaxies pursuing their paths. It could quickly cannibalize them, too, and grow to its present inflated proportions.

The regular clusters may have started as large groups of ordinary galaxies. Pulled together by one another's gravitation, they settled into a more compact group—and as they came closer together, each one ripped off gas from its neighbours. The loose gases fell into a huge pool which filled the centre of the cluster. But now the galaxies in the cluster found themselves moving, not through empty space, but pushing through the gas pool within the cluster. The stars in a galaxy are not affected unduly by the tenuous gas and, since the bulk of a galaxy is made up of stars, it looks as though the galaxy sails serenely through. But any gas *within* the galaxy will be stopped dead when it plunges into the pool. The galaxy's gas will be left behind to mix with the gas already in the cluster, while the galaxy is left as a gas-bereft star skeleton.

Elliptical galaxies are the gas-free type of galaxy—so here is an obvious explanation for the preponderance of ellipticals in the regular clusters. (It is not so easy to explain how ellipticals which are not in clusters have been swept so clear of gas, though, and theories range from collisions between spirals, to ellipticals simply using up all their initial gas in their first burst of star-formation.) Within the clusters, there is also a weird hybrid type of galaxy. These *lenticulars* are discs of stars only, looking for all the world like spirals stripped of gas. They are, in fact, likely to be exactly that. Lenticulars were once galaxies like our own, which had the misfortune to drift into the cluster's pool of gas.

The gas in clusters should be hot—at temperatures of some 100 million degrees—hot enough to emit huge quantities of X-rays. The Einstein Observatory has studied dozens of these clusters, and has spotted the gas pools as predicted, sitting at the centre of each one. The total quantity amounts to about one-tenth the mass of the galaxies in the cluster.

It has long been suspected that clusters contain far more invisible matter than this. In a typical cluster, the galaxies are moving so fast that their mutual gravitational pull cannot contain the individual galaxies. The clusters should have broken up long ago, and their galaxies dispersed into space.

The solution must be that the cluster is heavier than the sum total of its galaxies. Something invisible is spread out through it, its gravitation restraining the speeding galaxies. This *missing mass* is estimated to be 5 to 10 times heavier than the galaxies we see. Although the evidence for it is strong, some astronomers do doubt its existence for, over the decades, better measurements of galaxy motions have consistently reduced the amount of missing mass needed, and it is just possible that it is our measurements that are at fault. To common sense, it certainly seems absurd that most of the matter in a cluster should be totally invisible, to all types of telescopes, and shows itself only by its gravitational effects.

But more preposterous ideas have been mooted before, and have turned out to be correct. The idea that the Earth moves around the Sun appeared to be even more ridiculous when it was first proposed. It is quite on the cards that most of the matter in the Universe *is* completely invisible. The stars and galaxies we see are only a minor constituent: they cannot even govern their own motions. The gravitation of an invisible ring-master keeps them in check, whether they are whole galaxies bound together in clusters, or stars in the disc of rotating spiral galaxies.

If it does exist, no one knows what the vast amount of invisible matter can be. Since it is spread out it cannot consist of very heavy single objects, such as invisible galaxies or giant black holes. On the other hand, it cannot be distributed in a form as fine as the dust particles in space. If it were, it would form a haze that would block off light from beyond, just as fine particles of mist are more opaque than an equal amount of water condensed into larger rain drops.

A huge number of snowballs, planets—or even bricks—would do the trick. But it is almost impossible to understand why most of the Universe's matter should have gone to form these, rather than to make up stars. Cambridge astronomer Martin Rees has suggested a more likely picture. In the very early history of the Universe, most of the gas formed into stars which lived very quickly and died as supernovae. Their cores were left as black holes. They were a few times heavier than the Sun, like the black hole in the Cygnus X-1 star partnership. Rees thinks that huge numbers of black holes could constitute the invisible missing mass.

It is also possible that the subatomic particles called neutrinos are to blame. As we saw in chapter 6, neutrinos are ghostly

Galaxies of the local group	Distance (light years)	Type	Size (light years)	Mass (millions of Suns)
Milky Way and companions				
Milky Way Galaxy	—	spiral	100,000	150,000
Large Magellanic Cloud	170,000	spiral/irregular	26,000	10,000
Small Magellanic Cloud	205,000	irregular	16,000	1500
Ursa Minor	210,000	dwarf elliptical	4000	0.1
Leo I	750,000	dwarf elliptical	3000	4
Fornax	610,000	dwarf elliptical	10,000	20
Andromeda Galaxy and companions				
Andromeda Galaxy (M31)	2,200,000	spiral	170,000	300,000
M32	2,200,000	elliptical	7000	4000
NGC 205	2,200,000	elliptical	14,000	8000
NGC 147	2,200,000	dwarf elliptical	8000	9000
NGC 185	2,200,000	dwarf elliptical	9000	10,000
M33	2,350,000	spiral	60,000	20,000
Other galaxies				
NGC 6822	1,500,000	dwarf irregular	6000	1000
Carina	500,000	dwarf elliptical	3000	1?
IC 1613	2,400,000	dwarf irregular	15,000	400
GR8	3,000,000	dwarf irregular	1000	10

particles that are very difficult to catch; but theory predicts that a huge number were generated in the birth of the Universe, and are still around today. Physicists have always believed these elusive particles should have no measurable effect on the Universe, particularly as they are thought to have zero mass—if you could stop one, it wouldn't register any weight on the scales. But recent experiments in both the United States and Russia have suggested that neutrinos *do* have some mass. Although it is less than one-ten thousandth of the weight of an electron, there are so many around that they would outweigh all the ordinary matter in the Universe! So perhaps the invisible Universal ring-master consists of neutrinos held together by their own gravitation, and these elusive particles are what control the matter on the large scale. Galaxies would then just be the bright lights that mark the layout of neutrinos in the Universe.

No one is putting all his money on this theory yet, because the critical experiments have still to be confirmed. Whether neutrinos are involved or not, the missing matter remains one of the most intriguing mysteries in astronomy.

THE END OF THE UNIVERSE

Cosmologists look at the whole Universe as an enormous fabric of space-time. They regard clusters of galaxies, not as milling swarms, but simply as position markers in the Universe. And what intrigues them is that all the

Types of normal galaxy

Type	Proportion	Composition	Range in mass (millions of Suns)	Range in brightness (millions of Suns)
Elliptical	60%	stars only	0.1–3,000,000	0.01–100,000
Spiral	30%	90% stars 10% gas	1000–300,000	100–30,000
Irregular	10%	75% stars 25% gas	10–30,000	3–1000

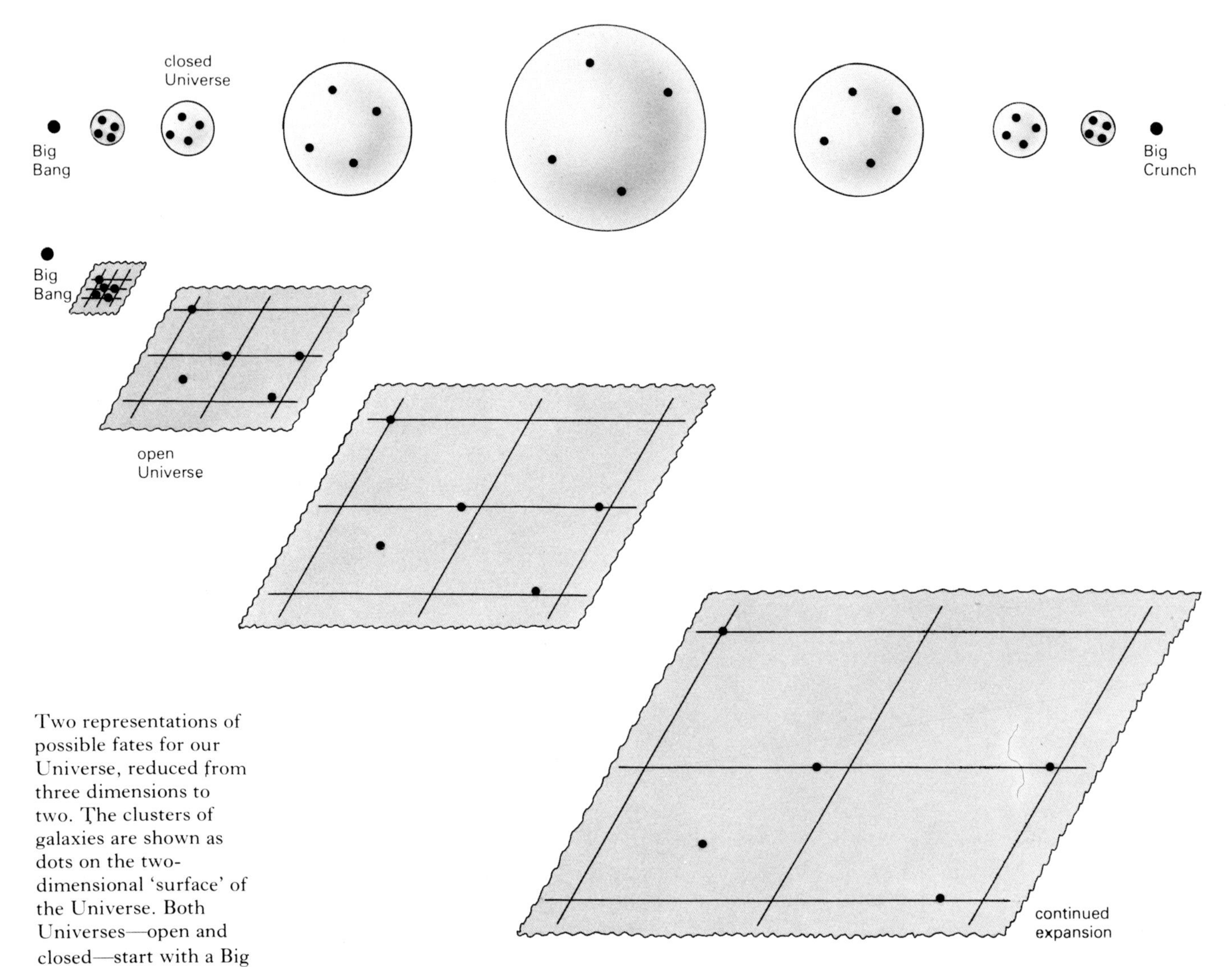

Two representations of possible fates for our Universe, reduced from three dimensions to two. The clusters of galaxies are shown as dots on the two-dimensional 'surface' of the Universe. Both Universes—open and closed—start with a Big Bang (*left*). A closed Universe (*top*) is represented by an expanding sphere, in which the clusters of galaxies move apart until the Universe reaches a maximum size. At this point, the outward flight of the clusters is just balanced by their combined pull of gravity. Gravity triumphs, and the Universe starts to collapse, ending in a 'Big Crunch'. An open Universe, in which there is insufficient matter to halt the expansion, is represented by a flat sheet of paper (*below*). As time passes, the clusters of galaxies move steadily apart from one another, and the Universe becomes increasingly empty.

markers are moving apart from one another: the Universe is expanding.

Albert Einstein pointed out that it is wrong to think of the clusters of galaxies flying apart like shrapnel from a bursting shell, that we must think of space-time itself expanding. Each cluster is merely a marker, tied to its own local piece of the fabric. As the Universe expands between them, the markers are carried apart willy-nilly, like floats on an ocean current.

Space stretches in a uniform way, so the speed at which two clusters of galaxies are carried apart depends on the amount of space between them—that is, on their distance apart. So the speed of recession between any two clusters is proportional to their separation. Two clusters 200 million light years apart are moving away from each other twice as fast as two clusters separated by 100 million light years. (The speeds associated with different distances are discussed later.)

From our viewpoint in the Local Group, we see other clusters moving away from us and, in accord with Einstein's theory, the speed of the galaxy is proportional to its distance from us. This does not mean that our Galaxy is at the 'centre' of the Universe. An astronomer on any other galaxy would see precisely the same effect: to him, our Galaxy and its Local Group would seem to be shooting away. The important point is that every cluster of galaxies is at a fixed point in Einstein's space-time; and the amount of space between them is continually growing.

In fact, the Universe does not have a centre. Einstein's results assure us of that—and also show that it is almost impossible to talk meaningfully about the 'size' of the Universe. The results are difficult to describe in non-mathematical terms, but basically there are two possibilities. One is that the Universe is infinitely large, and it goes on for ever in space in all directions. In that case, any place in the Universe can call itself 'the centre', because all points are infinitely far from the 'edge'. In fact, the idea of a 'centre' has no meaning in this kind of Universe, which is called an *open Universe*.

The alternative is even more difficult to visualize. A *closed Universe* is looped in the fourth dimension, so that it closes on itself. An analogy is the bending of the two-dimensional surface of a sphere in the third dimension. A two-dimensional creature living on a flat surface would be aware only of directions in the surface: he could not envisage the third dimension perpendicular to the surface. He would expect that a journey in a straight line would simply take him farther and farther from home. But we, as three-dimensional creatures, can see that if he lives on the surface of a sphere, his journey will take him round the globe. Eventually, he would find himself outside his own back door—to the two-dimensional voyager's great surprise, because he never deviated from a straight line in the sphere's surface. Only a knowledge of the third dimension could explain his situation.

Similarly, our Universe could curve in the fourth dimension, making a *hypersphere*. A journey in a straight line, out past nearby galaxies and then distant clusters of galaxies, would eventually bring us to our Milky Way again, coming from the opposite direction from the way we left. Such a closed Universe is limited in size: it contains only a certain finite number of galaxies. But it has no edge; and so it also has no centre. No particular galaxy can call itself the centre of the Universe. Making a comparison with the two-dimensional Universe of a sphere's surface, we could argue that this Universe *does* have a centre—namely the middle of the sphere. But this point does not lie in the two-dimensional Universe, which is the sphere's surface, but off in the third dimension. Similarly, one could say that our Universe does have a centre if it is closed—this centre is not anywhere in the Universe, however, but lies somewhere else in the fourth dimension.

To be fair, some mathematicians have suggested variations on Einstein's theme, Universes which do have a centre, but they are not popular at present. Tests within the solar system, of the binary pulsar and the double quasar, have all confirmed Einstein's view of space-time and gravitation; and his mathematically simple theories of the Universe itself explain very well the cosmological observations—particularly those that relate to the birth of the Universe in the Big Bang.

Space is continually expanding, and carrying the clusters of galaxies apart. The simple theory predicts a bleak future—for astronomers at least—when other clusters are so far off that we see only our local neighbours, like Andromeda, in space. But the real Universe is more complicated. The clusters of galaxies are not just indicators of position in space-time, they contain a lot of matter, and that matter has a gravitational field. If their pull on one another is strong enough, they can stop the expansion of space.

It is quite easy to work out how much matter is needed to slow down the expansion to an eventual, grinding halt. And the clusters are just not heavy enough, since their matter comes to only 3 percent of the required amount. But here the missing mass becomes crucial. If most of the mass in a cluster is invisible, then the galaxies contain a lot more mass than the amount they are credited with. Even on the extravagant side, though, it is unlikely that it outweighs the visible matter more than 10-to-1. So we still have only 30 percent: even the missing mass is insufficient to brake the expansion. But this is getting pretty near the required amount. Could there be matter elsewhere that could contribute the extra? Many clusters are arranged into larger *superclusters*—each some 200 million light years across. These superclusters are not bound together: the expanding Universe will break them up into individual clusters. But there may be matter lurking between the clusters in a supercluster.

Many X-ray astronomers thought so. They picked up radiation coming from beyond our Galaxy, radiation which was attributed to hot gas in superclusters. If this interpretation had been correct, then the gas in superclusters would indeed have been more than enough to brake the universal expansion. But the new Einstein Observatory has now found that

50 million light years away, the Virgo Cluster is the nearest large cluster of galaxies. Its 3000 members form the nucleus of a vast Supercluster, which includes our small and insignificant Local Group near its outskirts.

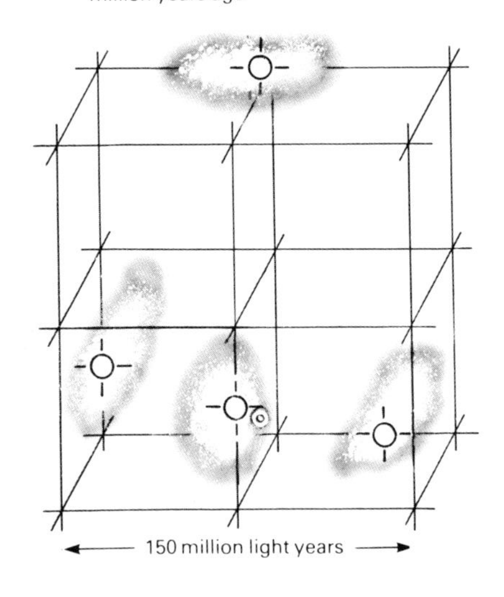

This representation of our 'local' region of the Universe, out to a distance of a few hundred million light years, shows how the superclusters have been moving apart with the general expansion of the Universe. The latter is indicated by the imaginary grid lines at the Big Bang (zero size), when the Universe was half its present age, and today. The gravitational inpull of the galaxies and clusters within the superclusters has made them grow at a slower rate than the Universe itself; although the superclusters are enlarging, it is not at the same rate as the grid. The effect on our Local Group of galaxies is shown in the inset. (*below left*)

We are moving away from the centre of the local supercluster at 1000 kilometres (600 miles) per second, but this is not as fast as we should be moving if we were 'anchored' to our own piece of space-time. Relative to this, we are effectively moving *towards* the supercluster centre (shown by the arrow).

most of this radiation comes from individual, very distant quasars: it seems there is no gas in the superclusters after all.

Confirmation that the missing mass is no larger than one would expect has come more recently from studies of our own Supercluster. Our modest Local Group of Galaxies is part of a supercluster centred on the huge Virgo Cluster. To see what these measurements mean, imagine that you have been transported to the Virgo Cluster, and are looking out at the Local Group. From Virgo, you would see the Milky Way Galaxy and its neighbours being carried away on the universal expansion, at the rate of 1000 kilometres per second (620 miles per second). But fast though this is, it should be an even higher speed. Space-time in the region of the Local Group is moving away from the Virgo Cluster at 1500 kilometres per second (950 miles per second).

There is no doubt that the huge Virgo Cluster is well tied to its own fabric of space-time. These observations indicate that the galaxies of the Local Group are being retarded and, instead of travelling in the flow of space-time in the vicinity, the gravitational pull of the Supercluster is trying to hold them back. Put into figures, it turns out that the Supercluster contains about 10 times as much invisible matter as galaxies—very much like

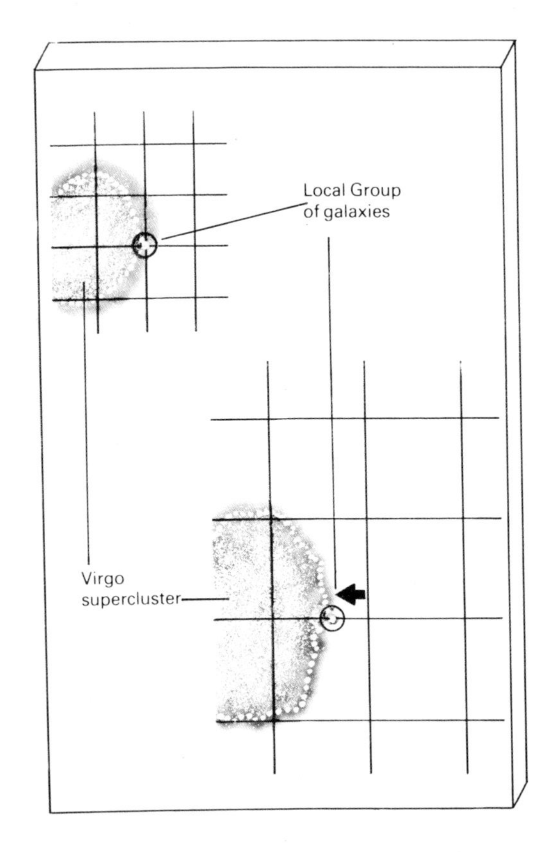

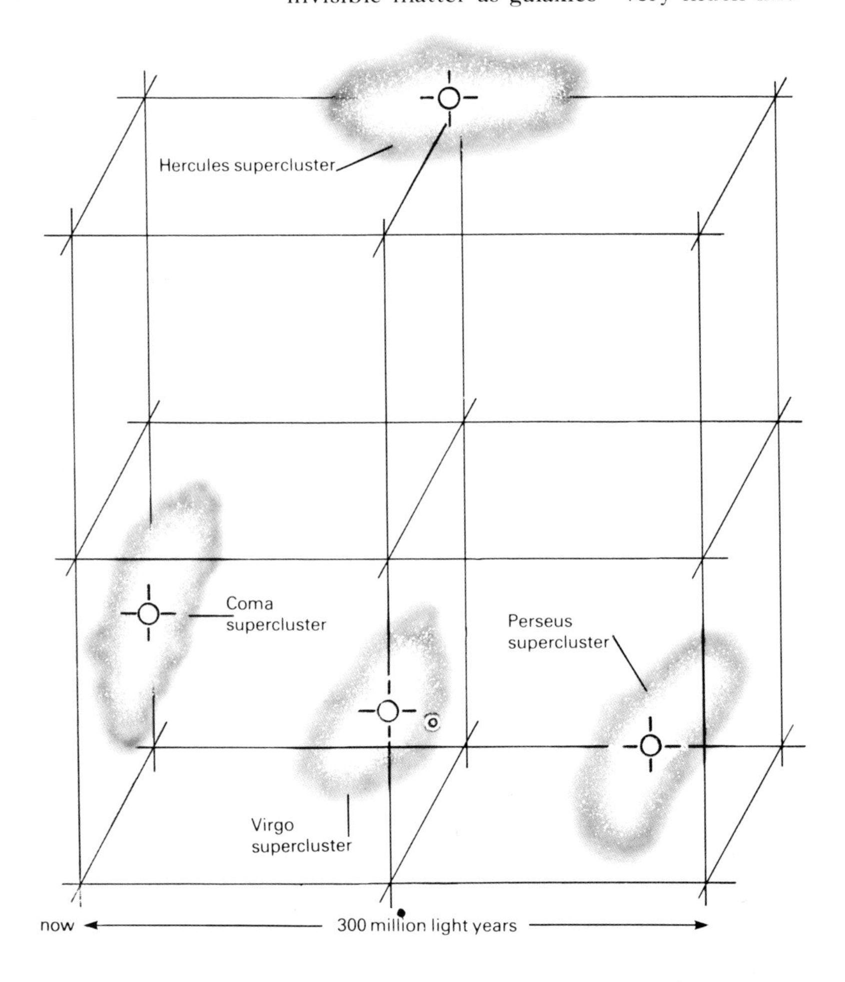

ordinary clusters. So even when we consider the largest concentration of matter in the Universe, the superclusters, we only reach 30 percent of the density of matter needed to stop the expansion.

These measurements are not yet universally accepted. Some astronomers still believe that the Local Group is actually more or less at rest in its local space-time fabric. In this case, though, there would be even less matter in the Supercluster and so much less mass to retard the Universal expansion. There is always the possibility that we may discover more missing mass · hidden in different forms in the Universe, so the issue is not entirely resolved. But now that we are sampling the largest concentrations of matter, it does seem that we can account for all the mass in the Universe, whether visible or 'missing'. And this leads to the conclusion that the Universe will expand for ever. This, incidentally, means that our Universe is open, infinite in size, because a closed Universe would eventually collapse again. As time goes by, the clusters of galaxies will move apart; the stars in each galaxy will die; the once-glowing galaxy will become the dark graveyard of cooled-off white dwarf stars, dead pulsars and necrophiliac black holes.

The very far future is uncertain. Possibly all the star-corpses in a galaxy will come together to make each galaxy a huge black hole. Physicists predict that at some even more distant date—so far ahead that the number of years involved would run to a couple of lines of text—these black holes will 'evaporate' away to nothingness. In this odd process, the black holes' energy will end up as radiation.

The most recent advances in physics now predict, however, that one of the most fundamental of the subatomic particles is unstable. The proton, an ingredient of all matter, is thought to be radioactive. Though it takes a long time to get around to it, any proton will eventually break up. The time involved is far, far longer than the present age of the Universe: a matter of some 10,000,000,000,000,000,000,000,000,000,000 years, but it is shorter than the time for the remains of a galaxy to collapse into a single black hole. Again, the result is the decay of matter, leaving just radiation.

In this longest-term of all predictions, new discoveries in physics may well change our ideas yet again. But all scientists agree that the Universe is running down. The fire of stars will not last for ever; eventually all the hydrogen fuel in every galaxy will run out. The stars will die. Life will die. And, eventually, matter will die. The Universe will contain only radiation. In its enormously extended space, this will be the longest of radio waves, and so thinly spread out that the faintest of the stars we see today would be a blazing beacon to the cold, black, dead Universe of the distant future.

One of the two giant elliptical galaxies (*above*) at the heart of the 5000-strong Coma Cluster of galaxies. Several large clusters are dominated by supergiant galaxies, believed to have grown to their present size by 'cannibalizing' other galaxies passing nearby.

The Hubble Constant

The expansion of the Universe tells not just of its future but also of its past. Galaxies must have been closer together long ago. Astronomers can backtrack their motions into the distant past when the space between them was more compact. Going back far enough, there must have been a time when the galaxies were touching; before that there were no separate galaxies, just a gas filling the entire Universe. And tracing back further, the gas must have been increasingly hotter and denser. This hot, dense fireball of gases exploded to become our Universe. Astronomers call it the Big Bang.

In principle, it is very easy to put a date to the Big Bang simply by tracing galaxy motions backwards—using the clusters of galaxies as markers in the expansion flow of space. The huge cluster that lies in the constellation Coma Berenices is some 350 million light years away. Its redshift tells us that the Coma Cluster is travelling away from our Galaxy at a rate of 7000 kilometres (4000 miles) per second: changing this speed into astronomers' units, the cluster's distance is increasing by 1 light year every 43 years. To cover the 350 million light years that separate us from the Coma Cluster would take $350,000,000 \times 43 = 15,000,000,000$ years. Fifteen thousand million years ago, this giant cluster must have been right on top of us.

The same figure of 15,000 million years comes up for all clusters of galaxies. One which lies at twice the Coma Cluster's distance is moving away twice as fast; so it has taken exactly the same time to cover the distance from us to it. All the matter in the Universe was heaped together 15,000 million years ago: the moment of the Big Bang.

American astronomer Edwin Hubble first found the law relating speed of recession of galaxies (redshift) to their distance from us.

The Coma Cluster is 350 million light years away, and its speed is 7000 kilometres per second. So every million light years in the distance to a cluster means a speed increase of 20 kilometres per second. A cluster of galaxies 100 million light years away is travelling at $100 \times 20 = 2000$ kilometres per second; a cluster at 1000 million light years at $1000 \times 20 = 20{,}000$ kilometres per second. (The formula has to be changed slightly at very large distances, because the speed can never exceed the speed of light.)

This figure of 20 kilometres per second of speed for every million light years of distance is known as the *Hubble constant*. It is a general law describing how the Universe expands, and applies to all clusters of galaxies in the Universe. In fact, to calculate the date of the Big Bang, we do not need to consider any particular cluster: it can be calculated from the value of the Hubble constant. But one of the major disputes in astronomy in recent years has been the actual value of this important constant and, as a result, astronomers have been uncertain about the actual date of the Big Bang.

To find the Hubble constant, just measure the speed at which a cluster like Coma is receding, and divide by its distance. Repeat for other clusters: they should give more or less the same answer, so average them to get the best result. It is simple to measure the speed, because it comes directly from the redshift of the light from galaxies in the cluster. But plumbing the distance to far-off clusters is a difficult problem. By looking at individual stars in nearby galaxies, we can be fairly certain how far it is to galaxies like Andromeda, but this method takes us only a few million light years into space. To find clusters outside our local Supercluster, we need to penetrate a hundred times farther.

Different astronomers use different methods, but most are based on the apparent brightness of individual galaxies. If all galaxies contained the same number of stars, it would be easy for they would be ready-made 'standard candles' of the same intrinsic brilliance. Far-off galaxies would be dimmed simply by distance, so a comparison with nearby galaxies would immediately reveal their distances. As it is, galaxies vary tremendously from one to another, and so we must use indirect methods to assure ourselves that particular galaxies are of the same type, and can indeed be used as standard candles for determining distances.

And there are conflicting results. In a mammoth investigation in the late 1960s and early 1970s, Allan Sandage and Gustav Tammann at the Hale Observatories in California found the Hubble constant to be 17 kilometres per second per million light years. This makes the Universe 18,000 million years old.

The great Andromeda Galaxy (*above*) is so close that large telescopes on Earth can pick out individual stars in its spiral arms. By studying its most luminous stars, and in particular the changes in light of its Cepheid variables, astronomers have established that this galaxy lies 2.2 million light-years away.

But not all astronomers agreed. Although no one had galaxy measurements as good as Sandage and Tammann's data, there were indications that all the clusters are closer to us: the Hubble constant would then have a larger value, and the Universe would be younger (because the clusters would have less far to travel). In the mid-1970s, Gerard de Vaucouleurs of the University of Texas at Austin undertook a similar massive investigation and found that the clusters were only just over half the distances that Sandage and Tammann had assigned to them. The Hubble constant was then 30 kilometres per second per million light years, and the Universe only 10,000 million years old.

Followers of Sandage and Tammann did not accept the new results, and the controversy has simmered ever since. The age of the Universe is one problem left unresolved, but more directly it affects the distances to all far-off galaxies. As a result, we do not know accurately the intrinsic brightnesses of most galaxies; nor do we know their sizes. In all publications on galaxies and the Universe at large, astronomers have to preface their calculations by saying what value of the Hubble constant they are using. And if two researchers at a conference are using different values, then listeners must convert one set of results to the other distance-scale before comparing them!

The huge radio galaxy 3C 236 is 1600 million light years away, and 17 million light years in size—if we use Sandage and Tammann's value for the Hubble constant. On the de Vaucouleurs scale, it is a lot closer, at 900 million light years, and is only $9\frac{1}{2}$ million light years across—although it still

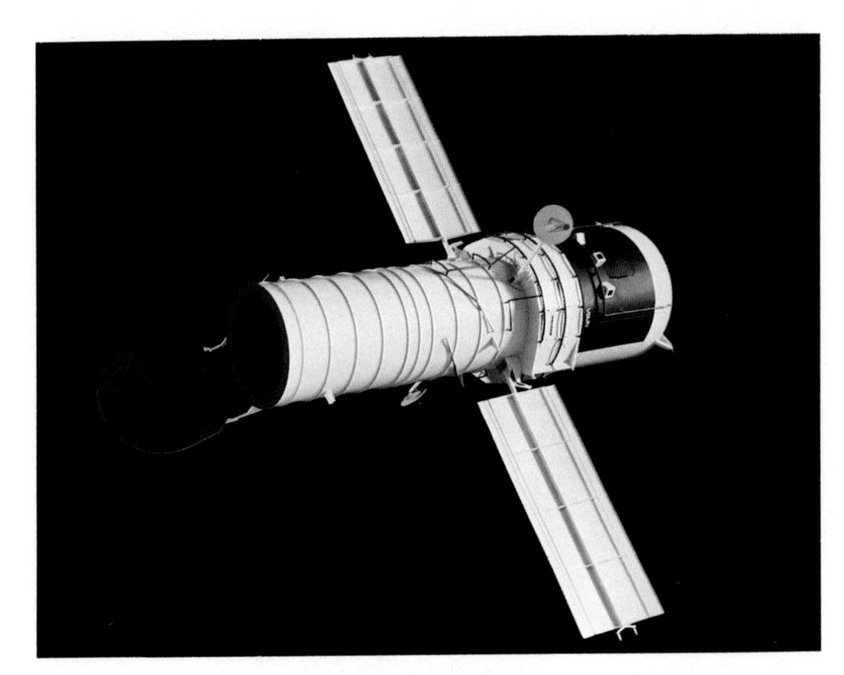

The Space Telescope (*above*) is a 2.4 metre optical telescope due to be carried into orbit by the US Space Shuttle in 1984. Sited above our polluted and turbulent atmosphere, it will be able to detect objects 50 times fainter than even the largest Earth-based telescopes, allowing more precise measurements of the distances to remote galaxies.

ranks as the largest known radio galaxy because all the others are reduced in proportion. (In this book, all figures have been calculated using an intermediate value for the Hubble constant, 20 kilometres per second per million light years, corresponding to an age of 15,000 million years for the Universe.)

A third group of cosmic surveyors, also in the United States—John Huchra, Marc Aaronson and Jeremy Mould—have recently devised a new technique to spot intrinsically-similar galaxies lying at different distances. Their results, published in 1980, come out remarkably close to de Vaucouleurs', especially when they look at the most distant clusters. Oddly enough, they get a different result for the Hubble constant when they restrict themselves to just the nearest giant cluster, the Virgo Cluster. This value is nearer to the Sandage and Tammann Hubble constant. Huchra, Aaronson and Mould explain the discrepancy by saying that our Galaxy and the Virgo Cluster are not being carried apart quite as fast as we would expect from the general expansion of the Universe. The natural interpretation is that the gravitational pull of the Virgo Cluster is affecting our Local Group (including the Milky Way); it slows down the separation velocity, and so the Hubble constant you work out from the Virgo Cluster is wrong. That is why it is important to look outside our local Supercluster before we can be sure that we are getting the correct Hubble constant for the Universe as a whole.

The Hubble controversy is by no means over, but more astronomers are now plumping for a larger value—a smaller separation between clusters of galaxies, and a younger Universe, born about 10,000 million years ago. Despite the growing evidence for this age, from the Hubble constant measurements, there is a problem with the Universe being so 'young'. The best measurements show that the oldest stars in our Galaxy are 15,000 million years old.

We cannot easily measure the ages of individual stars, but we can date clusters. All the stars in a cluster are born at the same time; the heaviest live for the shortest time, and die first. By investigating what kind of stars are left in a cluster, it is possible to date their time of birth—and some appear to pre-date the Universe itself. This cannot be true.

Those who prefer to think the Universe is 18,000 million years old naturally point to the old star clusters as evidence that they are right. The growing band who prefer a younger, 10,000 million-year-old Universe conclude that our theories of stars are not as accurate as has been assumed; that the oldest star clusters are quite a bit younger than current theory tells us, and that star theorists have some serious thinking to do.

THE HOT BIG BANG

The problems with the Hubble constant, and with 'stars older than the Universe' should be settled by another decade of research—and in particular by one new telescope. It is already designed, and should start observing by the mid-1980s. In view of astronomers' high hopes, it seems rather small: with a mirror 2.4 metres (95 inches) across, it is outclassed by a dozen telescopes in the world. The crucial difference is that this one will not be 'in the world': the Space Telescope will be orbiting high above our heads.

Out there, a telescope can receive light unhindered by our atmosphere; the images will be crisper, for the light will not be blurred by air currents. And the Space Telescope will be free from the background light scattered in the atmosphere, background light that normally fogs a photograph and prevents us from seeing the faintest possible objects.

The Space Telescope will help all branches of astronomy, from showing up clouds on our neighbouring planets to revealing the details of quasars. It will be able to pick up individual stars in the galaxies of the Virgo Cluster, and so put a precise distance on the nearest cluster of galaxies. Perhaps most important of all, it will be able to look out farther into the remotest depths of space. And because the depths of space are a time-machine, they will show us the beginnings of our Universe.

The reason is simple. Although light (and other radiations) travels fast, it takes time to get from one point to another. The sunlight shining through the window at present left the Sun eight minutes ago—so I am seeing the

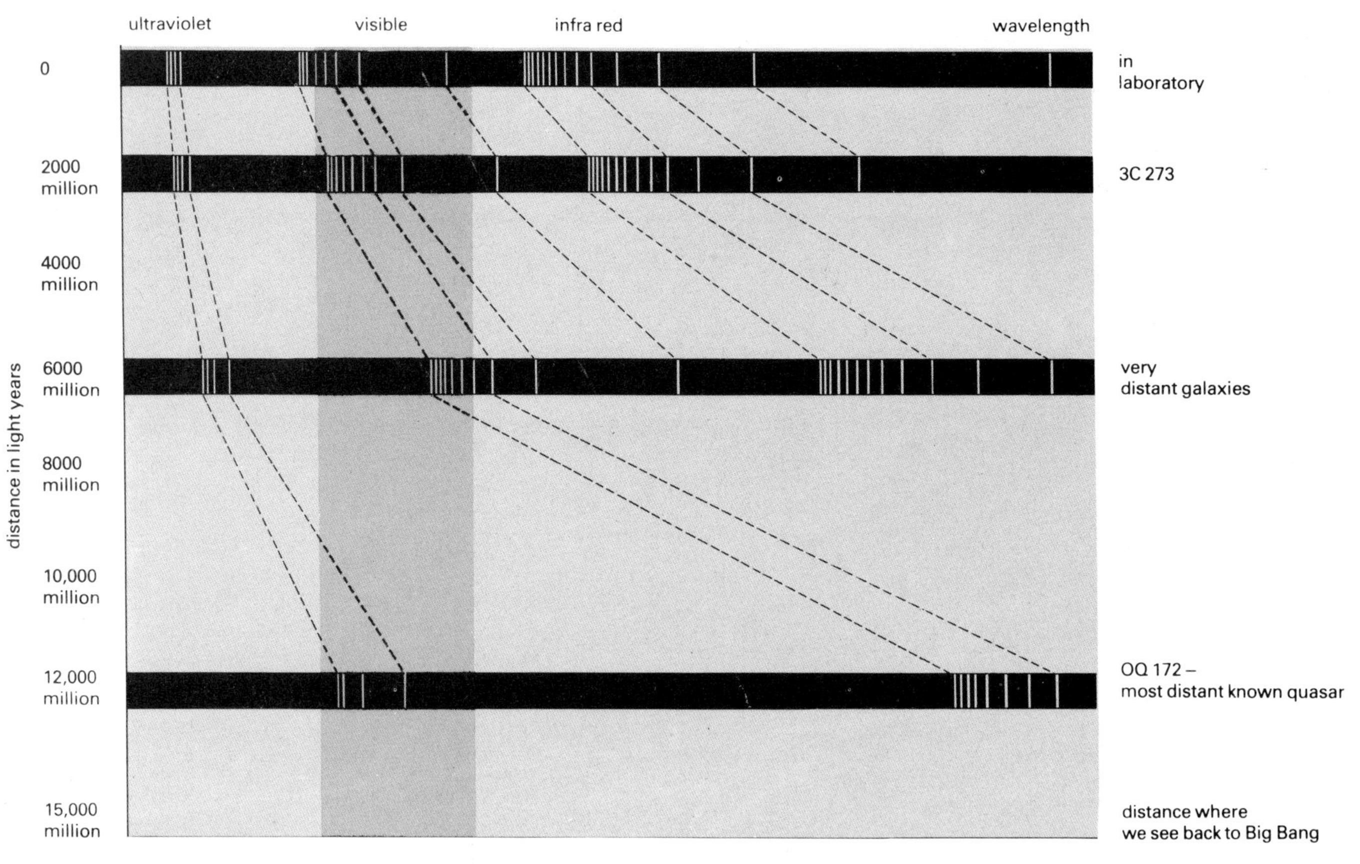

The wavelengths of light from distant objects are stretched by the expansion of the Universe—they are 'redshifted' to longer wavelengths. Here the spectral lines from hydrogen are shown for objects at different distances, starting with gas in the laboratory on Earth. The wavelengths are stretched in proportion to their original wavelength, so longer lines are moved further, but the ratio of wavelength increase to original wavelength (the redshift) is the same for any line from a particular object. In the most distant quasars, lines that were originally emitted as ultraviolet reach Earth as visible light. At a distance of 15,000 million light years, we are looking back in time to the Big Bang, and any radiation from the exact moment of the Big Bang would be shifted to infinitely long wavelengths.

Sun as it *was* eight minutes ago, not as it *is* now.

We see the Andromeda Galaxy as it was two million years ago—as Man was appearing on Earth. Our view of the most distant quasar, OQ 172, is a snapshot of a quasar taken effectively 12,000 million years ago, only 3000 million years after the birth of the Universe. In other words, as we see it the world of OQ 172 is much more the world just after the Big Bang than it is the Universe of today.

This far out in space—and back in time—there are more quasars about. It is a fact that ties in nicely with the idea that galaxies are more 'active' when they are younger; before a galaxy has settled down, there is more turbulent gas about to fuel a central black hole, and so give birth to a quasar.

Beyond the quasars, we should be seeing back to the time when galaxies themselves formed out of the all-pervading gas clouds racing apart from the Big Bang. The gas clouds must have been slightly irregular. There were pockets where the gas was denser and regions where it was more tenuous. As space expanded between the gas atoms, some of the dense patches stayed together, held in by their own gravitation. Space within these bound clouds could not expand; but it expanded between them where the gas was more tenuous, to separate the clouds from one another. Each cloud became a cluster of galaxies.

The details of galaxy formation from these clouds is still disputed. The clouds presumably broke up into smaller patches, each of which became a galaxy; within each, the gas jostled into nebulae where the first generation of stars was born. But beyond these vague generalizations, there are many theoretical variations. Russian scientists believe that the original gas clouds collapsed in one direction first, to make a thin pancake of gas, which then broke up into galaxies. It has even been suggested that the stars came first, forming directly from the Big Bang gases, and then clumped together as galaxies, and the galaxies into clusters. Martin Rees has proposed that most of the gas very quickly turned into stars which rapidly 'burned out' to leave most of the Universe's matter as individual black holes scattered through space—the missing mass. This then dictated where the rest of the gas should go to make up galaxies.

If we could see out far enough, we should see right back to the time when the galaxies formed. In fact, the first bursts of star formation should make the young galaxies much more luminous than the galaxies of today, so making them easier to see. Unfortunately, their distance means that they are too dim to be picked up by present-day Earth-based telescopes. The Space Telescope should detect these youthful galaxies—and tell us just how galaxies did begin.

However it happened, the galaxies must have formed about half a million years after the Big Bang. During the preceding phase of the expanding Universe, it was filled with gases—

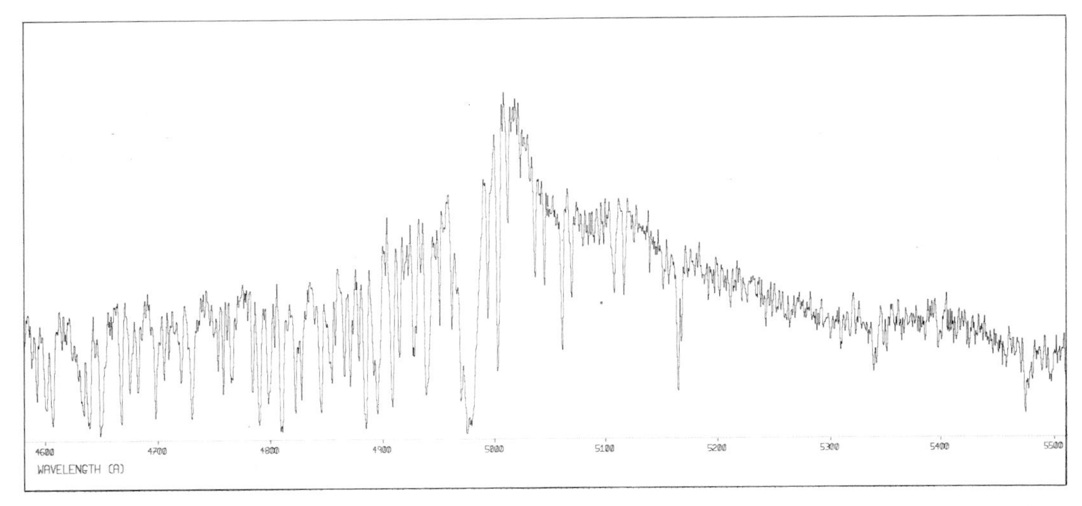

Spectrum of the distant quasar QO420–388 (*right*) recorded by the IPCS (described on pages 148–9). The broad peak is a Lyman-alpha emission line from hydrogen gas, which is emitted as ultraviolet radiation and here stretched to visible wavelengths by the redshift effect. (Lyman-alpha is the longest-wavelength of the left-hand group of hydrogen lines shown (*left*) When the quasar's light passes through gas clouds in intergalactic space, some radiation is absorbed by atoms, at a wavelength determined by the gas cloud's redshift.

Because any gas cloud must lie nearer to us than the quasar, its absorptions must occur at lower redshifts, and so absorb light of a shorter wavelength than one would associate with the quasar itself—that is, to the left in a spectrum. The huge number of dips to the left of Lyman-alpha peak in this spectrum show that the quasar light has passed through a vast number of intergalactic hydrogen clouds. Quasar light can thus act as a probe of the constitution of the Universe at large.

tremendously hot to begin with, but cooling with the expansion. The gas produced radiation appropriate to its temperature, just like the light from a star's surface. But the Universe was infinite in size, even then, so the radiation had nowhere to go. It just kept travelling through the gases. In fact, that radiation is still around today. It reaches us from the depths of space beyond the quasars, from beyond the forming galaxies, from almost 15,000 million light years away, where we see the Universe as it was only half a million years after the Big Bang.

Two American radio astronomers, Arno Penzias and Robert Wilson, picked up the radiation by accident in 1965. Although the early Universe was exceptionally hot, the waves of radiation have been so stretched by redshift that they reach us as long-wavelength radio waves. Provided we 'look out' far enough, whatever direction we look in the sky, we can see a region as it was just after the Big Bang. For this reason, the radio background arrives equally from all directions.

It started from a gas at a temperature of thousands of degrees Centigrade but it has now been so stretched that it is equivalent to radio waves from a very cold gas, at some $-270°$C, which is $3°$ above the absolute zero of temperature. It is called the 3 K background radiation (K being the Kelvin temperature measured from absolute zero).

This temperature ought to be exactly the same in all directions in space, because we are looking back to the same kind of gas. But astronomers flying a tiny radio telescope in a U-2 spyplane—to get above the worst of Earth's atmosphere—have discovered that the background is slightly 'warmer' in the direction of the constellation Leo. The only reasonable explanation is that our Galaxy is not fixed absolutely to its local fabric of space-time and that, relative to the Universal background, it is moving towards Leo. No one is quite certain why we should be adrift like this, but it could be that the Virgo Cluster is pulling on us. Virgo is the next-door constellation to Leo, so the directions agree —and, as we saw earlier, there is independent evidence from studying the redshift of nearby galaxies that Virgo is indeed holding us in its gravitational embrace.

A second surprise of the background radiation is its strength in the infrared region of the spectrum. Although its radio strength fits well the radiation expected from heated gas in the early Universe, the infrared measurements show it to be weaker than expected. There is no agreed explanation for this; but the measurements are tricky, and must be repeated before we should start rewriting our favourite theories of the Big Bang.

Background radiation did, however, kill off an alternative theory for the Universe—the Steady State. This said that the Universe had no beginning. It expands forever and, while the clusters of galaxies move apart, new galaxies spring into existence between them. So on the broadcast scales, the Universe always looks much the same, and always has done. But this fails to explain why there are more quasars far off in space, or back in time, and there is no way the Steady State theory can produce a background of radio waves.

One or two Steady State proponents have clung on, however, and its strongest advocate, Sir Fred Hoyle, has come up with another version which differs both from the Steady State and the Big Bang. He refuses to believe that the Universe is really expanding. Instead, all the clusters of galaxies stay the same distance apart, but they are increasing in mass all the time, and this mass increase causes the redshift. Mathematically in fact, the two pictures are very similar, although they sound so different. According to Hoyle, the background radiation is scrambled up from a previous phase of the Universe—a time when all masses were negative. But the Big Bang is *the* accepted view: it explains most of what is known about the Universe, making the fewest possible assumptions, and while it *may*

15,000 million years
Present day

10,000 million years
Sun and solar system form

5000 million years
Many galaxies have active quasars at their centres

2000 million years
Most galaxies (including Milky Way) have formed: first stars shining, producing heavier elements

500 million years
Gas filling Universe begins to fragment and form galaxies

300,000 years
Electrons combine with protons and helium nuclei to make *atoms:* hydrogen and helium gas

10,000 years
Radiation becomes less important than matter

3 minutes
Protons and neutrons combine to make helium nuclei

1 second
Electrons and antielectrons (positrons) annihilate, leaving small residue of electrons

0.000 001 second
Quarks and antiquarks annihilate, leaving small residue of quarks: quarks team up to make protons and neutrons

0.000 000 000 000 000 000 000 000 000 000 000 01 second
Matter now predominates over antimatter (by one part in 100 million).

0.000 000 000 000 000 000 000 000 000 000 000 000 000 000 1 seconds
Farthest back we can calculate — limit set by uncertainty principle

0
Big Bang

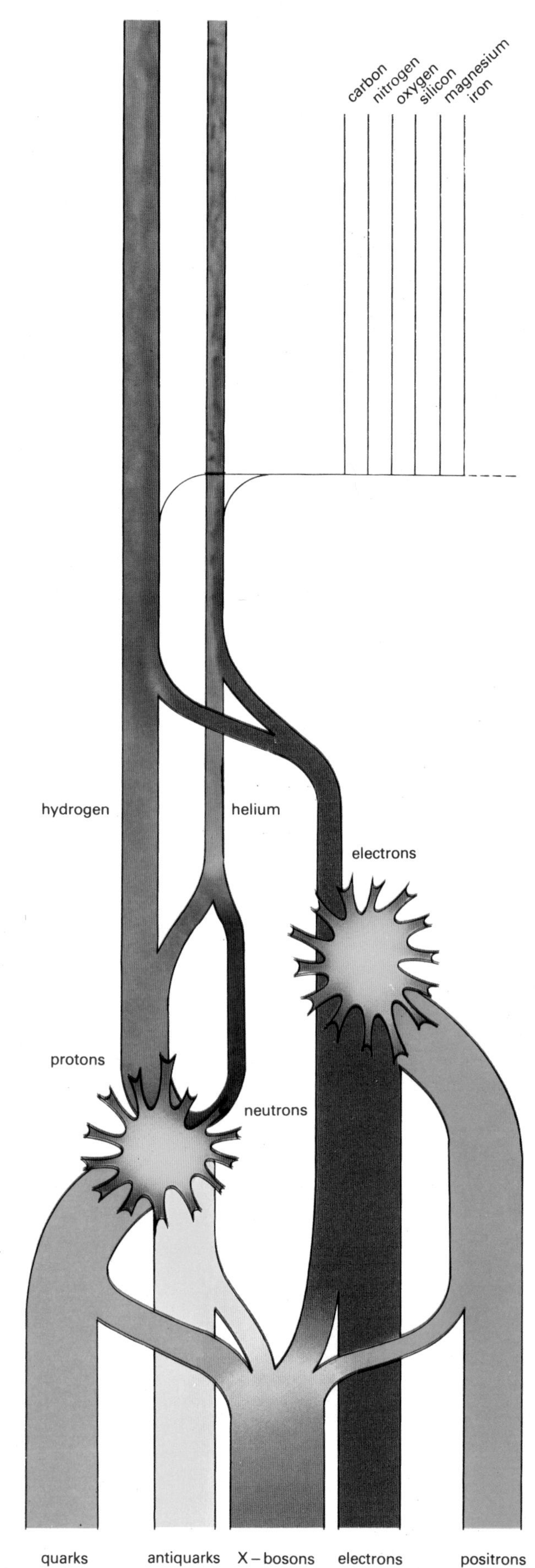

3°
3000°
neutrinos
1,000,000,000°
1,000,000,000,000°

distribution of matter

radiation (temperatures above absolute zero)

Potted history of the Universe from the Big Bang to the present day (*left*), according to the most widely-held theories. Events become more uncertain as we calculate back to the earliest moments. The constitution of the Universe (left-hand columns) was largely determined in the first three minutes, apart from the heavier elements which were made later in stars. The initially uniform distribution of gases eventually broke up into galaxies (centre column). Radiation and neutrinos still survive from the Big Bang (right-hand columns); the radiation from the early days has now cooled to a weak radio wave background; the present-day importance of neutrinos is still uncertain.

eventually turn out to be wrong, there is no good evidence for discarding it now.

The Big Bang theory comes in two forms: the Hot Big Bang and the Cold Big Bang. In the hot version, the Universe began by being *extremely* hot. Mathematically speaking, it was infinitely hot at the instant of the Big Bang itself, and ever since it has been cooling down gradually. The 'hot' gases which formed into galaxies only contained the residual heat from the inferno of the Big Bang.

On the other hand, the Universe could have started off cold. Just before the galaxies formed, something happened to heat up all the gas. If most of it formed into the stars which burned out to form the black holes of the missing mass, they must have blazed out a huge amount of light and heat during their brief lives. Perhaps this heated up the cool gases expanding from a Cold Big Bang, up to the thousands of degrees that the background radiation tells us they had.

The Hot Big Bang is the current favourite. At first it seems that little can be said about its hot gases, except that they expanded and cooled down until they broke up into galaxies. But physicists can calculate a lot of what must have happened, on the basis of experiments with the ultimate particles that make up matter: subatomic particles, electrons, neutrinos and quarks. And locked up at this time is one of the greatest mysteries of the Universe. *Why* is there matter here at all?

Within the first one-millionth of a second after the Big Bang, matter as we know it did not exist. There were no atoms, just subatomic particles. And in the inferno of heat and pressure, there was antimatter, too: each of the subatomic particles has its exact opposite. For every type of quark there is an antiquark; for each neutrino an antineutrino; for every electron an antielectron (usually known as a positron). A particle and its antiparticle will annihilate each other when they meet, in a burst of radiation. And conversely, intense radiation, like gamma rays, can *create* a particle-antiparticle pair—by sacrificing its own energy.

Just after the Big Bang, radiation was so intense that it created a whole jostling sea of particles and antiparticles. There is every reason to expect equal amounts of each. By the time the Universe was a couple of seconds old, though, all the quarks and antiquarks should have been annihilated, and all the electrons and positrons, too. At this time, there was still a lot of radiation, but it was too weak to create new particle-antiparticle pairs. But if all the matter and antimatter were annihilated, the Universe should be empty save for radiation. There should be no matter left to make up atoms, to constitute the gases that could eventually turn into galaxies.

Since there *are* stars, there is matter in the present-day Universe, so it is obvious that all the matter was not wiped out. (And there is no evidence for naturally-occurring antimatter in the Universe at present to compensate.) There must have been an imbalance between matter and antimatter in the first moments of the Universe that left some matter around after the great annihilation. Calculations show that after the Big Bang there must have been 100,000,001 quarks for every 100,000,000 antiquarks, and 100,000,001 electrons for every 100,000,000 positrons. So *almost* all the matter did disappear. Only one particle survived for every 100 million original particles.

The imbalance between matter and antimatter was small; but why was there any imbalance at all? The current favourite explanation is a new theory of the fundamental subatomic particles, which supposes that both quarks and antiquarks are fragments from another, heavy kind of subatomic particle. This *X-boson* existed even earlier after the Big Bang. As it broke down into quarks and antiquarks, the split would be slightly uneven—rather as a pencil balanced on its point is perfectly symmetrical, but must eventually fall in just one direction. So the content of the Universe had to end up unbalanced. The theory may or may not be correct; experiments on Earth to recreate the X-boson will eventually decide. If it is correct, it affects our view on the end of the Universe, as well as the beginning, for this theory also predicts that protons have only a limited lifespan, so that all matter will eventually decay into radiation.

And what of the Big Bang itself? Simply working back in time, cosmologists say that the Universe started in a 'primeval atom', which was infinitely compressed and infinitely hot. (It contained all of space, incidentally, so it did not 'explode' in the normal sense, out into empty space; in its explosion, the primeval fireball *created* space between all its particles of matter, and that creation continued between the clusters of galaxies today.)

That is too simple, perhaps. Taking present theories at face value, there is a moment before which we cannot calculate what was going on. There is no way we can measure the state of the Universe before then either. The portcullis falls because of a fundamental limitation on our knowledge of very fast events, called the Uncertainty Principle. For the Universe itself, this limit prevents us from knowing anything about its very earliest moments—even in principle. The curtain comes down when we try to think of events happening before our Universe was 0.000 000 000 000 000 000 000 000 000 000 000 000 000 000 1 second old. For everyday purposes, that is as close to nought as possible, but in the early Universe it does represent a crucial difference. At this time, the temperature was still seeringly hot—but not *infinitely* hot—and the matter had spread out, so while it was incredibly condensed, it was not packed to infinite density.

As far as scientists are concerned, this was the moment of the Universe's birth. Our knowledge can only tell us that, at its birth, the Universe was a very hot, very dense, and rapidly exploding mass of matter, antimatter and radiation. Of its brief gestation, and of its moment of 'conception' we know nothing.

Other universes

There is no scientific answer to what happened *before* the Big Bang, nor to *why* those hot gases of the primeval fireball exploded to make up our Universe—but it is possible to speculate in a philosophical way.

If our Universe contains enough matter to halt its expansion, it will eventually slow to a standstill and then start contracting until all the galaxies run together in a 'Big Crunch'. From this, the matter may bounce back in another Big Bang to recreate another Universe. On this view, the Universe follows recurrent phases. It continually passes through cycles of expansion and contraction, with a Big Crunch/Big Bang marking off one cycle from the next. The Universe could thus last for ever; 'our Universe' is just one cycle between one particular Big Bang and the succeeding Big Crunch, out of an infinite series of others stretching both backwards and forwards in time. What was happening before the Big Bang, then, was the collapsing Universe of the previous cycle, headed for its doom in the Big Crunch which rebounded as the Big Bang of our Universe.

Appealing though this idea is, the observations now seem to show that there is not enough matter about to stop the present expansion. So our Universe has a one-way ticket to indefinite expansion—and demise. In this case, it cannot be part of an infinite cycle of expansion and contraction, and we must accept that the Universe did *begin* in the Big Bang. Scientists with a religious bent have suggested that the primeval fireball had always existed before this, from the infinite past, and that its explosion 15,000 million years ago was triggered by a divine Hand. Or, as a variant of this view, perhaps nothing existed before. The exploding fireball was created supernaturally to become our expanding Universe: on the most enormous scale possible, God said, 'Let there be light.'

Although one cannot rule out divine intervention, it is not necessary to invoke it. Space was created in the Big Bang; perhaps *time* was, too. We generally think of time as something rolling inexorably by, from infinite

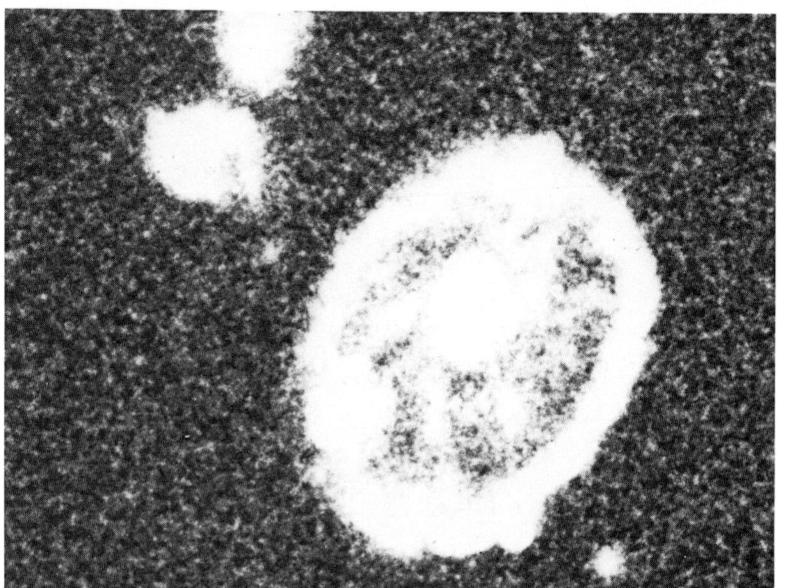

Enigmas of the Cosmos: three of the many peculiar galaxies which still challenge astronomers. NGC 1344 (*top*) is an elliptical galaxy surrounded by mysterious shells of stars. Some astronomers think that a collision between two spiral galaxies leading to sudden starbirth may explain its strange appearance. The 'Cartwheel galaxy' (*middle*) is thought to have evolved in a similar way: a small galaxy (*left*) may have passed through a normal spiral, shocking its outermost gas into forming a ring of stars. Stefan's Quintet (*bottom*) was once believed to be a small, ordinary group of galaxies. However, the spiral galaxy NGC 7319 has a very much lower redshift than the other members of the group, which may mean that it is simply a nearby galaxy lying in the same direction as the others, or that it is a genuine group member—but one whose redshift derives from something other than the expansion of the Universe.

past to infinite future, but it is closely linked to space, according to relativity. Suppose that time itself started in the Big Bang, just as space did, and is now unrolling. Time measurement is then something absolute: not just measuring intervals—from noon to noon, say—but measuring from a definite beginning, like measuring the length of string from a bobbin, or temperatures from nature's coldest point, absolute zero.

On this view, there was not a 'before the Big Bang' because time did not exist until after the Big Bang happened. It is as meaningless as winding back onto a bobbin, keeping on winding and expecting to find a negative length of string. There need not be a *cause* for the Big Bang. It was a unique event, the only unique event in the history of the Universe, and perhaps it was unique, too, in being the only event which happened without anything causing it. The Universe's conception just happened; at that happening itself, neither time nor space existed. The event of the Big Bang brought both time and infinite space into existence, as well as matter.

THE RULES OF THE UNIVERSE

In whatever way the Universe's explosion was 'triggered', right from the start it had rules built into it. These described the kind of forces which would exist—gravitational, electric, magnetic, nuclear—and the kinds of subatomic particles which would be stable and would eventually make up the matter in the Universe. The reason for these particular rules is puzzling—and controversial. One thing is certain, though: if the rules were slightly different, life could not exist in the Universe. Despite its rather esoteric air, this question of 'rules' is probably the strangest enigma of all.

The rule which allowed some matter to survive the matter-antimatter annihilation is obviously vital. It allowed matter to exist after the Big Bang, and so build up stars, planets and life. So, less evidently, is the fact that the Universe has always been expanding equally in all its parts. There is no obvious reason why different parts of the Universe should not have exploded at different rates from the Big Bang, but if they had the early gases would have tried to even out the irregularities, and in the process would have created so much radiation that galaxies could not have formed.

A critical balance between the nuclear forces and the electric forces ensure that stars have nuclear reactions which turn hydrogen into helium gradually. As a result, the Sun has continued to shine steadily for billions of years, so giving life the chance to arise on Earth. An imbalance one way would have meant that all the hydrogen turned to helium in the Big Bang, so there would be no Sun-like stars at the present time—and no hydrogen about to make up water and other molecules essential to life. The other way, and atomic nuclei would not be stable: no atoms could exist (except hydrogen)—no carbon, no oxygen, no iron—and so no planets, and no life.

Cosmologists have drawn up a startling list of such 'coincidences' that allow life in the Universe. Our existence depends critically on a whole variety of rules operating at levels too microscopic for us to readily appreciate.

From a religious point of view, it is easy to argue that God designed the Universe intentionally so that life could evolve and flourish in it. But that is perhaps too easy an explanation. Scientists have always tried to explain nature without invoking gods, and have succeeded in understanding thunder without invoking Thor, sunshine without Helios. And cosmologists have devised an even more audacious explanation for the orderliness of the Universe: The Universe is bound by these rules *because we are here*.

This bold *anthropic principle* starts out by assuming that there are more universes than ours. Although ours stretches out in all directions in space, and back in time at least to the Big Bang, there could be others very like ours off in different dimensions (see chapter 9). Suppose that each of these universes starts off in the mêlée of a Big Bang by adopting its own selection of rules. These could be completely different from those pertaining in our Universe: different kinds of 'matter', forces totally unlike gravitation or the electric force that we know. Most universes will end up in absolute chaos: unstable atoms, no galaxies, no stars, no planets, no life. Only a very tiny proportion will happen to start off with a suitable set of rules for building up an orderly universe. And only in these is there any chance that life can evolve, and eventually come to study the universe around.

So in the huge number of possible universes, there are just a few where life can exist: and these must necessarily be universes which are orderly, where conditions are suitable for life. Seen in this multi-universe perspective, our Universe is no longer so remarkable. It just happens to be one of the few where life *can* develop. So the orderliness of the Universe is intimately linked to the fact that we are here to observe it. It is exaggerating only a little to say the Universe is orderly *because* we are here.

The consciousness of modern scientists has advanced now to embrace not just the totality of our Universe but also to unknowable universes beyond our own. No longer is man a creature of the Earth, or our Galaxy, or even our Universe. In imagination at least, he is exploring universes beyond.

The spiral galaxy (*above*) is very much a twin of our own Milky Way: to astronomers it represents all the problems of understanding spiral structure in galaxies, starbirth and stardeath, and the formation of galaxies themselves from the embers of the Big Bang.

These crop up in different contexts in modern science. In discussing black holes, we saw how Einstein's equations can describe universes on other dimensions—and possibly even allow 'bridges' which connect one universe to another. If the Universe is cyclic, repeating its expansion and contraction over and over again, there have been other universes before ours, and there will be others to follow, after ours has collapsed in the total destruction of a future Big Crunch.

In one interpretation of quantum theory, our Universe is all the time splitting up into a plethora of different universes. Our consciousness follows just one path in all this splitting. In simple terms, you may decide to get married and settle down to have a family; but there is another universe, just as real, where your decision was to stay single, and where your children never exist.

So mysteries of space and time extend far beyond our own Universe. We do not know whether these other universes are 'real'—whatever that may mean—or whether we can reach them. But it is only a few decades since Man even began to understand the size and structure of this Universe.

There is plenty of time. The Sun will continue to shine for another 5000 million years. If we consider the development of our civilization from the Stone Age to the present day as one hour, then present technology began to take shape in the Industrial Revolution only a couple of minutes ago, and we have 50 whole years until the Sun dies.

After the Sun's demise, there will still be other stars blazing in our Galaxy for billions of years thereafter. By the time our Galaxy has become a mere charnel-house of star-corpses, Man may indeed have found other ways to seek out other Universes—and to travel to them. As our Universe sinks into a cold, black death, our descendants may be elsewhere, and elsewhen. They may be warming themselves in the remnant fires from another Big Bang. And when this universe in its turn drags to a lingering death, they will find another universe among the infinite variety nestled in the multi-dimensional complexity of ultimate reality.

Even if man finally tracks down and explains all the puzzles of our home Universe, the day may dawn when our descendants are poised to explore the infinite mysteries of new, unimaginable, universes beyond.

INDEX

Note: *numbers in italics refer to illustrations and captions*

Pictures supplied by:
Anglo-Australian Observatory: 96&97 (D. F. Malin), 150/1, 167, 178
Ardea: 74T (Peter Green), 74CL (P. Morris), 74BL, 75
Association of Universities for Research in Astronomy: 20, 90TR
Arecibo Observatory: 108B (National Astronomy & Ionosphere Centre, Cornell University), 109L&R ('The Search for Extraterrestrial Intelligence' by Carl Sagan & Frank Drake/Scientific American/Lyn Cawley
Bavaria: 70/1
Stephen Benson: 116
Heinrich Bevann: 62/3 (published by B. Heezan & M. Thorp, N.Y.)
Dr. Alec Boksenberg/University College: 148T, 149TL
Brookhaven National Laboratories: 85
California Institute of Technology: 43
Bruce Coleman Ltd: 74CR
C.S.I.R.O.: 83T
Erstes Physikalisches Institute: 134/5
European Space Agency: 67B
Foy and Bonneau: 39T&C
Mike Freeman: 164BC
Hale Observatories: 39B, 72, 90, 161BR, 164BL, 164BR, 169, 170
Reproduced by kind courtesy of Stephen Hawking: 140
John Hillelson Agency: 24TR (Dr. George Gerster)
Vincent Icke/University of Minnesota: 136/7
Institute of Geological Sciences: 24C, 24B, 58, 60, 74BR
Kitt Peak National Observatory: 24TL, 176B
Lowell Observatory: 50T
Mats Wibe Lund: 63BR
McLean Observatory: 20/1 (Dr. Leo Conolly)
Mullard Radio Astronomy Observatory: 119
P. Morris: 115
M.R.A.O.: 122T
N.A.S.A.: 23, 32, 33, 37, 38, 44, 45, 50B, 54, 63CR, 67T, 90TL
Novosti Press Agency: 25, 76/7
Photri: 6, 28, 87(NASA), 117, 130, 135TR (Centre for Astrophysics)
Royal Astronomical Society: 86
Royal Greenwich Observatory: 94BR, 95BL, 156
Royal Observatory, Edinburgh: 14 (from original UK Schmidt plates by Dr. Malin) 89, 127 (UK Schmidt Telescope Unit), 176TL
The Science Museum: 22/3
Science Photo Library: 37(NASA), 51(NASA), 131, 154, 158/9
John S. Shelton: 63TR
Space Frontiers: 55 (NASA), 83B, 109B (NASA), 122B (NASA), 147B (NASA)
Spectrum: 71
Sterrewacht Leiden Huygens Laboratorium: 155, 162
Alan Stockton/Institute for Astronomy, Hawaii University: 151
University of London: 173

Artwork supplied by:
David Hardy: 36/7, 111, 120/1
Bernard Fallon: 66
Bill le Fever: 9, 16/7, 43, 48/9, 52/3, 56, 59, 79, 80, 94TL, 95TR
Mike Freeman: 10/1, 112/3, 139
Ben Mayer: 164T
Andy Miles: 27, 30, 34, 41, 46/7, 73, 82, 99, 102, 103, 104/5, 106, 107, 124, 126, 132/3, 142, 147T, 152, 153, 157, 160, 161TL, 166, 168, 172, 175
Mick Robinson: 112TL
Tom Stimpson: 12/3, 65, 84/5
Jon Wells: 100/1
Martin Woodford: 62

Artwork visualized by:
Barry Jackson

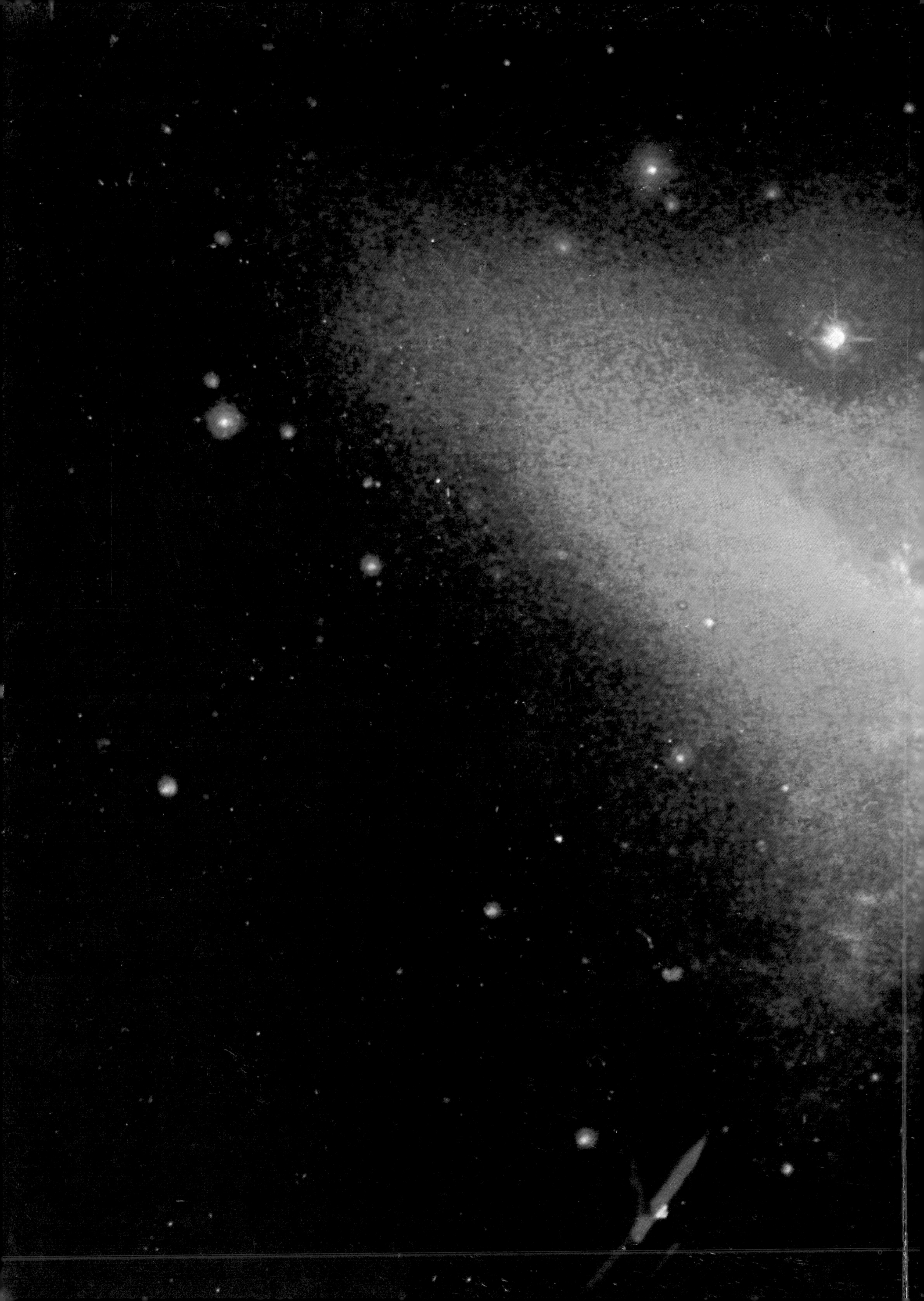